Ulrich Poth | Reinhold Schwalm | Manfred Schwartz

Acrylic Resins

Poth/Schwalm/Schwartz: Acrylic Resins

ISBN: 978-3-86630-857-2

Cover: BASF SE

Ulrich Poth, Reinhold Schwalm, Manfred Schwartz
Acrylic Resins
Hanover: Vincentz Network, 2011
European Coatings Tech Files
ISBN 3-86630-857-4
ISBN 978-3-86630-857-2

Please ask for our book catalogue
Vincentz Network, Plathnerstr. 4c, 30175 Hannover, Germany
T (202) 684-6630, F (202) 380-9129
E-mail: books@american-coatings.com, www.american-coatings.com

Layout: Vincentz Network, Hanover, Germany

ISBN 3-86630-857-4
ISBN 978-3-86630-857-2

European Coatings TECH FILES

Ulrich Poth | Reinhold Schwalm | Manfred Schwartz

Acrylic Resins

Poth/Schwalm/Schwartz: Acrylic Resins

ISBN: 978-3-86630-857-2

Preface

Acrylic polymers are essential products for various industrial application fields. Particularly, they play important roles as binders (acrylic resins), dispersion resins and polymer thickeners in the coatings industry.

Since acrylic resins were launched onto the coatings market, they have distinguished themselves by meeting various quality requirements and replacing other resins. Acrylic resins are used in eco-friendly coating systems, for example. Thanks to their broad range of properties and to ongoing developments within this product class, they also serve as ingredients for many different coating systems.

The present book presents an overview of the production, properties and application of acrylic resins and their distinctive features. It covers acrylic resins prepared by solution polymerization and emulsion polymerization, as well as reactive acrylic resins which form films by radiation curing. Besides a general description of these three product classes, there are chapters dealing with particular properties for the diverse application fields.

The goal of the book is to convey the latest knowledge about acrylic resins in solvent-borne and water-borne systems, and for radiation curing in an understandable and descriptive manner. Given the breadth of applications for acrylic resins, the book discusses the different chemical and physical aspects of the production methods and the related application properties. There are numerous literature citations for further reading on specific issues.

The book is aimed at students and newcomers to the field of coatings technology, and also at well-versed experts in coatings applications and other related industries. The essential background information herein will underpin decisions concerning the choice and use of acrylic resins.

Münster, Ludwigshafen, São Paulo, March 2011
Ulrich Poth Dr. Reinhold Schwalm Dr. Manfred Schwartz

Poth/Schwalm/Schwartz: Acrylic Resins

ISBN: 978-3-86630-857-2

Contents

1 Definitions

Ulrich Poth

Acrylic resins for coatings are binders (polymers) composed mainly of esters of acrylic acid or methacrylic acid. By analogy with the salts of inorganic acids, the esters are called acrylates or methacrylates. The IUPAC systematic name for acrylic acid is prop-2-enoic acid while that for methacrylic acid is 2-methylpropenoic acid.

Formula 1.1: Esters of acrylic acid and methacrylic acid

Although acrylic esters and methacrylic esters have quite different properties, polymers made from either are called acrylic resins. Numerous products contain mixtures of both esters.

Binders are the film forming components contained in all coating systems. They are so-called because of their ability to wet pigments and to bind coating layers to substrates.

The term **resin** is derived from natural resin (rosin), which historically were used as a binder for coatings. The term alludes to the physical state of the products. Resins consist of polymers or oligomers that exhibit glass-like behaviour. Physically, they are liquids that have extremely high viscosities, i.e. they are solidified melts. But there are also binders which have the appearance of being true liquids at ambient temperatures.

Esters of acrylic acid and methacrylic acid are distinguished by the reactivity of their double bonds. Among others, the presence of these double bonds renders the esters amenable to polymerization, which may be initiated either by free-radicals or by ions.

Poth/Schwalm/Schwartz: Acrylic Resins

ISBN: 978-3-86630-857-2

Acrylic resins for coatings fall into two groups:

The first comprises the **polyacrylates**, which are prepared by polymerizing acrylic or methacrylic esters via their double bonds.

Polyacrylates are made by various polymerization processes. The choice of polymerization process critically determines the resultant properties of the acrylic resin.

One process is **solution polymerization**. In this process, the polymers are prepared in organic solutions, which may be used directly in coatings formulations (see Chapter 3). In addition, such polymers can be transformed into secondary aqueous dispersions or into powder coatings resins. Some binders for the aforementioned applications are produced by **bulk polymerization**, or **pearl polymerization** .

Another process is **emulsion polymerization**, which is employed in the production of primary aqueous acrylic dispersions (see Chapter 4).

Esters of acrylic acid and methacrylic acid which act as building blocks for polymers are called **monomers**. Further building blocks capable of forming polymers in conjunction with acrylic esters and methacrylic esters are called **comonomers**.

The second group of acrylic resins for coatings comprises **acrylic** or **methacrylic ester resins** which still contain **double bonds**. These binders are called reactive acrylic resins. Addition or condensation reactions are employed to incorporate the acrylic or methacrylic ester into polymer or oligomer molecules. The resultant binders are capable of forming films by polymerization after application, and are notable for their resistance properties. The polymerization reactions are initiated mainly by energy-rich radiation (e.g. UV light), which yields three-dimensional crosslinked macromolecules.

These reactive acrylic resins, which form films through polymerization of the double bonds of acrylic or methacrylic esters, are classified and named for the reactions by which they are incorporated into polymers or oligomers (see Chapter 5).

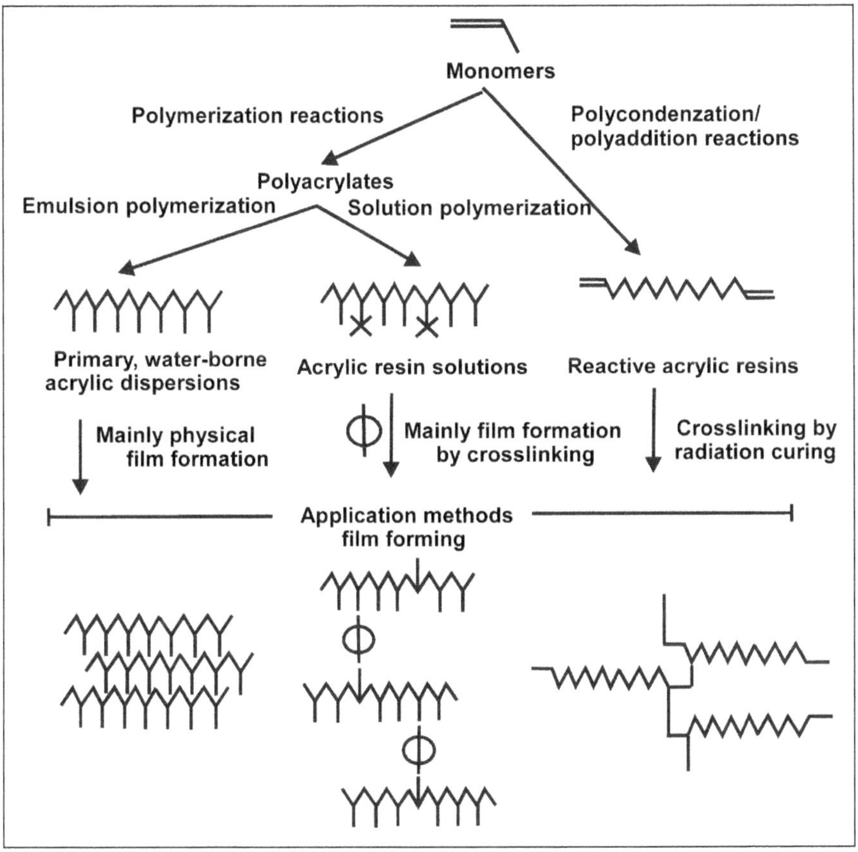

Figure 1.1: Classification of acrylic resins for coatings

2 General composition and structure

Ulrich Poth

The composition and structure of acrylic resins determine the application properties of the acrylic binders prepared from them.

2.1 Free-radical polymerization

2.1.1 Polymerization reactions

Polymerization reactions at the double bonds of acrylic monomers [1–6] may be initiated by free-radicals or ions. The most important and common reaction is free-radical initiation. The initiators employed (peroxy and azo compounds) decompose spontaneously into free-radicals when the temperature is increased (initiator reaction). The decomposition rate is influenced by the type of initiator and the temperature.

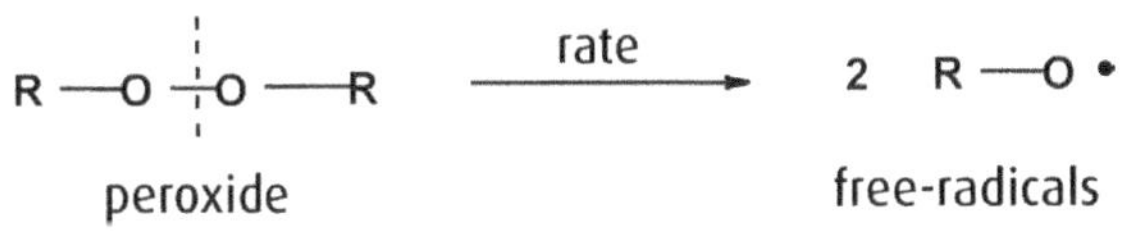

Formula 2.1: Initiator reaction (e.g. for a peroxy compound)

The resultant free-radicals react with the π-electrons of the double bonds on the monomers. The monomer molecule forms a new single bond and a single electron, i.e. a new free-radical. This is the start reaction in free-radical polymerization.

Poth/Schwalm/Schwartz: Acrylic Resins

ISBN: 978-3-86630-857-2

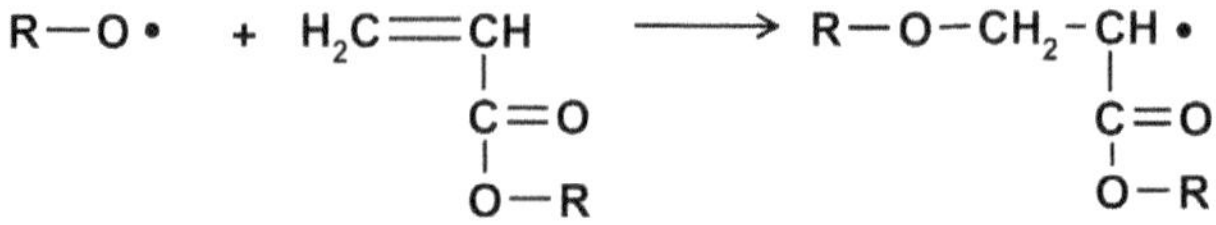

Formula 2.2: Start reaction

The planar double-bond system becomes a tetrahedral molecule of lower bond energy. Consequently, a substantial amount of energy is released in the form of heat (exothermic reaction). As the activation energy of the free-radical reaction is relatively low, the free polymerization enthalpy of the monomer reaction is markedly negative at -60 to -80 kJ/mol (examples: acrylic acid -75 kJ/mol, methyl methacrylate -58 kJ/mol at 298 K).

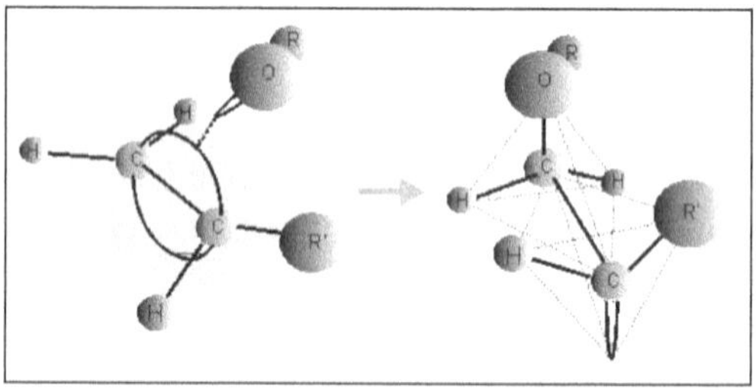

Figure 2.1: Model of the reaction at the double bond

The free-radical monomer formed by the start reaction can then add a further monomer molecule, to which it transfers its free-radical nature. This reaction is repeated as long as monomer molecules are available. Such a reaction is called chain propagation.

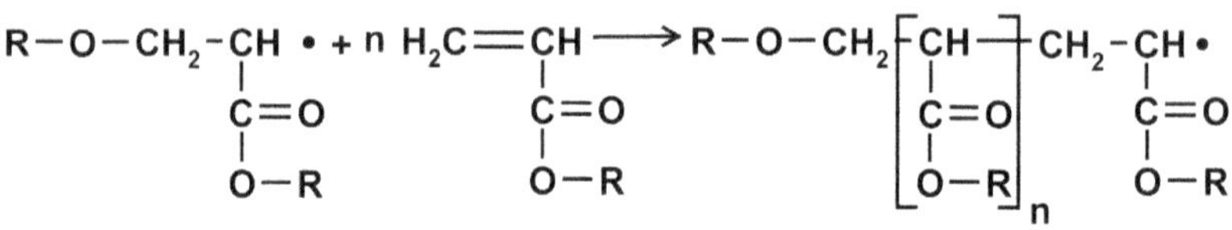

Formula 2.3: Chain propagation

The free-radical chains may grow until two free-radicals collide, forming a single σ-bond. The second free-radical may be another free-radical chain or an initiator free-radical. This chain-termination reaction is called recombination.

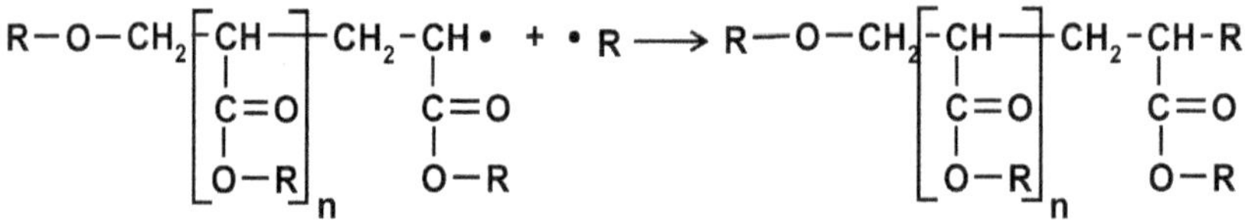

Formula 2.4: Chain-termination reaction by recombination

Propagation may also be terminated by the reaction between a free-radical at the end of a chain and movable atoms or atom groups. This reaction, which creates a chain end and a new free-radical on the partner molecule, is called chain transfer. On the one hand, the reaction partner might be another chain with active hydrogen atoms. In that case, the free-radical formed on this chain would act as the starting point for renewed chain propagation and lead to the generation of branched polymer molecules. On the other hand, chain transfers also take place by reaction between free-radical chains and other molecules in the reaction mixture, e.g. with solvent molecules. In turn, these free-radicals, formed on the partner molecules, can initiate new polymer chains. Some compounds are ideal for chain-transfer reactions, e.g. mercaptans. Such compounds are called regulators; they are added in tiny amounts to control chain propagation.

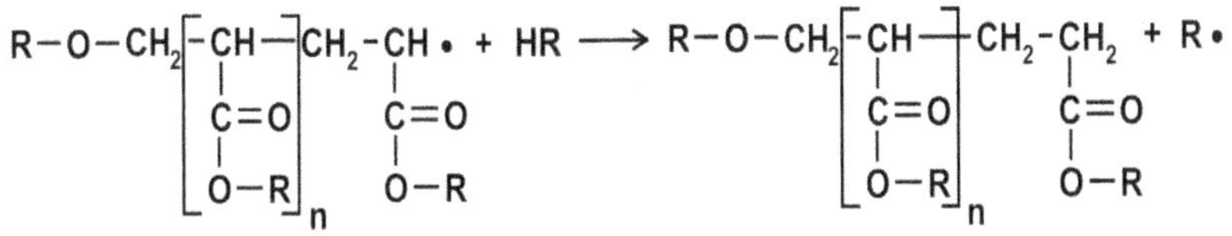

Formula 2.5: Chain termination by chain transfer

An alternative chain-termination reaction is disproportionation. This occurs when the ends of two free-radical chains react by transferring a hydrogen atom. The driving force for this reaction is the high energy content of free-radical molecules. In a disproportionation reaction, two neighbouring molecular states of lower energy are created. Of the two free-radicals, one becomes a saturated chain end and the other, a chain end with a new double bond.

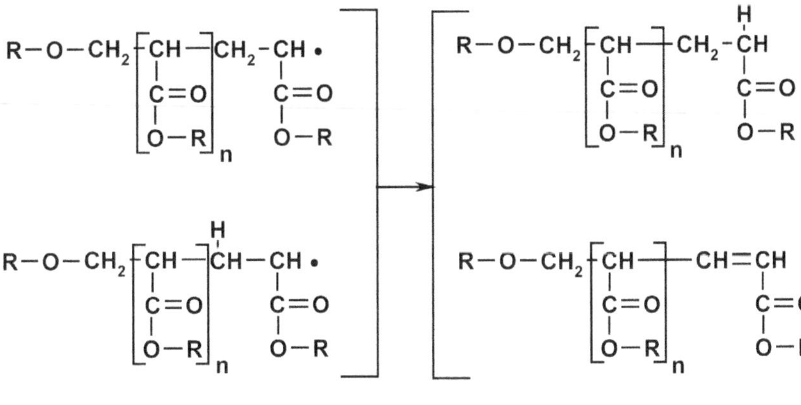

Formula 2.6: Chain termination by disproportionation

The polymer molecule bearing the new double bond at the end of its chain can act as a macro-monomer. If it is incorporated into a further chain propargation, the second way to form branched polymer molecules is created.

The relative proportions of the various chain-termination reactions depend on the type of polymerization process and the reaction conditions. The predominant chain-termination reactions are believed to be recombination with other polymer molecules or chain transfer to solvent molecules or regulators. These favour the generation of linear polymer molecules. However, it must be remembered that linear acrylic polymers are composed of extensively coiled molecules. The formation of branched acrylic polymers, by chain transfer to other polymer chains or by disproportionation, confers markedly different application properties on such acrylic polymers in coating formulations.

The various polymerization processes and the chosen reaction conditions can give rise to side reactions. These exert a marked influence on the properties of the polymer molecules and coating systems prepared from them (see description of polymerization processes).

When reactive acrylic resins undergo film formation – usually under the influence of high-energy radiation (see Chapter 5) – the reactions involved are the same as those described above. For the most part, the reactive acrylic resins employed contain molecules bearing two or more double bonds. Thus, it is possible for each molecule to participate in more than one chain-polymerization process. That is how polymer molecules become crosslinked in the film matrix.

2.1.2 Kinetics of free-radical chain polymerization

The initiator reaction is a first-order decomposition reaction that yields two free-radicals [7]. The reaction rate (v_1) varies with the concentration of initiator (c_I) and the temperature. The rate of the initiator reaction is shown in the following Formula 2.7 [2]:

$$v_1 = \frac{2 \cdot dn_I}{dt} = k_1 \cdot c_I$$

Formula 2.7: Velocity of initiator reaction

The velocity of start reaction (v_2) varies with the concentration of initiator free-radicals ($c_{R\bullet}$), the concentration of monomers (c_M) and, of course, the temperature.

$$v_2 = \frac{-\,dn_{R\bullet}}{dt} = k_2 \cdot c_{R\bullet} \cdot c_M$$

Formula 2.8: Velocity of start reaction

Comparison of velocity constants for initiator and start reactions show that the constant of start reaction $[k_2]$ is much larger than that of initiator reaction $[k_1]$. That is why the velocity of start reaction $[v_{st}]$ is ultimately determined exclusively bei the efficience of initiator decomposition $[k_1]$.

$$k_1 \ll k_2$$

$$k_1 = k_{st}(T)$$

$$v_{st} = k_{st} \cdot c_I$$

Formula 2.9: Comparison of velocity constants

After a very short time, the chain propagation reaction reaches equilibrium with respect to chain termination. The rate of chain termination (v_{ct}) varies with the concentration of chains bearing free-radicals $(c_{RnM\bullet})$.

$$v_{ct} = \frac{-\,dn_{RnM\bullet}}{dt} = k_{ct} \cdot c^2_{RnM\bullet}$$

Formula 2.10: Velocity of chain termination

At equilibrium, the change in concentration of chain free-radicals $(c_{RnM\bullet})$ is zero. The rate of chain propagation (v_p) is then equal to the rate of chain termination (v_{ct}).

$$\frac{dn_{RnM\bullet}}{dt} = 0$$

$$v_{st} = v_{ct}$$

$$k_{st} \cdot c_I = k_{ct} \cdot c^2_{RnM\bullet}$$

Formula 2.11: Equilibrium

The propagation velocity (v_p) is then given by:

$$v_p = k_p \cdot c_M \cdot c_{RnM\bullet}$$

$$v_p = k_p \cdot c_M \cdot \sqrt{c_I \cdot k_{st} / k_{ct}}$$

Formula 2.12: Propagation velocity

The molar propagation velocity (n_M) varies with the propagation constant (k_P) and the monomer concentration (c_M) as a function of time.

$$n_M(t) = k_p \cdot c_M$$

Formula 2.13: Molar propagation velocity

Examples of kinetic data for monomers are presented in the Table 2.1 for methyl acrylate and methyl methacrylate [02].

Substitution of the data for methyl acrylate into the propagation rate formulas above yields a figure of 165 · 10^{-5} mol/(l s), i.e. 20,380 molecules per second. For methyl methacrylate, the corresponding propagation rate computes to 17 · 10^{-5} mol/(l s), i.e. 3,120 molecules per second. For a polymer with an average molecular weight of 10,000g/mol, the average time needed to form single molecules is 0.007s in the case of acrylic monomer, and 0.03s in the case of methacrylic monomer.

Table 2.1: Kinetic data for monomer examples

Parameters (at 60°C)	Symbols [dimensions]	Monomer	
		methyl acrylate	methyl-methacrylate
Initiator concentration (azo-isobutyronitrile)	c_I [10–3 mol/l]	6.1	6.1
Monomer concentration	c_M [mol/l]	9.75	8.50
Constant (start reaction)	k_{st} [· 10–6 l/(mol s)]	4.6	4.6
Constant (termination reaction)	k_{ct} [· 106 l/(mol s)]	4.3	9.3
Constant (propagation reaction)	k_p [· l/(mol s)]	2090	367
Propagation rate	v_p [· 10–5 mol/(l s)]	165	17
Molecular rate	n_M [mol/sec–1]	20,380	3120

In other words, esters of acrylic acid polymerize 10 times as fast as esters of methacrylic acid. That is why acrylic resins bearing reactive double bonds consist mostly of esters of acrylic acid (see Chapter 5).

In contrast to polycondensation reactions, where the molecules are formed over many hours, individual molecules of acrylic polymers prepared by free-radical initiated polymerization process are formed in fractions of seconds. Furthermore, again in contrast to polycondensation, free-radical-initiated polymerization is not an equilibrium process. Therefore, the molecular-weight distribution as a function of the average molecular weight obeys statistical laws for the given reaction conditions. Naturally, however, the polymerization reaction depends on the energy conditions. The *Gibbs-Helmholtz* equation is valid.

$$\Delta G = \Delta H - T \cdot \Delta S$$

Formula 2.14: Simplified version of the Gibbs-Helmholtz equation

This equation states that a reaction takes place only if the free reaction enthalpy (G), which is the difference between the value of the total reaction enthalpy (H) and the product of the temperature (T) and the reaction entropy, is negative. As the temperature rises, the value of the free reaction enthalpy becomes smaller and smaller. If the value reaches zero or the free reaction enthalpy becomes positive, polymerization of monomers is no longer possible. This temperature limit is called the ceiling temperature. For example, the ceiling temperature for methyl methacrylate is 373K (200°C, 328F). Generally, the ceiling temperatures for methacrylates are lower than those for acrylates. This relationship between polymerization efficiency and temperature is what makes low-temperature radiation curing of reactive acrylic resins so advantageous (see Chapter 5).

2.1.3 Influences on polymerization reactions

To summarize, polymerization reactions are induced by the use of initiators to raise the temperature and form free-radicals. Thereafter, all subsequent reaction steps (start reaction, chain propagation, and recombination) are highly exothermic. All polymerization processes must be technically capable of meeting these conditions. As a consequence, there is little scope for variation in production processes; ultimately, these restrictions influence the properties of the polymer products. Above a monomer-specific temperature, no polymerization reaction takes place.

As molecular propagation is not an equilibrium reaction, the polymerization obeys statistical laws [2]. That is the reason that molecular-weight distribution of acrylic resins prepared by free-radical polymerization is much broader than that of polymers prepared by polycondensation. Special techniques are available for achieving lower average molecular weights and narrower molecular-weight distributions. Some of the free-radically initiated polymerization processes also enable particularly large, non-crosslinked polymer molecules (average molecular weights up to more than 10^6g/mol) to be made.

As already mentioned, acrylic polymers prepared by the usual methods are mostly made up of linear molecules. Those molecules are extensively coiled, a fact which significantly influences the application properties of the resultant coatings, namely the solubility and viscosity of solutions, fastness of physical film formation (drying) of solutions and dispersions, wetting and flow during film formation, efficiency of crosslinking reactions. If, on account of the polymerization process conditions, the proportion of branched molecules increases, the tendency to coil decreases, the solution viscosity rises, and the functional groups become more amenable to crosslinking.

2.2 Monomers

2.2.1 Esters of acrylic acid

Currently, acrylic acid is almost exclusively made from propene by one-step or two-step catalytic oxidation process [8]. Acrylic acid esters are prepared by esterifying acrylic acid with the corresponding mono-alcohols in the presence of specific catalysts [9]. Enzymatic processes that consume less energy and have high yields have also become available [10]. Some esters are made directly. For example, ethyl acrylate can be prepared directly from acrylonitrile and ethanol in the presence of sulphuric acid, or from acetylene, carbon monoxide and ethanol [13]. In addition, acrylic esters of higher mono-alcohols can be prepared by transesterifying low-molecular esters.

In industry, acrylic acid esters for the production of polymers are formed with methanol, ethanol, n-butanol, iso-butanol, tert.-butanol, 2-ethylhexanol, n-dodecyl alcohol, and cyclohexanol, all of which are abundantly available. The other esters of acrylic acid play only a minor role. Chain length and the branching of alcohol side-chains exert a significant influence on the properties of acrylic resins and the resultant coatings (see Chapter 2.2.6).

2.2.2 Esters of methacrylic acid

Methacrylic acid is made by different processes. The most common utilises acetone und hydrocyanic acid, obtained via acetone cyanohydrin. The acetone cyanohydrin is dehydrated with sulphuric acid to methacrylamide.

Table 2.2: Physical data of industrially made acrylic esters [19–22]

Monomer	CAS No.	Mol. weight [g/mol]	Melting point [°C]	Boiling point [°C] (hPa)	Density [g/cm³] (°C)
Methyl acrylate	96-33-3	86.1	-75	80 (1013)	0.956 (20)
Ethyl acrylate	140-88-5	100.1	-72	100 (1013)	0.922 (20)
N-Butyl acrylate	141-32-2	128.2	-64	148 (1013)	0.898 (20)
iso-Butyl acrylate	106-63-8	128.2	-61	139 (1013)	0.890 (20)
tert.-Butyl acrylate	1663-39-4	128.2	-69	117 (1013)	0.885 (20)
2-Ethylhexyl acrylate	103-11-7	184.3	-90	91 (13)	0.887 (20)
Lauryl acrylate	2156-97-0	240 to 268	-8	200 (polym.)	0.87 (25)
Stearyl acrylate	4813-47-4	324	30	200 (polym.)	0.8545 (40)
4-tert.-Butylcyc-lohexyl acrylate	84100-23-2	210	-11	111 (2.7)	0.9399 (25)
3,3,5-Trime-thylcyclohexyl acrylate	86178-38-3	196.3	n. a.	40 (0.1)	n. a.
Dihydrodicyclo-pentadienyl acrylate	12542	204.3	-36	119 (5)	1.0697 (25)
Benzyl acrylate	2495-35-4	162.2	n. a	110 (11)	n. a
Phenylethyl acrylate	3530-36-7	176.2	n. a	102 (5)	n. a

The latter is then saponified either with water to form free methacrylic acid or with alcohols to yield different methacrylic esters [12].

Methacrylic acid is also obtained via methacrolein by oxidation of isobutene [18].

Propene and carbon monoxide yield isobutyric acid, which can be dehydrogenated to methacrylic acid.

Ethene and synthesis gas (hydrogen and carbon monoxide) combine to form propionaldehyde, which reacts with formaldehyde to yield a methylol compound that can be dehydrated and oxidised to methacrylic acid.

Polymers can be produced industrially from a swathe of different esters of methacrylic acid and mono-alcohols, namely esters of methanol, ethanol, n-butanol, iso-butanol, tert.-butanol, iso-decyl alcohol, iso-tridecyl alcohol, cyclohexanol.

Table 2.3: Physical data of industrially available methacrylic esters [19–22]

Monomer	CAS No.	Mol. weight [g/mol]	Melting point [°C]	Boiling point [°C] (hPa)	Density [g/cm³] (°C)
Methyl methacrylate	80-62-6	100.1	-48	100 (1013)	0.943 (20)
Ethyl methacrylate	97-63-2	114.1	-75	119 (1013)	0.917 (25)
iso-Propyl methacrylate	4655-34-9	128.2	n. a.	125 (1013)	0.885 (20)
N-butyl methacrylate	97-88-1	142.2	-75	163 (1013)	0.895 (20)
iso-Butyl methacrylate	97-86-9	142.2	-37	155 (1013)	0.886 (25)
tert.-butyl methacrylate	585-07-9	142.2	-48	136 (1013)	0.876 (20)
Neopentyl methacrylate	2397-76-4	156.2	n. a.	88 to 90 (24)	0.888 (25)
N-Hexyl methacrylate	142-09-6	170.3	-45	204 (1013)	0.888 (25)
2-Ethylhexyl methacrylate	688-84-6	198.3	-50	218 (1013)	0.885 (25)
N-Decyl methacrylate	3179-47-3	226	-44	155 to 156 (22)	0.876 (20)
iso-Decyl methacrylate	29964-84-9	226	-22	126 (10)	0.878 (25)
Lauryl methacrylate	142-90-5	254	-7	142 (4)	0.873 (20)
N-Tridecyl methacrylate	2495-25-2	268	n. a.	150 (10)	0.87 (20)
iso-Tridecyl methacrylate	94247-05-9	~ 270	n. a.	312 (1.13)	n. a.
Stearyl methacrylate	32360-05-7	332.3	n. a.	210 (6.7)	n. v.
Cyclohexyl methacrylat	101-43-9	168.2	n. a.	210 (1013)	0.936 (20)
4-tert.-Butylcyclohe-xyl- methacrylate	46729-07-1	224.3	n. a.	112 (1)	n a.
3.3.5-Trimethyl-cyclo-hexyl methacrylate	7779-31-9	210.2	-15	65 (0.2)	0.93 (25)
iso-Borneyl methacrylate	7534-94-3	222.3	-60	66 (0.6)	0.983 (25)
Benzyl methacrylate	2495-37-6	176.2	n. a.	232 (1013)	1.04 (25)
Phenyl methacrylate	2177-70-0	162.2	n. a.	95 to 100 (16)	1.052 (25)
Phenylethyl methacrylate	3683-12-3	190.2	n. a.	119 to 120 (11)	0.976

The esters of methacrylic acid with benzyl alcohol, p-tert.-butylcyclohexanol, norboneol, dicyclopentadienylic alcohol and fatty alcohols serve as specialty monomers.

The esters of methacrylic acid are prepared by standard esterification processes, while the esters of high-molecular alcohols may be also made by transesterification, e.g. starting with methyl methacrylate.

As is the case for the various acrylic esters, the type and size of the ester side-chains of methacrylic esters significantly influence the properties of the polymers and resultant coating systems (see Chapter 2.2.6).

As already mentioned, all esters of methacrylic acid polymerize at a much lower rate than esters of acrylic acid.

2.2.3 Functional monomers

Acrylic resins containing functional groups (see Chapter 3) and reactive acrylic resins containing double bonds (see Chapter 5) are made from monomers which bear various functional groups in addition to double bonds.

2.2.3.1 Hydroxyl-functional monomers

When just one hydroxyl group in a polyalcohol is esterified with acrylic acid or methacrylic acid, the resultant monomer still contains free hydroxyl groups. Industrially, such monomers are produced mainly by the addition reaction of epoxy compounds to acrylic acid or methacrylic acid. That method – conversion of ethylene oxide or propylene oxide with acrylic acid or methacrylic acid – is used to prepare the monomers 2-hydroxyethyl acrylate and 2-hydroxyethyl methacrylate, or 2-hydroxypropyl acrylate and 2-hydroxypropyl methacrylate, respectively [23].

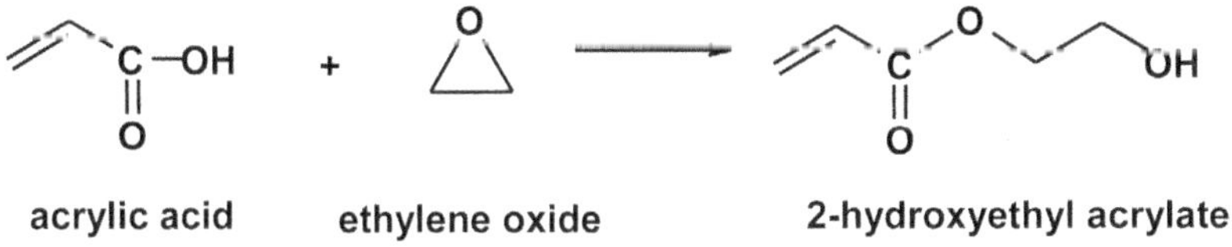

Formula 2.15: Production of 2-hydroxyethyl acrylate

Other hydroxy-functional monomers are prepared by partial esterification of polyols. One mole of acrylic acid and one mole of 1,4-butanediol yields butanediol monoacrylate.

Glycerol monoacrylate is a specialty monomer which is prepared from acrylic acid and glycidol.

Acrylic resins containing free hydroxyl groups are suitable for coatings which form films by chemical reaction. They are crosslinked with different reaction partners: polyisocyanates bearing free or blocked isocyanate groups, and amino resins bearing functional groups (methylol groups or methylol ether groups). The hydroxy-functional monomers differ in their structure and the reactivity conferred by the type of hydroxyl group. These differences determine their use in the production of acrylic resins.

2.2.3.2 Carboxy-functional monomers

Carboxyl groups are simply incorporated into acrylic resins by copolymerizing defined quantities of free acrylic or methacrylic acid with other monomers.

Like hydroxyl groups, the carboxyl groups lend themselves to crosslinking reactions. The preferred reaction partners are compounds containing epoxy groups.

Carboxyl groups in acrylic resins can be neutralized with alkalis or amines. The resultant carboxylate ions can form hydrophilic centres, so that it is possible to transform the polymers into water-borne colloidal solutions or into secondary dispersions which are suitable for water-borne coating systems.

Furthermore, certain quantities of carboxyl groups in acrylic resins exert a catalytic effect. This can be sufficient to effect reaction between hydroxyl groups and functional groups of amino resins (methylol groups, methylol ether groups). The crosslinking reaction is accelerated by the acidity of the carboxyl groups.

In addition, diacrylic acid (carboxyethyl acrylate) serves as a vehicle for incorporating carboxyl groups into special acrylic resins.

Other ways of introducing carboxyl groups into acrylic resins utilise the esters of a diol formed with acrylic acid or methacrylic acid and succinic acid or maleic acid, each of which ultimately bears one free carboxyl group per molecule.

2.2.3.3 Amino-functional monomers

Primary and secondary amines easily enter into an addition reaction with the double bonds of acrylic or methacrylic acid and their esters (Michael addition). Therefore stable monomer compounds are not formed. Initially, monomers bearing tertiary amino groups are formed which do not contain active hydrogen atoms. Such monomers are produced by converting tertiary-amino alcohols and acrylic acid or methacrylic acid into esters. The most important monomers in this class are N,N-dimethylaminoethyl acrylate and N,N-dimethylaminoethyl methacrylate.

However, there also exists a monomer containing a secondary amino group, namely N-tert.-butylaminomethacrylate. This monomer is stable only because steric hindrance by the tertiary-butyl group preventing a Michael addition reaction.

There are also monomers available which consist of quaternary ammonium salts, e.g. 2-(trimethyl ammonium) ethyl methacrylate chloride.

Polymers which have a significant content of amino groups can be neutralized by adding volatile acids. The resultant cationic carrier groups serve as a vehicle for preparing stable, aqueous colloidal solutions (comparable to the effect of anionically stabilized carboxylate ions).

Polymers bearing amino groups are noted for their outstanding adhesion on different surfaces (metals and plastic parts). They are also suitable for making paper and are used as ingredients for adhesives.

Monomers containing tertiary amines can act as catalysts. On one hand, these groups accelerate the reaction between hydroxyl groups and isocyanates (see Chapter 3). On the other, the tertiary amino groups force the crosslinking reaction of reactive acrylic resins by UV light (see Chapter 5).

2.2.3.4 Amide-functional monomers

Hydrolysis of acrylonitrile or methacrylonitrile leads to acrylamide or methacrylamide, respectively. Polymers with significant amounts of acrylamide or methacrylamide are highly polar, and are mostly used in the paper and textile industries. They are also suitable for water treatment.

Amide-functional monomers are also used in the manufacture of self-crosslinking acrylic resins for coatings applications (see Chapter 3). The resins are produced by making the amides react with formaldehyde and mono-alcohols, or the corresponding semi-acetals. These reactions can be performed in polymer analogeous reaction. There are also monomers that already consist the described modification, namely methylol acrylamide, N-(n-butoxymethyl) acrylamide, and N-(iso-butoxymethyl) acrylamide and the corresponding methacrylamides.

Other monomers available contain alkylated amides, e.g. N-ethylmethacrylamide and N,N-dimethylacrylamide.

In addition, there are (meth)acrylamides whose amide groups have been replaced by tertiary-aminoalkyl groups. Polymers containing such monomers are suitable for flocculants, paper coatings and textile finishing agents.

2.2.3.5 Epoxy-functional monomers

Of the various epoxy-functional monomers, glycidyl methacrylate (2,3-epoxypropyl-1-methacrylate) is made industrially. Acrylic resins containing specific quantities of glycidyl methacrylate can be crosslinked with compounds containing primary and secondary amino groups at ambient temperatures, or with compounds containing carboxyl groups mainly at elevated temperatures. The crosslinkers are polyamines or polycarboxylic acids or their derivatives. The most important application of acrylic resins containing the monomer glycidyl methacrylate is that of powder-coating resins.

Table 2.4: Physical data of industrially produced functional monomers [19–22]

Monomer	CAS No.	Mol. weight [g/mol]	Melting point [°C]	Boiling point [°C] (hPa)	Density [g/cm³] (°C)
1. Hydroxy-functional monomers					
2-Hydroxyethyl acrylate	818-61-1	116.1	< -65	60 (1.3)	1.11 (20)
2-Hydroxypropyl acrylate	25584-83-2	130.1	< -60	60 to 65 (1.3)	1.06 (20)
2-Hydroxyethyl methacrylate	868-77-9	130.1	< -60	87 (5)	1.08 (20)
2-Hydroxypropyl methacrylate	27813-02-1	144.2	< -58	87 (5)	1.03 (20)
Butanediol-1,4-monoacrylate	2478-10-6	144.2	-80	130 (27)	1.039 (20)
2. Carboxy-functional monomers					
Acrylic acid	79-10-7	72.1	13	141 (1013)	1.046 (20)
Methacrylic acid	79-41-4	86.0	16	161 (1013)	1.02 (20)
3. Amino-functional monomers					
N,N-Dimethylaminoethyl acrylate	2439-35-2	143.2	-92	70 (29)	0.9362
N,N-Dimethylaminoethyl methacrylate	2867-47-2	157.2		182 (1013)	
4. Amido-functional monomers					
Acrylamide	79-06-1	71.1	84	125 (33.3)	
Methacrylamide	79-39-0	85.1		110 (decomp.)	aprox.1.11 (20)
N-Methylol methacrylamide	923-02-4	115.1			
N-Butoxymethyl methacrylamide	5153-77-5	171.2		105 (0.4)	
N-iso-Butoxymethyl-methacrylamide	4548-27-0	171.2		127 (6.7)	
N-(3-Dimethylaminopropyl)- methacrylamide	5206-93-6	170.2		142 (4)	
5. Epoxy-functional monomers					
Glycidyl methacrylat	106-91-2	142.2	< -10	189 (1013)	1.08 (20)

2.2.4 Ether acrylates and methacrylates

Not only diols are partially esterified with acrylic acid or methacrylic acid, but also oligomeric etherdiols. If only one hydroxyl group is esterified, the resultant monomers contain an ether side-chain and a hydroxyl group. Examples are diethylene glycol acrylate and triethylene glycol methacrylate.

Glycol monoethers, too, are esterified with acrylic acid or methacrylic acid. Acrylates and methacrylates of methoxyethanol, ethoxyethanol, methyldiglycol and methyltriglycol are produced industrially. Furthermore, there are esters of acrylic acid with phenoxyethanol.

Polyethers are prepared by adding ethylene oxide or propylene oxide to monoalcohols or polyalcohols. There are also block copolymers which contain both cyclic oxides. The various polyethers may be esterified with acrylic acid or methacrylic acid. A number of monomers with ether side-chains of different chain lengths are available industrially.

Acrylic esters or methacrylic esters of such polyethers containing only one double bond per molecule serve as molecular building blocks for acrylic resins. The polyether side-chains perform a marked plasticizing effect. This increases with increase in length of the polyether chain. If the side-chains consist of polyethylene oxide, the polymers become distinctly hydrophilic. Acrylic resins with significant amounts of polyethylene oxide side-chains afford a way of preparing aqueous colloidal solutions or secondary dispersions featuring steric (non-ionic) stabilization.

Furfuryl and tetrahydrofurfuryl acrylate and methacrylate are ether monomers as well.

2.2.5 Polyunsaturated acrylic and methacrylic compounds

Polyunsaturated acrylic and methacrylic monomers result from the esterification of more than one hydroxyl group of a polyol with acrylic acid or methacrylic acid. As all the double bonds of such monomers are amenable to free-radical-initiated polymerization, the monomers build bridges between different polymer chains. The use of just relatively small quantities of such polyunsaturated monomers for copolymerization yields branched acrylic resins. However, larger quantities cause crosslinking of the polymers. This possibility is exploited in the preparation of aqueous and non-aqueous microgel dispersions.

The possibility of crosslinking is mainly employed in the production of reactive acrylic paint systems, mostly for radiation curing (see Chapter 5). Polyunsaturated monomers are also described there.

2.2.6 Comonomers

A series of unsaturated compounds easily copolymerize with esters of acrylic acid or methacrylic acid (see Chapter 2.2.7). Such compounds are called comonomers. They are added to the resin composition to achieve special properties.

Such comonomers are firstly derivatives of acrylic acid or methacrylic acid, mainly acrylonitrile and methacrylonitrile. Nitrile monomers perform outstanding adhesion properties on the resultant polymers, particularly on metallic substrates.

However, the most important comonomer is styrene (vinyl benzene). Also vinyl toluenes and the various methyl styrenes can copolymerize with esters of acrylic acid and methacrylic acid. Copolymers containing these aromatic monomers are noted for their hardness, and show rapid physical drying in film forming of solvent-borne paints.

Copolymerization of olefins with acrylic esters or methacrylic esters is not possible without further ado. This also applies to vinyl halides, vinyl ethers, and vinyl esters.

Also the copolymerization of acrylic esters or methacrylic esters with maleic acid or its derivates requires special proceedings.

By contrast, N-vinyl compounds copolymerize easily with esters of acrylic acid or methacrylic acid. Such comonomers, which are available industrially, are N-vinyl pyrrolidone, N-vinyl imidazole, N-vinyl caprolactam, and N-vinyl carbazole. These monomers perform excellent wetting and penetration on metal surfaces, plastic parts and paper. They are used to prepare polymers, but a much more important use is in reactive acrylic systems which are crosslinked by UV light (see Chapter 5). These monomers are more difficult to handle as the products solidify at ambient temperatures or higher.

Table 2.5: Physical data of industrially available comonomers [19–22]

Monomer	CAS No.	Mol. weight [g/mol]	Melting point [°C]	Boiling point [°C] (hPa)	Density [g/cm³] (°C)
Acrylonitrile	107-13-1	53	-83.5	77.6 (1013)	0.806 (20)
Methacrylonitrile	126-98-7	67.1	-35.8	90 to 92 (1013)	0.80 (25)
Styrene	100-42-5	104.2	-31	145 to 146 (1013)	0.909 (25)
p-Vinyl toluene	622-97-9	118.2	-34	170 to 175 (1013)	0.897 (25)
α-Methyl styrene	98-83-9	118.2	-24	165 to 169 (1013)	0.909 (25)
N-Vinyl imidazole	1072-63-5	94.1	78 to 79	192 to 194 (1013)	1.039 (25)
N-Vinyl pyrrolidone	88-12-0	111.1	13 to 14	92 to 95 (11)	1.04 (25)
N-Vinyl caprolactam	2235-00-9	139.2	35 to 38	128 (21)	1.029 (25)
N-Vinyl carbazole	1484-13-5	193.2	66 to 65	154 to 155 (1013)	1.085 (25)

Other special monomers contain fluorinated alkyl side-chains, e.g. hexafluorobutyl methacrylate. Such fluorine-containing monomers introduce special surface properties on the resultant polymers.

Surface-active polymers are also prepared by using derivatives of acrylic acid or methacrylic acid with siloxanes; e.g. trimethylsiloxyethyl methacrylate, polydimethylsiloxane methacrylate.

2.2.7 Copolymerization

When two monomers polymerize together, the probability that a further monomer 1 or monomer 2 will become attached to one free-radical end of the growing polymer chain of monomer 1 or vice versa depends on the concentration of the monomers and the velocity constants of the different addition reactions. There are four velocity constants ($k_{[1.1]}$, $k_{[1.2]}$ and $k_{[2.1]}$, $k_{[2.2]}$). If the velocity constants for the reaction between monomer free-radicals with molecules of their own kind are high ($k_{[1.1]}$ and $k_{[2.2]}$ >> ($k_{[1.2]}$ and $k_{[2.1]}$), no copolymerization will take place. The polymerization process leads to mixtures of two polymers. If the velocity constants of the mixtures are nearly equal ($k_{[1.1]} = k_{[1.2]}$ and ($k_{[2.1]} = k_{[2.2]}$), the outcome is ideal random copolymers whose composition depends solely on the monomer concentration.

If the velocity constants for the reaction between the different monomers are much higher than the velocity constants for that between monomers of the same kind ($k_{[1.2]}$ and $k_{[2.1]}$ >> ($k_{[1.1]}$ and $k_{[2.2]}$), the resultant copolymers have an strictly alternating composition, which is independent of the monomer concentration [2]. This copolymer is formed until one monomer is consumed totally.

The copolymerization behaviour can be determined from the differential change in monomer concentration over time as a function of the quotients of the velocity constants and the actual monomer concentrations. This is shown in Formula 2.16:

$$\frac{d\,c_{M_1}}{d\,c_{M_2}} = \frac{\dfrac{k_1 \cdot c_{M_1}}{k_2 \cdot c_{M_2}} + 1}{\dfrac{k_2 \cdot c_{M_2}}{k_1 \cdot c_{M_1}} + 1}$$

Formula 2.16: Calculation of copolymerization behaviour

Clearly, it is very difficult to calculate the copolymerization behaviour of just two combined monomers. The calculation becomes much more complex when more than two monomers are combined. Most acrylic resins consist of more than two monomers. Numerous trials are underway to quantify the copolymerization behaviour of monomers individually [24]. One method consists in fragmenting the velocity constants of the monomers into the effects of the resonance stability of the free-radicals (Q) and the polarity (e) of each monomer. Both terms are then expressed in terms of the values for styrene. For styrene, Q is defined as 1.0 and e as -0.8. Formula 2.17 shows how the Q and e values are calculated.

$$k_{1.2} = F_1 \cdot \underset{\uparrow}{Q_2} \cdot \underset{\uparrow}{e}^{(e_1 \cdot e_2)}$$

Formula 2.17: Calculating the Q and e values

Plotting the Q and e values for the various monomers yields the following diagram (Figure 2.2):

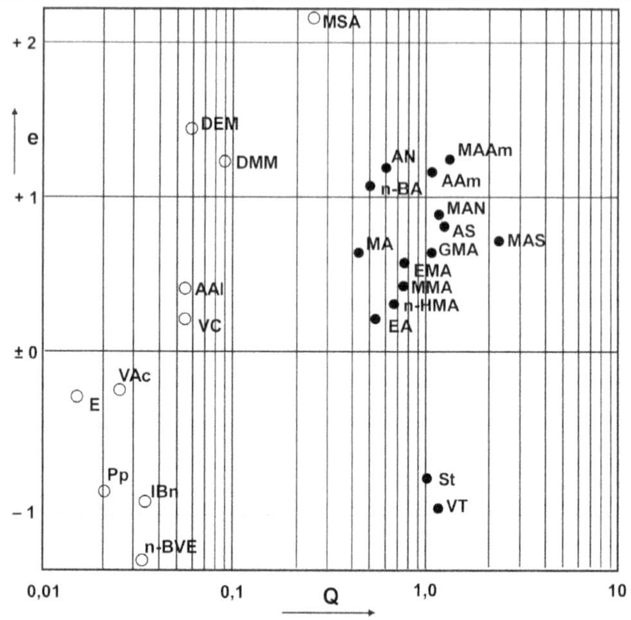

E	= ethene	St	= styrene	MA	= methyl acrylate
Pp	= propene	VT	= vinyl toluene	EA	= ethyl acrylate
IBn	= iso-butene			n-BA	= n-butyl acrylate
VAc	= vinyl acetate	AN	= acrylonitrile	MMA	= methyl methacrylate
VC	= vinyl chloride	MAN	= methacrylonitrile	EMA	= ethyl methacrylate
n-BVE	= n-butylvinyl ether	AS	= acrylic acid	n-HMA	= n-hexyl methacrylate
AAl	= allyl alcohol	MAS	= methacrylic acid	GMA	= glycidyl methacrylate
DMM	= dimethyl maleate	AAm	= acrylamide		
DEM	= diethyl maleate	MAAm	= methacrylamide		
MSA	= maleic anhydride				

Figure 2.2: Q-e diagram

The Q-e diagram makes it possible to estimate the tendency of different monomers to copolymerize efficiently or not. Generally, monomers with nearly comparable values for Q and e will yield highly random copolymers. For most acrylic resins, a random distribution of monomers along the polymer chain is desirable.

2.2.8 Characterization of monomers

2.2.8.1 Glass transition temperature

There are different ways to quantify the properties of acrylic monomers with respect to the influences of the different side-chains. The most important is the glass transition temperature (T_g). The glass transition temperature of polymers is the temperature at which the molecule composite changes from a glassy state into an elastic state. In the glassy state, the molecules of polymers are highly associated – coiled – and have high resistance to mechanical deformation. As the temperature rises, the molecules start to become mobile. In the elastic state, molecules respond to mechanical influence, but return to their former configuration as soon as it ceases. Finally, at even higher temperatures the molecules start to uncoil and will no longer return to their former composite state. The polymer becomes plastic and transfers into a melt. The temperature at which the transition from the glassy state to the elastic state occurs is influenced by the degree to which the polymer molecules are associated. The molecular associations themselves are due to physical chain-to-chain interactions (e.g. due to polar groups) and the stiffness of the chains.

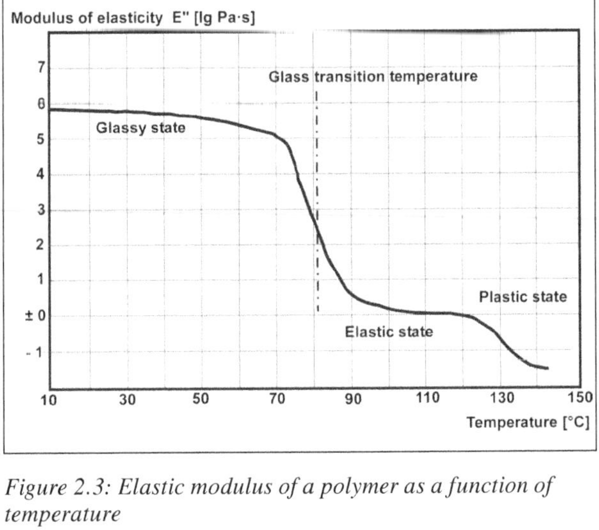

Figure 2.3: Elastic modulus of a polymer as a function of temperature

The glass transition temperature can be determined in various ways. The most informative of these is dynamic thermo-mechanical analysis (DMTA). It measures the response of polymer films to periodic mechanical forces as a function of temperature. The polymer response is plotted as a chart of elastic modulus (E'', storage modulus) over temperature. Figure 2.3 shows the change in elastic modulus of a non-crosslinked polymer with change in temperature. The temperature at the point of inflection between the glassy and the elastic states is the glass transition temperature (T_g).

Another method for determining glass transition temperatures is dynamic scanning calorimetry (DSC) [26]. The precise value of the glass transition temperature varies with the measurement method and the measuring conditions. Above a certain molecular weight, the value of the glass transition temperature is independent of the molecular weight and is influenced only by the structure and building blocks in polymer chains.

Generally, for acrylic resins: the stiffer the polymer chain is, the shorter the side-chains are; further, the more polar the building blocks are, the higher is the glass transition temperature. Figure 2.4 shows the glass transition temperatures of homopolymers of different monomers, ordered by the number of C atoms in the side-chains.

The chart shows that methacrylic esters confer much higher glass transition temperatures than the corresponding acrylic esters. The reason is that the methyl group – located on the same C atom as the carboxyl group – restricts the mobility of the chain. This contrasts with the behaviour of methyl side-chains in the case of polyester building blocks (e.g. neopentyl glycol).

The glass transition temperatures show a marked decrease with increase in the length of the side-chains. The effect is more pronounced for methacrylic esters than for acrylic esters. The difference is 95°C for the two methyl esters and only 37°C for the two n-hexyl esters.

Surprisingly, the glass transition temperature of relatively long-chain, linear acrylic esters increases with increase in chain length. The reason is that the longer linear side-chains can associate physically (waxy behaviour of long, linear aliphatic chains). This increase is more pronounced for acrylic esters than for methacrylic esters, because the methyl group on the polymer chain acts as a spacer. Branched side-chains generally confer higher glass transition temperatures than the corresponding linear side-chains. Aromatic groups in side-chains (e.g. styrene) generate relatively high glass transition temperatures due to the association effect of π-electron systems in the aromatic ring structure.

On account of their polarity, hydroxy-functional monomers lead to higher glass transition temperatures than ester monomers having the corresponding number of C atoms. However, butanediol-1,4-monoacrylate generates very low glass transition temperatures. Other polar monomers, such as acids, amides and nitriles, yield very high glass transition temperatures due to the additional scope for molecular association of their polar groups.

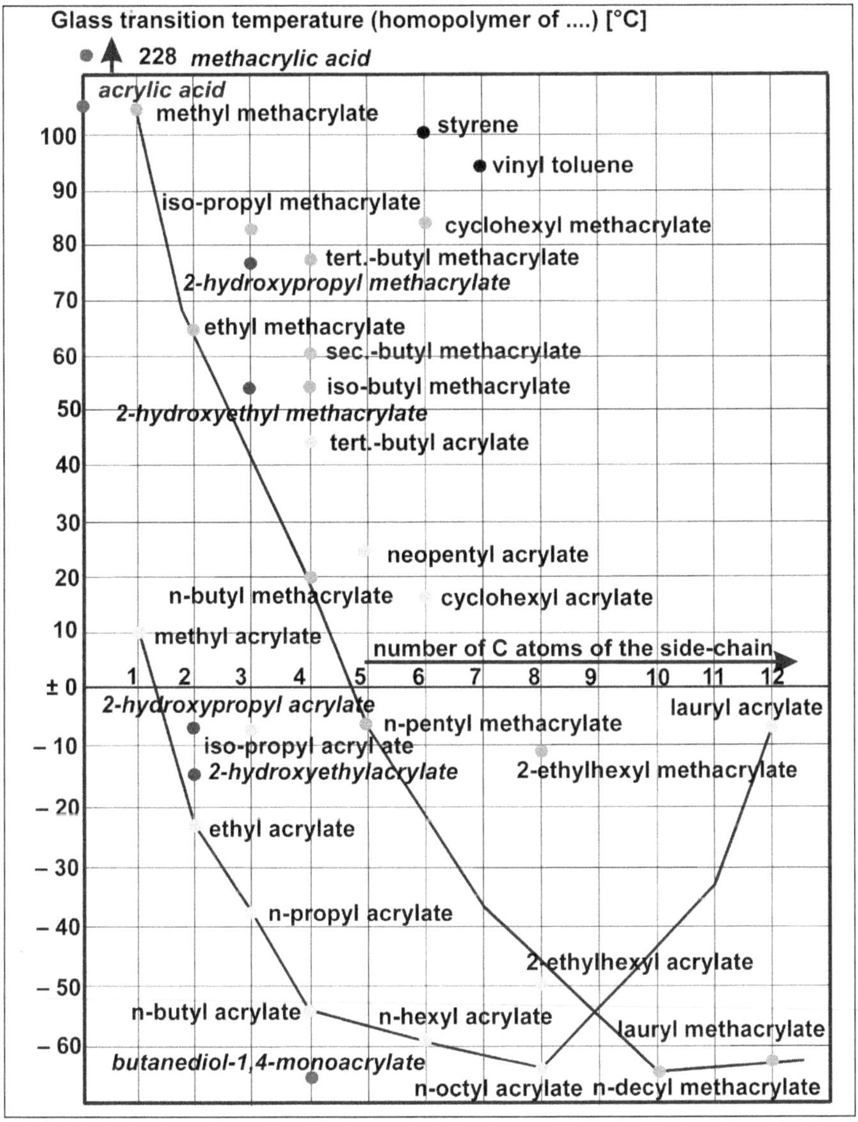

Figure 2.4: Glass transition temperature as a function of the number of C atoms on the side-chains

As most acrylic resins consist of a couple of different monomers, the question arises as to the glass transition temperatures of monomer mixtures. The answer is that the glass transition temperature of copolymers is determined by the mass fractions of the different monomers.

However, the resultant glass transition temperature is not the linear mean value of the mass fractions of monomers and their individual glass temperatures. The calculation is based on reciprocal mean values, as shown in Formula 2.18.

$$\frac{1}{T_{G\ co\text{-}polymer}} = \sum_{i=1}^{n} \frac{m_i}{T_{G\ monomer\ i}}$$

Formula 2.18: Glass transition temperature of copolymers

Figure 2.5 shows the glass transition temperature of polymers containing different quantities of styrene and n-butyl acrylate. The chart shows measured glass transition temperatures (determined by DSC) and curves of the linear and reciprocal mean values.

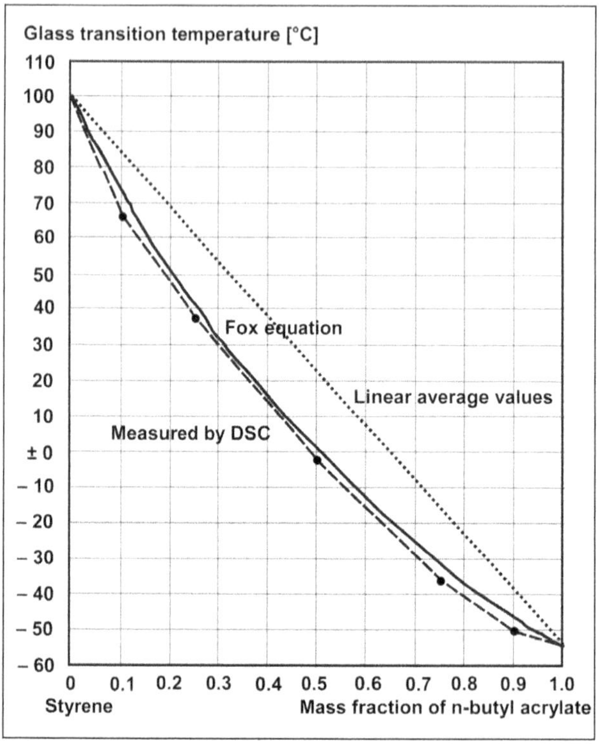

Figure 2.5: Glass transition temperature of different copolymers of styrene and n-butyl acrylate

It is clear that even small fractions of n-butyl acrylate lower the glass transition temperature of styrene copolymers significantly, as they keep the rigid styrene chains apart. Conversely, small fractions of styrene in n-butyl acrylate copolymers are unable to associate and hence unable to increase the glass transition temperature. Mathematically, this behaviour is described relatively well by reciprocal mean values. There are some exceptions, e.g. as when molecule segments interact with each other.

The value of the glass transition temperature represents a great deal of polymer properties. High glass transition temperatures confer high mechanical resistance at ambient temperatures, resulting in polymers of high hardness. A related effect is high diffusion density, which leads to better chemical and solvent resistance. However, a corollary of a high glass transition temperature is low flexibility, and such polymers are brittle at ambient temperatures. The brittleness cannot be compensated until the elastic component of the flexibility has been improved, e.g. by crosslinking reactions. (Flexibility is defined as the sum of plasticity and elasticity.)

Low glass transition temperatures signify high flexibility, as conferred by the plasticity component, but also low hardness and less chemical and solvent resistance. Flexible polymers possess better scratch resistance, because the polymer surface can escape the mechanical stress and because of the ability to compensate surface damage by virtue of their special thermoplasticity (cold plastic flow).

Crosslinking reactions raise the glass transition temperatures of polymers. Optimum crosslinking is associated with better flexibility (elastic component of flexibility), better mechanical strength, and chemical and solvent resistance.

2.2.8.2 Material properties

The nature of the side-chains on monomers in **acrylic resins** influences not only the balance of hardness and flexibility. Acrylic resins containing monomers with **short side-chains** are relatively polar polymers. The production of stable solutions therefore requires polar solvents. This also applies to acrylic resins which contain significant fractions of functional monomers and to monomers containing aromatic side-chains. Generally, acrylic polymers containing mainly methacrylic derivatives are more polar than those containing acrylic derivatives.

Acrylic polymers containing **long aliphatic side-chains** are particular non-polar, and so are readily soluble in non-polar solvents. In addition, the low polarity leads to optimum application behaviour, better spraying properties, efficient wetting of different surfaces, optimum flow and levelling during film-forming, and to more homogeneous films from aqueous dispersions.

Acrylic polymers containing long aliphatic side-chains have significantly lower viscosities in organic solutions and in melts.

Acrylic resins consisting mostly of methacrylic monomers usually yield higher solutions viscosities than resins made from the corresponding acrylic monomers.

Monomers containing cycloaliphatic compounds confer special effects. Although the glass transition temperatures are relatively high – near those of the corresponding aromatic compounds – and although high hardness values are achieved, the solution viscosities are considerably lower than those of polymers containing short side-chains or aromatic compounds. Cycloaliphatic monomers generate optimum balance of hardness and resistance on one hand, but low viscosity and sufficient flexibility on the other. Unlike the aromatic building blocks, cycloaliphatic monomers generate permeability to UV light, a fact which is an advantage for clear coat formulations. However, these monomers are more expensive than other products.

Functional monomers are introduced in acrylic resins mainly for crosslinking reactions. However, they can additionally act as carrier groups for water-borne secondary dispersions. Furthermore, they support wetting and adhesion and some have a catalytic effect on various crosslinking reactions, as mentioned above.

Monomers containing ether side-chains confer higher plasticity than the corresponding monomers with alkyl side-chains. Polymers containing long side-chains of polyethylene oxides offer scope for preparing non-ionically stabilized water-borne secondary dispersions. The ether groups in such polymers are not resistant to UV light.

2.2.9 Handling of monomers

Acrylic monomers and comonomers can polymerize spontaneously under catalytic influences and at elevated temperatures. For this reason, they are stabilized for storage and transportation. The additives for stabilization are reducing compounds. Historically, hydroquinone was the most important stabilizer. Currently, the most common product is methyl hydroquinone, due to its broader solubility. Other suitable stabilizers are hydroquinone monomethyl ether (4-methoxyphenol), 4-tert.-butyl catechol and 2,5-di-tert.-butyl-4-methylphenol (BHT, Ionol).

The optimum quantity of added stabilizer depends on the monomer type and transport destination (e.g. greater quantities are needed for shipments to the tropics). The quantity varies from 15 to 500ppm. Monomers containing acids, amides, nitriles, or hydroxyl groups require more stabilizer than simple ester monomers. Methacrylic monomers require less stabilizer than acrylic monomers.

As monomers are highly reactive compounds, there are health risks associated with handling them. These vary substantially from monomer to monomer [19–22].

For example, N,N-dimethylaminoethyl acrylate is classified as very toxic (T+), but N,N-dimethylaminoethyl methacrylate is classified "only" as harmful and irritant (X_n and X_i).

The following monomers are classified as toxic (hazard symbol: T): 2-hydroxyethyl acrylate, 2-hydroxypropyl acrylate, acrylonitrile, methacrylonitrile, acrylamide, N,N-dimethyl acrylamide, N-iso-butoxymethyl acrylamide and a polyethylene glycol methyl ether acrylate.

The monomers acrylic acid, methacrylic acid, 2-hydroxyethyl acrylate, 2-hydroxypropyl acrylate und N,N-dimethylaminoethyl acrylate are classified as corrosive (hazard symbol: C). The low-boiling esters of acrylic acid and methacrylic acid, the acids themselves and the nitriles are classified as highly inflammable or extremely inflammable (hazard symbols: F and F+).

Acrylic esters containing side-chains of up to four C atoms, and 4-hydroxybutyl acrylate (butanediol-1,4 mono acrylic ester) are classified as harmful (X_n); in addition, 4-hydroxybutyl acrylate is defined as dangerous for the environment (N).

All derivatives of acrylic acid and methacrylic acid are classified as irritant (X_i).

All esters of acrylic acid with side-chains of up to eight C atoms and all esters of methacrylic acid with side-chains of up to four C atoms, and also most of the polyunsaturated monomers, esters of polyols, are deemed sensitising. Derivatives of methacrylic acid are less hazardous than the derivatives of acrylic acid. This becomes particularly apparent for the functional monomers of the both groups.

In summary, the appropriate safety regulations must be observed when monomers are handled. Acrylic polymers have to be checked to avoid that the amounts of residue free-monomers do not exceed the given limiting values. For the use of reactive acrylic resins, free-monomers are placed into circulation. Application of these coating systems requires the adherence of particular safety regulations.

2.3 Production processes – generally

The processes for producing acrylic polymers are designed to efficiently dissipate the heat of polymerization (enthalpy) and to guarantee product reproducibility.

By reproducibility here is meant the molecular weights and molecular-weight distributions and the optimum random distribution of monomers along the copolymer chains.

Suitable processes are the mass polymerization, the suspension polymerization, the emulsion polymerization, and the solution polymerization.

Reactive acrylic resins are produced by reactions of functional monomers with different oligomers, without reaction of double bonds. These processes employ the common methods of polyaddition and polycondensation (see Chapter 5).

2.3.1 Bulk polymerization

For bulk polymerization, monomer compositions are mixed with suitable initiators and heated to the requisite polymerization temperature. The exothermic energy of the polymerization must be compensated by external cooling equipment. The resulting polymers have a relatively broad molecular-weight distribution, and very high molecular weights may be achievable. Total conversion of all monomers is difficult to accomplish. As the reaction is highly exothermic, only small batches are produced. It is more easily to prepare bulk polymerization product in thin layer. Plastic parts are produced in that way, e.g. the socalled acrylic glass, consisting of poly methyl methacrylate. This polymerization process is preferred if the target is to prepare 100 % polymers, e.g. for solvent-free secondary dispersions or resins for powder coatings – but the aforementioned restriction has to be taken in mind.

One variant of bulk polymerization is incomplete polymerization in a continuous process. In this, monomer mixtures and suitable initiators are fed at relatively high temperatures through a tube reactor or a flow-through container, where they are partially polymerized. Free monomer is then separated from the solid polymer by distillation. The residual monomers can be recycled. The process conditions must be selected so as to guarantee constant polymer composition and reproducible molecular weights and molecular-weight distributions. The risk of side reactions at high temperatures must be compensated. The polymerization temperatures must, of course, be lower than the ceiling temperature of the monomers. Principially, this process affords polymers of lower molecular weight and narrower molecular-weight distribution than those of common bulk polymerization. This socalled SGO process (solid grade oligomer) can be used to make acrylic resins for high-solid coatings [29] which are dissolved in appropriate solvents. The products are also suitable for preparing solvent-free, water-borne secondary dispersions.

2.3.2 Suspension polymerization

In suspension polymerization [30], monomer mixtures are converted into an aqueous emulsion and a free-radical polymerization process is started by initiators which are soluble in organic solvent. The emulsion consists of rather large particles. To prevent fusion of the particles, protective colloids (e.g. polyvinyl alcohol) and specific emulsifiers are added. The particle size and the particle size distribution depend on the shearing conditions (type of stirrer and shear rate).

Suspension polymerization is mostly performed with monomers that generate high glass transition temperatures.

The process is principially the same as that of bulk polymerization, but takes place in individual droplets. The surrounding water readily absorbs the heat of polymerization. When polymerization is complete, the resultant polymer particles are filtered and washed thoroughly. The target is to remove as much of the protective colloids and emulsifier as possible. The polymer particles look like pearls and are mostly delivered as such. Most of these polymers have relatively high molecular weights and a broad molecular-weight distribution, just like bulk polymers. The resultant products are mainly used for the production of physically drying coatings.

2.3.3 Solution polymerization

In solution polymerization, the monomers and organophilic initiators are reacted in organic solvents. The monomers as well as the polymers are soluble in the selected solvents, which absorb the reaction enthalpy. The resultant solutions are mainly the form in which the acrylic resins are shipped. Consequently, the selected solvents must meet the application requirements of the resultant acrylic coatings (see Chapter 3).

2.3.4 Emulsion polymerization

In emulsion polymerization, monomer mixtures are dispersed in water with the aid of suitable emulsifiers. The polymerization reaction, which takes place at elevated temperatures, is initiated by water-soluble initiators and occurs in micelles. The outcome is stable aqueous dispersions which are used for wall paints and house paints (for craftsmen and for the do-it-yourself sector), and for adhesives, printing inks, and textile treatments (see Chapter 4).

2.4 Literature

[1] J. C. Bevington: Journ. Chem. Soc. (1956)
[2] Bruno Vollmert: Grundriss der Makromolekularen Chemie, E. Vollmert Verlag, Karlsruhe (1979)
[3] P. J. Flory: Principles of Polymer Chem.; Cornell Univ. Press; Ithaka (1986)
[4] K. Matyjaszewski, T.P. Davis: Handbook of Radical Polymerization, J. Wiley & Sons, N. Y. (2002)
[5] B. Tieke: Makromolekulare Chemie, Wiley-VCH (2002)
[6] H. G. Elias: Makromoleküle, Wiley-VCH (2002)

[7] A. M. North: The Kinetics of Free Radical Polym.; Pergamon Press; Oxford (1965)
[8] F. T. Maler, W. Bayer: Encyclopedia Chem. Process Des. 1 (1976)
[9] D. J. Hadley, E. M. Evans: Propylene and its Derivates, J. Wiley & Sons, New York (1973)
[10] Specified in DE 102004033555 and US 2006084779, BASF SE
[11] E. H. Riddle: Monomeric Acrylic Esters; Reinhold; New York (1954)
[12] H. Rauch-Puntigam, T. Völker: Acryl- und Methacrylverbindungen. Springer, Berlin (1967)
[13] W. Reppe 1939 at Röhm & Haas
[14] D. J. Hucknell: Selective Oxidation of Hydrocarbons, Academic Press, London (1974)
[15] Acetoncyanhydrinprozess 1933 at Röhm & Haas, Darmstadt
[16] J. W. Crawford: Chem. Abstr. 28 (1934)
[17] Specified in GB 405 699 (1934) from ICI
[18] Isobutene process: J. W. Nemec, W. Bayer: Acrylic and Methacrylic Polymers, Encycl. Polym. Science and Engineering; J. Wiley & Sons, New York (1985)
[19] Technical data sheets from Evonic (Röhm)
[20] Technical data sheets from BP
[21] Technical data sheets from BASF SE
[22] Technical data sheets from International Speciality Chemicals (Bisomer)
[23] Specified e.g. in DE 1568487 (1966) from Bayer
[24] Alfrey and Price, derivation of Arrhenius equation
[25] W. F. Hemminger, H. K. Cammenga: Methoden der Thermischen Analyse. Springer Verlag Berlin
[26] Höhne, G. Hemminger, W. and Flammersheim, H.-J. (1996): Differential Scanning Calorimetry – An introduction for Practioners. Springer-Verlag Berlin
[27] Fox equation
[28] H. F. Mark: Encyclopedia of Polymer Science and Engineering, J. Wiley & Sons, London (1985)
[29] specified in US 6552144 (2000) from Johnson Polymer
[30] C. E. Schildknecht und I. Skeist: Polymerisationsprozesses; Polymerisation in Suspension J. Wiley & Sons, London (1977)

3 Solution polymerization products

Ulrich Poth

3.1 Definition

This chapter discusses acrylic resins which are made by polymerization in organic solution. Most organic polymer solutions produced in this way serve directly as resin solutions for all kinds of solvent-borne coating materials.

They may also be used in the production of water-borne coatings. In that event, the solvent used for the polymerization process also serves as co-solvent for the aqueous formulation. Or, after the resin has been transformed into the aqueous phase, the solvent is distilled off, to yield a solvent-free water-borne system. The aqueous colloidal solutions or secondary dispersions of resins are stabilized by ionic or non-ionic carrier groups. Acrylic resins prepared by solution polymerization can also be converted into secondary aqueous dispersions by adding emulsifiers. In most of these cases, the process solvent is distilled off. Combinations of both processes are possible, too.

Removing solvent by distillation also affords a route to solid resins. Special solid resins are used for powder coatings.

3.2 History of acrylic resins made by solution polymerization

From the very beginning, it was believed that the use of solvents would be advantageous to the production of acrylic resins for coatings. First, the solvent is able to absorb and dissipate the heat of exothermic polymerization reactions for exploitation in reflux cooling. The second advantage is, the resultant solutions can be used directly in the preparation of coatings.

Acrylic resins dissolved in organic solvents were initially used for paint systems that dry physically. They were in competition with coatings based on cellulose nitrate. The one particular advantage they offered was superior weatherability.

Poth/Schwalm/Schwartz: Acrylic Resins

ISBN: 978-3-86630-857-2

At that time – in the 1940s – the fact that the attainable molecular weights of solution polymers were relatively low due to the regulating effect of the process solvents was considered a drawback. Only when it became clear in the early 1950s that very high molecular weights arose from crosslinking reactions **after application** of functional acrylic paints did acrylic resins for coatings became more important. In those days, the prime application of acrylic resins was for stoving enamels, predominantly in the U.S.A. Acrylic monomers were less expensive there than in Europe, where the dominant class of resin for that application was that of alkyd resins. The first commercially important acrylic resins for coatings prepared by solution polymerization were paints based on resins containing etherified methylol acrylamides. Such resins can self-crosslink at elevated temperatures. On account of the films' particularly good chemical resistance, the products were used to make coatings for household appliances (washing machines, refrigerators), so-called "white goods". In the 1960s, the market for automotive OEM (original equipment manufacturing) finishes in the U.S.A., mainly topcoats, was dominated by stoving enamels based on hydroxy-functional acrylic resins and melamine resins as crosslinkers. In the 1970s, acrylic resins made the great breakthrough in two coatings application fields in Europe. Automotive OEM clear coats were now formulated on acrylic resins and melamine resins because they offered better weatherability than their alkyd counterparts. And, automotive repair topcoats were increasingly reformulated to products based on hydroxy-functional acrylic resins which were crosslinked with polyisocyanate adducts (two-component coatings). Meanwhile in the U.S.A., the aromatic solvents in common use were (and still are) considered critical due to their photolytic effect in the atmosphere. They had to be replaced [1]. This spurred the development and application of, among others, acrylic resins in the form of non-aqueous dispersions (NADs) in aliphatic hydrocarbon solvents. These have the additional advantage of allowing higher application solids, a fact which stems from different thinning behaviour on the part of dispersions. Since the late 1970s, there had also been numerous attempts in Europe to reduce or avoid emissions of volatile organic compounds (VOCs) from coatings during application processes. First to come onto the market, in the 1980s, were so-called high-solid acrylic resins. Their purpose was to yield paints with the maximum-possible application solids. Such products dominate the market of industrial coatings to this day. Special solid acrylic resins for high-solid coatings are made by a continuous high-temperature polymerization process [2]. At the same time, intense research was being done on new ways to crosslink acrylic resins. However, even now, the well-established crosslinking principles are still in use, even though there has been progress in this regard.

Although the principles behind the production of water-borne secondary dispersions of acrylic resins have been widely known since the advent of acrylic polymers, the field of application is still limited. The main problem with such dispersions is that they have difficult application behaviour.

Since the early 1990s, powder coatings based on acrylic resins made by solution polymerization are in development. On account of their typical properties, these powder coatings are mainly used for weatherable topcoats and clear coats. However, due to the relatively high costs of the raw materials and production process, such powder coatings are not widely used. An additional constraint is that powder coatings must be applied in relatively thick layers if they are to yield smooth and glossy coating films. But other solventless coating materials are launched onto the coatings market have gained ground in several application fields, namely reactive acrylic resins, which are liquid, 100% systems based on unsaturated oligomers and acrylic monomers, and are cured by radiation, mainly UV light. They are enjoying continuous growth rates in the coatings market (see Chapter 5).

3.3 Solution polymerization process

3.3.1 Influence of the process on the properties of acrylic resins

As already mentioned, in the past, solution polymerization was criticised for the fact that only relatively small molecular weights are attainable, compared with those of bulk polymerization or suspension polymerization. In the meanwhile, this supposed disadvantage is now prized as a specific advantage. However, there are other advantages to solution polymerization. Conducting solution polymerization by feeding of monomer mixture over a period of time under suitable reaction conditions yields molecules of relatively uniform size and an optimum statistical distribution of the monomers along the polymer chain. This benefits the application properties of the resultant acrylic resins in numerous ways. These are to be found in the production and application of paints (viscosity, solids content, application behaviour, e.g. optimum sprayability, flow and levelling), as well as in the film properties (optimum crosslinking, hardness, flexibility, chemical resistance, and weatherability). Although a great deal of effort has been expended on trying to achieve such properties by other processes, solution polymerization offers favourable physical preconditions for achieving such molecules, including the possibility of maximum optimization.

3.3.2 Production procedure

When acrylic resins were first made by solution polymerization, the one-pot technique was employed. In this, all the ingredients – monomers, solvent and initiator

– are mixed together in the reactor and heated. This approach naturally yields a strongly exothermic reaction which is hard to compensate by reflux cooling. The resultant products are relatively inconsistent and reproducibility is poor.

The process was soon changed to the now-common feed technique. This commences with the addition of solvent to the reactor, which is heated to the chosen polymerization temperature. The monomer mixture and the initiator (in solution) are charged into separate feed tanks. Both are then charged in parallel and uniformly into the reactor. In the older method, the reaction temperature was the reflux temperature. Here was found that the products were much more uniform and reproducible if the reaction temperature was constantly kept below the boiling temperature of the solvent. Usually the process was terminated – some considerable time after the feeding of monomers and initiator solution – by adding a special quantity of initiator. Here was found that it is proved advantageous to keep adding the initiator solution continuously in excess to the monomer feed. This, too, promotes the formation of uniform molecules over the entire process, as documented impressively by GPC analysis. Better results were also obtained here, by feeding the initiator solution some minutes before the monomer. Obviously – following that procedure – there is a formation of an optimum radical concentration just in time. Although single, polymer molecules are formed in fractions of seconds, as mentioned in Chapter 2, this ensures that the entire polymerization process is spread out over time. As a first result, the exothermic reaction can be controlled much better than in other methods. This production method is shown schematically in Figure 3.1 and is described in various literature proceedings [3].

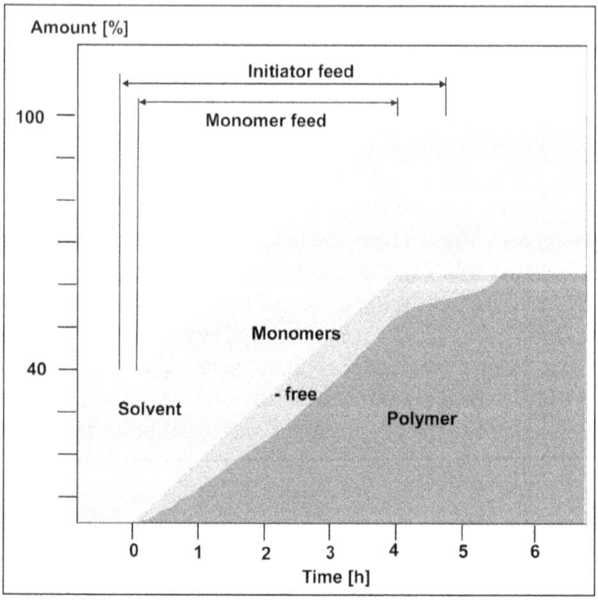

Figure 3.1: Optimum operation of solution polymerization process

The chart shows the preparation of an acrylic resin in the form of a solution containing 60% solids. The amount of free monomer was determined by GPC analysis of samples taken over a certain period of time

– polymerization within the samples is stopped by adding inhibitors. The exposure does not take the amount of initiator into consideration.

Of much greater importance than the fact that the exothermic reaction is distributed over a period of time is that a kind of stationary state is generated. That is the main reason why homogeneous polymer molecules are formed. It is even possible principially to keep the absolute concentration of free monomer constant over time. Due to the increasing amount of polymer, the relative concentration of monomers decreases, but nevertheless GPC analysis shows virtually no change in molecular mass distribution or mean value. The reason may be that the viscosity of the reaction mixture increases with increasing amount of polymer molecules. It has been reported that the rate of the termination reaction decreases with rise in viscosity, a fact which leads to high molecular mass values. This effect [4] is observed mainly in bulk polymerization, but it is possible that it applies to solution polymerization, too. The effect of the decreasing of free monomer content, which yields a lower propagation rate, is compensated by a rise in viscosity, which leads to a lower termination reaction velocity.

The third advantage of the continuous-feed process is that the conditions guarantee an optimum distribution of monomer composition along the polymer chains. If the monomer concentration is kept relatively small and virtually constant during the feed process, the probability that all monomers will react is greater than in other production processes. Slower-reacting monomers also have a better chance of being incorporated into polymer chains. Only if the polymerization parameters (Q and e) differ extensively (e.g. in the case of vinyl ethers) can such monomers accumulate in this process. In that event, special production conditions need to be chosen in order that the optimum copolymerization behaviour will be achieved. Such conditions could be a gradient feed process (different feed rates) or adding the "slower" monomers to the solvent at the beginning of the process.

3.3.3 Influence of the process conditions

The conditions for calculating the chain-propagation rate (see Formula 2.12 in Chapter 2) can be used to derive the principal parameters which influence solution polymerization, while disregarding the material factors of monomer, solvent and initiator. Those are the polymerization temperature (T), the initiator concentration (I), the quantity of solvent as represented by the corresponding expected quantity of polymer (P, acrylic resin), and the feed time (t). The quantitative effects of these influences were studied by by a statistical experimental design. [5] using an example of an acrylic resin formulation [6]. Experts should be familiar with these parameters. Target of the study was to analyse the values and variations in the parameters and primarily their interaction. Common limits for the experimental design were selected:

- Initiator concentration (I), 1.0 to 2.5% related on monomer mixture as 100%, monomer mixture
- Polymerization temperature (T): 100 to 160°C
- Solvent amount or polymer solids content (P): 50 to 80 wt.%
- Feed time (t) of monomer mixture: 0 to 5 hours.

The resultant acrylic resin solutions were tested for their solids content, viscosities, and molecular masses (GPC) and some application parameters were determined and evaluated. Large differences were found for the molecular masses and viscosities. This is of particular relevance for the formulation of low-VOC paint systems, e.g. high-solid coatings. As expected, the molecular masses and viscosities decrease substantially with increasing initiator concentration (I). At higher initiator amounts, the effect subsides. Later experiments showed that the influence of initiator amount depends heavily on the type of monomer. For example, monomer compositions containing high levels of methacrylic esters react much more extensively at higher initiator concentration than those containing high quantities of acrylic esters.

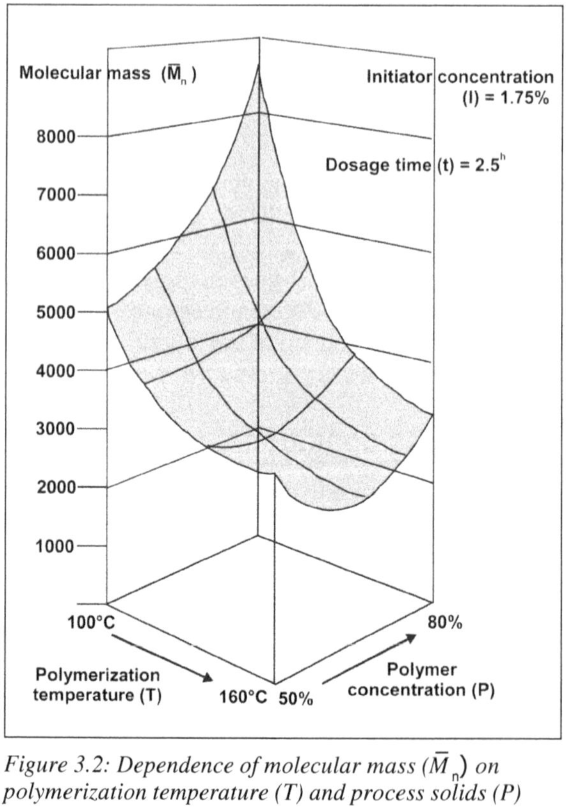

Figure 3.2: Dependence of molecular mass ($\bar{M}_n$) on polymerization temperature (T) and process solids (P)

Increasing polymerization temperature (T) yields a substantial decline in molecular mass and viscosity. At very high temperatures, the viscosity may increase again, mainly at higher solids content (P). This rise may be due side-reactions (chain transfer, transesterification, disproportionation). Nevertheless, high polymerization temperatures are selected, one reason being to lower the quantity of initiator and thus save on costs. Sometimes, the chosen temperatures exceed the boiling temperature of the solvent. However, increasing the pressure allows the process to be run at really high temperatures, without solvent boiling or the need for adjusting the reflux temperature.

The feed time (t) exerts a major influence, especially at low values, i.e. very short times. Due to the high amount of free-monomer in such cases, particularly large polymer molecules are then formed. However, after a certain feed time, the changes in molecular weights and viscosities are no longer significant.

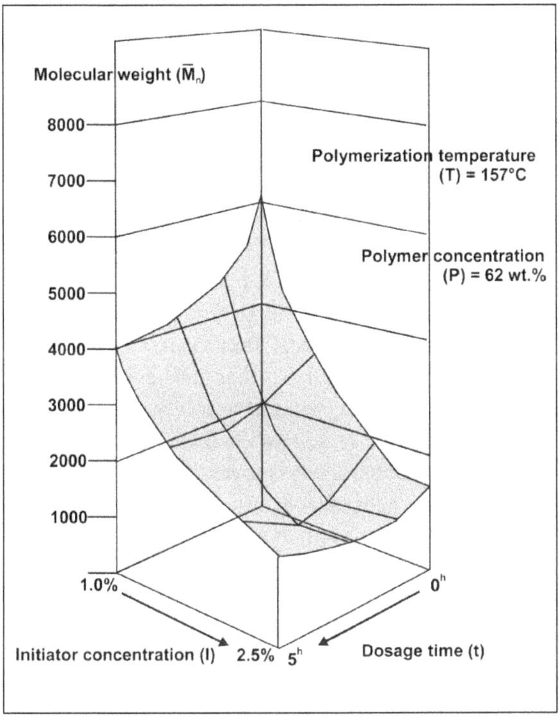

Figure 3.3: Dependence of molecular mass ($\bar{M}_n$) on initiator concentration (I) and feed time (t)

Remarkable is the effect of the amounts of solvent or the desired solid contents of acrylic resin, respectively. On account of chain-transfer reactions, an increasing solvent content effects an albeit moderate, yet continuous decrease in molecular mass and viscosity of the resultant acrylic resins. Further experiments showed that the effect does not subside. More and more process solvent leads to continuously decreasing molecular masses. For high-solid applications such solvents must, of course, be distilled off.

Figures 3.2 and 3.3 show the interpretation of the results of the experimental design, the dependence of the molecular mass ($\bar{M}_n$, number average, GPC analysis) on polymerization temperature (T) and polymer solids content (P) (Figure 3.2), and on initiator concentration (I) and feed time (t) (Figure 3.3). The other parameters are each kept constant.

3.3.4 Alternatives to solution polymerization

Although solution polymerization offers optimum preconditions for the production of acrylic resins for solvent-borne coatings, alternatives exist for special applications.

As long as acrylic resins were used for solvent-borne paints, which dry solely physically, it is preferable for the resins to have relatively high molecular masses and broad molecular mass distributions. Solution polymerization affords such resins only in a few cases. It is possible to choose unsaturated compounds to act as solvent which may be incorporated at least partly into the polymer composition, e.g. cyclopentadiene. Otherwise, suspension polymerization must be chosen. However, there are methods that utilise the monomers themselves as solvent. The polymerization is carried out at relatively high temperatures, but not to completion. The polymer is separated from residual free monomer by distillation, and the residual monomers are recycled to the process. This is a continuous process [7]. Primarily, this process was used to make thermoplastic acrylic resin (TPA), wax-like polymers, and hotmelt adhesives. Meanwhile it is possible to prepare acrylic resins with functional groups for crosslinkable coating systems. Such polymers are useful for the production of powder coatings and of solvent-less, water-borne secondary acrylic dispersions. The resultant acrylic resins are also suggested for the formulation of high-solid (low-VOC), solvent-borne paints.

3.4 Composition of acrylic resins, influences on properties

The raw materials employed in solution polymerization affect both the production process and the application properties of the resultant resins. Besides the monomers, the polymerization initiators, regulating agents, and process solvents exert a major influence.

3.4.1 Influence of monomer types

As mentioned before, acrylic derivatives polymerize at a much higher velocity than methacrylic derivatives. Therefore, all other conditions in the aforementioned process being kept constant, acrylic resins based on acrylic monomers would have significantly higher average molecular masses than their methacrylic counterparts. Although the viscosity of polymer solutions varies with the molecular mass – there is a direct mathematical correlation between viscosity and the weight-average molecular mass [8] – the viscosity of an acrylic resin solution based on methacrylic monomers would be lower than that of one based on acrylic monomers, although the latter have a much lower molecular masses. The reason lies in the much lower chain mobility of methacrylic polymers, because the methyl group on the side-chain opposite the carboxyl group gain sterical hindrance.

As the glass transition temperature also depends on chain mobility and molecular association, the viscosities of the resin solutions usually emulate the glass transition temperatures. However, there are some important exceptions. Some monomers generate relatively high glass transition temperatures, but fairly low solution viscosities: tert.-butyl esters, cyclohexyl esters, isobornoyl esters of acrylic acid or methacrylic acid, where the effect is more pronounced in the case of acrylic esters. The side-chains mentioned above contribute through their structure to the increase in glass transition temperature, but by influencing the solubility, and leading to relatively low solution viscosities. Such monomers are preferred for acrylic resins intended for high-solid paints. Remembering that the values of the glass transition temperatures follow the same order as physical hardness, it must be added that high glass transition temperatures also generate high diffusion densities. Consequently, coating films made from acrylic resins with high glass transition temperature gain better solvent resistance and chemical resistance.

Long, aliphatic side-chains on the monomers result in plasticity to the resultant polymers and generate flexible coating films. They are beneficial in respect of solubility, low solution viscosities, substrate and pigment wetting, flow and levelling, and high gloss. However, due to their lower physical hardness, combined with lower diffusion density, films made from acrylic resins containing such monomers are less resistant to mechanical impact, chemicals and solvents. These are all reasons why industrial acrylic resins very often consist of a mixture of monomers of distinctly different glass transition temperatures – to produce the optimum balance of film properties.

Functional monomers follow their glass transition temperatures with regard to the physical properties of their films, but offer scope for crosslinking. Significant differences exist in the crosslinking behaviour of functional groups. This is particularly apparent in the case of hydroxy-functional monomers. The crosslinking rate of hydroxy-functional monomers increases markedly in the following order: 2-hydroxypropyl methacrylate, 2-hydroxypropyl acrylate, 2-hydroxyethyl methacrylate, 2-hydroxyethyl acrylate, 4-hydroxybutyl acrylate (butanediol-1,4-monoacrylate). This order comes about from the differences in primary and secondary hydroxyl groups, the different mobility of polymer chains (acrylic monomers versus methacrylic monomers), and the exposed primary hydroxyl group of the last monomer.

Nitrogen-functional monomers mainly confer optimum metal adhesion of polymers containing them.

The various types of crosslinking are described separately in Chapter 3.5.

3.4.2 Initiators

Just two classes of initiators are used almost exclusively for free-radical initiation of solution polymerization: azo compounds and peroxy compounds. Both types decompose spontaneously into free-radicals at elevated temperatures. The radicals start the polymerization reactions (see Chapter 2.1.1). The individual products differ in the velocity at which they decompose as a function of temperature. The dependence is described by the half-life temperature over time, i.e. the temperature and time at which half of the initiator compound has decomposed into free-radicals (see Figure 3.4 [9]). The yield of free-radicals at a given temperature naturally influences the rate of polymerization and the value of the molecular mass of the acrylic resin, too.

For a given polymerization temperature, the initiator type is chosen on the basis of its half-life time. The differences in decomposition rates are influenced by the ligands on the initiator. Decomposition is boosted by bulky alkyl groups (e.g. tertiary alkyl), aromatic groups, carboxyl derivatives (esters, anhydrides), and carbonic acid esters.

Azo compounds have relatively high decomposition rates. The most widely known compound is azo iso-butyronitrile (AIBN, 10-hour half-life temperature 64°C [10]). It decomposes into iso-butyronitrile radicals and nitrogen. In certain conditions, some of the iso-butyronitrile can react by recombination to form tetramethyl succinic dinitrile, which is toxic. Therefore, the current preference is for azo isovalero nitrile (AVN), which offers comparable decomposition conditions (10-hour half-life temperature 66°C). The primary and secondary decomposition products of AVN are less toxic.

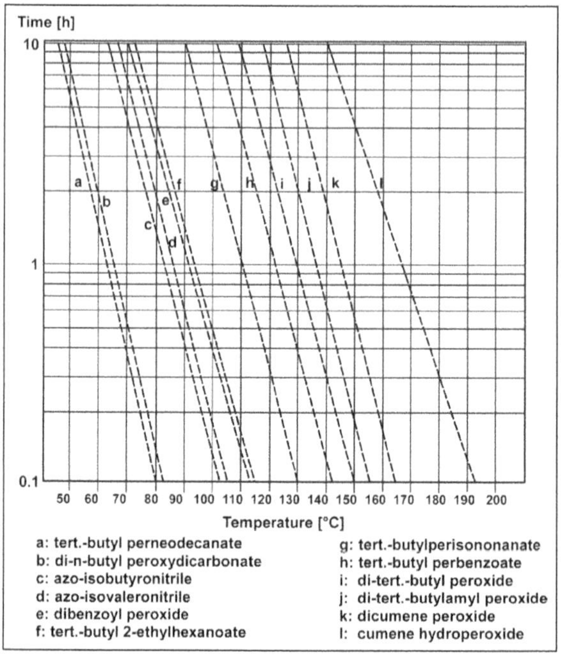

Figure 3.4: Half-life times of initiators as a function of temperature

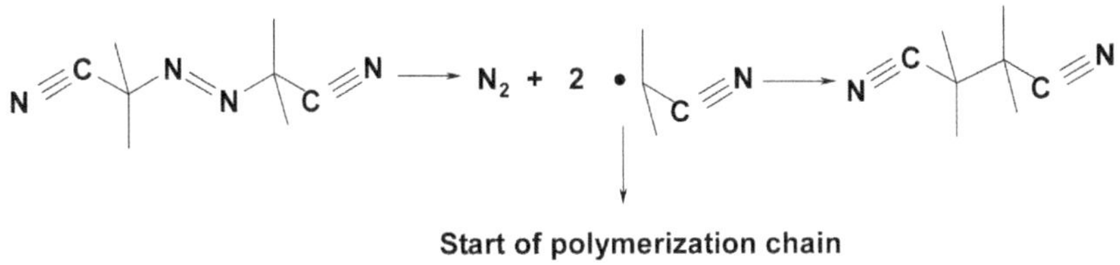

Formula 3.1: Reactions of azo iso-butyronitrile

These azo compounds are crystalline solids, and they must be dissolved or dispersed in a solvent for use. AVN is much more soluble than AIBN, which is soluble in polar solvents in only small quantities. The advantage of the solid initiators is that they have much better storage stability (maximum storage temperature is 24°C) than other initiators of comparable decomposition rates. A further advantage is that azo compounds have very little potential for oxidation, compared with peroxy compounds. Consequently, there is less risk of discolouration or dissociation reactions.

A particularly reactive species is the **peroxy dicarbonates**, with a 10-hour half-life temperature of about 40°C [10]. They therefore have to be transported and stored at relatively low temperatures.

In the past, benzoyl peroxide [10] was used as polymerization initiator. Benzoyl peroxide is a sparingly soluble crystalline solid, and has a 10-hour half-life temperature of 71°C. It is decelerated by adding some water. As it is a peroxy anhydride, it decomposes into carboxy radicals, which in turn decompose into phenyl radicals. All these radicals can initiate polymerization reactions, but they can also recombine to diphenyl or trigger chain-transfer reactions with mobile hydrogen atoms to yield free-radicals and benzene. Because of the possibility of generating such toxic compounds and its poor solubility, benzoyl peroxide has already been widely replaced.

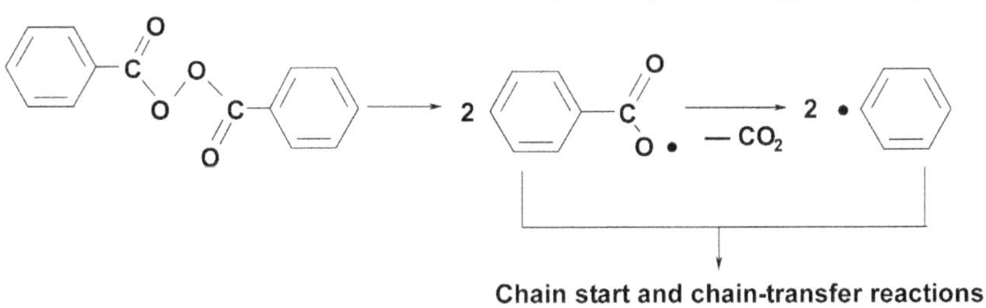

Formula 3.2: Reactions of benzoyl peroxide

The various industrially available **peresters** span a large area of reactivity, depending on the acid group [10]. Tert.-butyl perneodecanate is a liquid and has a low 10-hour half-life temperature of 46°C. It must be stored at temperatures below -10°C. Widely used until now, tert.-butyl perbenzoate, with a 10-hour half-life temperature of 104°C, can also generate phenyl free-radicals which may be able to form benzene (see Formula 3.3).

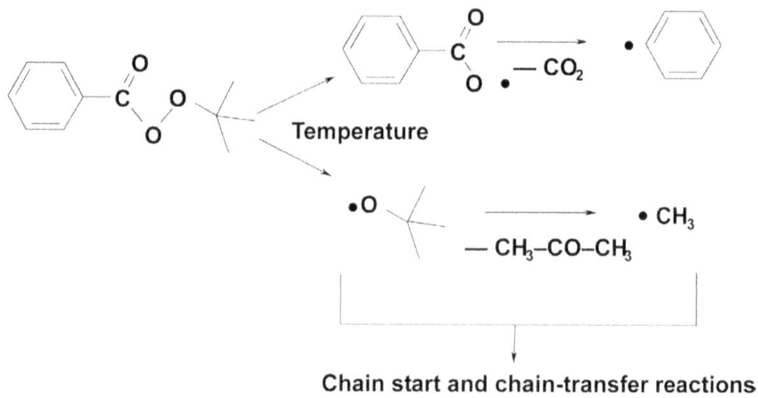

Formula 3.3: Reactions of tert.-butyl perbenzoate

Consequently, current practice is to use peresters of branched monocarboxylic acids prepared by oxo synthesis. However, tert.-butyl per-2-ethyl hexanoate, with a 10-hour half-life temperature of 74°C (maximum storage temperature: 19 to 20°C) is also under discussion, as 2-ethylhexanoic acid is suspected of being teratogenic. It is expected that tert.-butyl perisononanate (tert.-butyl per-3,5,5-trimethylhexanoate), with a 10-hour half-life temperature of 100°C and a maximum storage temperature of 25°C, will replace the other perester initiators.

Perethers decompose only at higher temperatures in sufficient yield of free-radicals. The best-known product is di-tert.-butyl peroxide, with a 10-hour half-life temperature of 125°C and a maximum storage temperature of 40°C, which is often used because, among other reasons, it is less expensive.

Di-tert.-butyl peroxide decomposes into tert.-butoxy radicals. Secondary decomposition yields acetone, and methyl radicals. Methyl radicals are highly reactive and can readily trigger chain-transfer reactions, followed by grafting reactions. To avoid such side reactions, it is recommended to use the corresponding tertiary-amyl compounds [11]. Di-tert.-amyl peroxide has a 10-hour half-life temperature of 118°C and a maximum storage temperature of 30°C. During the secondary decomposition reaction, tert.-amyloxy radical, which is generated first, forms acetone and an ethyl radical. Ethyl radicals are less reactive than methyl radicals.

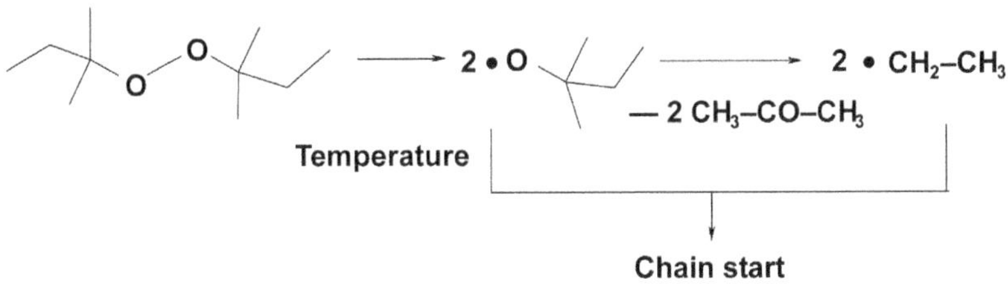

Formula 3.4: Reactions of di-tert.-amyl peroxide

Dicumyl peroxide is a crystalline solid with a 10-hour half-life temperature of 116°C and a maximum storage temperature of 30°C. Decomposition leads first to cumyl radicals, which decompose to acetone and phenyl radicals, with the same toxicity risks described above.

Hydroperoxides react relatively slowly. The most widely used type is the relatively inexpensive cumene hydroperoxide, which has a 10-hour half-life temperature of 140°C and a maximum storage temperature of 40°C. However, cumene hydroperoxide, too, generates phenyl radicals during secondary decomposition, with the attendant risk of the generation of toxic benzene by chain transfer. The decomposition of hydroperoxides can be accelerated by redox agents. This possibility is rarely used for polymerization reactions, but more for crosslinking unsaturated resins (UP resins).

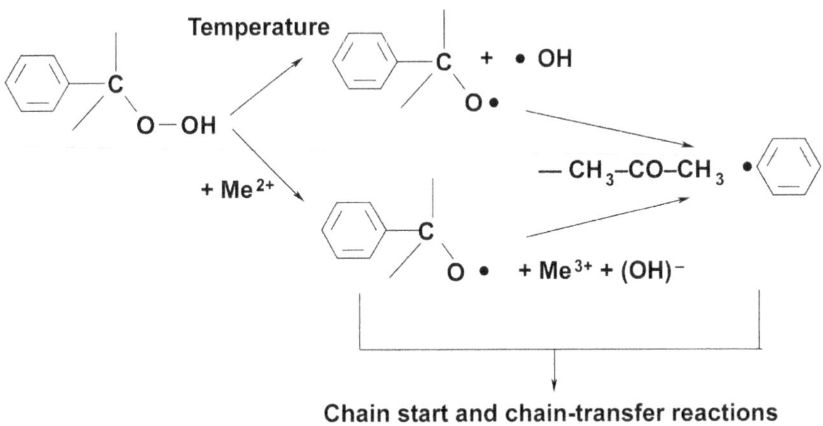

Formula 3.5: Reactions of cumyl hydroperoxide

When choosing initiators for a defined polymerization temperature, it makes sense to select products with corresponding half-life temperatures of 3 to 10 hours. Other decision criteria are the solubility of the initiator, available storage conditions and, of course, the raw material costs.

3.4.3 Regulation agents

Regulation agents are products which effect chain transfer easily, even when added in a tiny amount. The regulation agents for solution polymerization are mainly sulphur compounds: thiols and thiocarbonic esters.

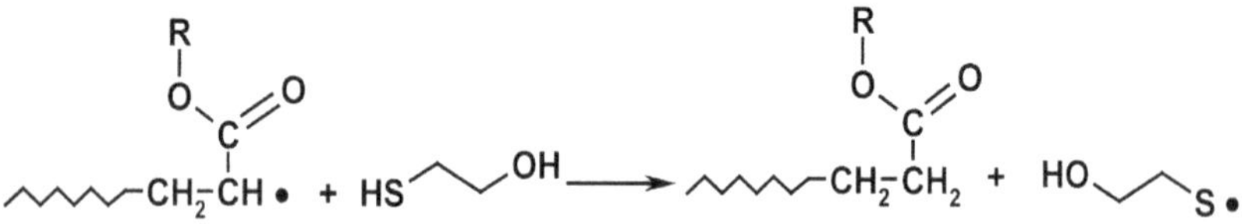

Formula 3.6: Chain-transfer reaction by a regulation agent

Thiols (mercaptans) have an unpleasant odour and readily release hydrogen sulphide. Therefore, compounds are chosen which are higher molecular or which for other reasons have lower vapour pressure at ambient temperatures, e.g. dodecyl mercaptan or mercapto ethanol. Regulation agents stop chain propagation and start a new polymer chain by a transfer reaction. Both these reactions significantly decrease the molecular masses of polymers. Radicals of regulation agents, as well as radicals of initiators, always form chain ends. Using regulation agents or initiators with additional functional groups affords a way of making polymer chains with terminal functional groups. Such polymers can be used to prepare other polymers, e.g. polyacrylic diols as building blocks for polyurethanes.

In many cases, the regulation effect of process solvents is used to avoid the unpleasant odour of regulation agents, which despite intensive efforts cannot be totally suppressed.

3.4.4 Process solvents

From a process point of view, solvents serve to absorb the heat of reaction and to lower the viscosity of the reaction mixture. The amounts of solvent control the monomer concentration during the process. From a material point of view, process solvents act as chain-transfer agents and by virtue of their solvent power. Numerous studies have been conducted to quantify solvent power (solvent parameters [12]). However, coating experts tend to be very pragmatic about their choice of solvents. Whereas a good solvent in the physical sense is one which interacts very effectively with polymer molecules to form solvates by opening up the molecule coils – a fact which results in maximum solution viscosity – experts tend to look at a solvent's thinning effect. Any solvent which acts as a thinner is actually a poor solvent in the physical sense. Of course, for a solvent to perform its task, it must be capable of forming stable colloidal solutions. These are the reasons that most coating formulations consist of blends of solvents of different solvent power (good solvents, poor solvents and thinners).

The combination of solvating and thinning effects versus the regulating effects of solvents is studied in an experimental design using solvents which are suitable for coatings. The solvents concerned reflect the different chemical classes of solvents: aromatic hydrocarbons, esters, alcohols, glycol ethers, glycol ether esters, and ketones. Aliphatic hydrocarbons and terpene hydrocarbons, the residual solvent classes, are unsuitable for acrylic resins. In the study, a model acrylic resin is made by polymerization in each of the various solvents (8), under the same polymerization conditions. The target solids content was 67 wt.%. All the resultant were are then thinned with the various solvents to 50% solids, so that the final 64 solutions consist of all possible solvent mixtures in the ratio 1 : 1. All solvents are process solvents and also thinning solvents (cross-trials). The differences in the viscosity of solutions of same solvent composition – the first solvent acting as a process solvent and the second as thinner and vice versa – allow to estimate the extent of regulation effected by transfer reaction, and the extent of thinning conferred by all solvents.

For the chosen solvents, this approach yields the following list, arranged in order of decreasing regulation power: C_3 alkyl aromatic solvent (Aromatic 100), methyl iso-butyl ketone, xylene, C_5 alkyl aromatic solvent (Aromatic 150), iso-butanol, methoxy propyl acetate, methoxy propanol, n-butyl acetate.

The order of decreasing thinning effect is: methyl iso-butyl ketone, n-butyl acetate, methoxy propyl acetate, methoxy propanol, xylene, iso-butanol, C_3 alkyl aromatic solvent (Aromatic 100), C_5 alkyl aromatic solvent (Aromatic 150).

This shows that alkyl aromatic solvents have the best regulation effect. These are widely available and relatively inexpensive. In some parts of the U.S.A., they are banned because of their photolytic effect. They contain compounds (xylene, cumene, trimethylbenzene) which are restricted in use by limitation of mass amounts. When used as solvents delivered in paints, they must be labelled as harmful (X_n) and dangerous for the environment (N). They are suitable for formulating coatings provided that they comply with VOC regulations. Another result of the study is that polar compounds are the best solvents for achieving thinning effects. However, that has long been state of the art.

In reality, methyl iso-butyl ketone emerges as the best solvent from the study as it combines high regulating efficiency and thinning behaviour. For that reason, for a long time in the U.S.A., ketones were preferred for process solvents and for paint formulations. However, they could not be used on the European market because customers rejected their pronounced ketone odour. In the meanwhile, some α-ketones are suspected of being harmful to health and are being replaced by other solvents.

3.5 Types, properties and application of acrylic resins

The biggest field of application for acrylic resins made by solution polymerization is solvent-borne paints, of course. However, products made by solution polymerization also lend themselves to the production of water-borne secondary acrylic dispersions and powder coatings.

3.5.1 Acrylic resins for solvent-borne coatings

Solvent-borne coatings based on acrylic resins are nearly exclusively used in paints that form films by crosslinking. The most important crosslinking reactions are those of methylol acrylamides (self-crosslinking) and the reaction between hydroxyl groups and amino resins or polyisocyanates that contain free or blocked isocyanate groups. Other crosslinking reactions have so far played a minor role.

When acrylic resins were first developed, the resin solutions were used to make paints that dry solely physically (thermoplastic acrylic resins, TPAs). Nowadays, such products are only used in special cases.

3.5.1.1 Thermoplastic acrylic resins

From the mid-1920s on, industrial coatings were predominantly formulated with a cellulose nitrate binder, which was combined with plasticizers or, later, with alkyd resins. These paints were distinguished by ease of application, fast physical drying at ambient temperatures and good pigment loading (high brilliance). Their disadvantages included poor weatherability, a tendency to yellow, and relatively low resistance to solvents and chemicals. The mid-1930s saw the launch of stoving enamels for automotive OEM paints, which were based on alkyd and amino resins and which confer much better resistance. In the 1940th were developed th thermoplastic acrylic resins for coatings that form films at ambient temperatures.

These can be produced by both solution and suspension polymerization. The products mainly contain methyl methacrylate, or its combination with other methacrylic esters for adjusting the glass transition temperature. The paints may also contain plasticizers; they dry by physical means only. They are notable for rapid initial drying at ambient temperatures, relatively good weatherability, and do not yellow. On account of those properties, thermoplastic acrylic resins were used for automotive repair finishes, mainly in the U.S.A. However, these failed to match the application properties of cellulose nitrate combinations, or their high pigment loading (brilliance). Thermoplastic acrylic resins are therefore mainly used for physically drying clear coats.

A further disadvantage is that coatings based on simple thermoplastic acrylic resins do not adhere very well on metal surfaces. Much better adhesion on metal results from modifying such acrylic polymers with certain amounts of methacrylic acid or, even better, with aminoalkyl acrylic esters (see Chapter 2.2.3.3), methacrylamide or methacrylonitrile [13]. Paints containing such copolymers are used for metal coatings, even for can-coatings, e.g. adhesion coatings for crown caps.

Another field of application for thermoplastic acrylic resins is impregnation of mineral substrates. They serve here as a primer sealer for concrete and plastered masonry, prior to application of the wall paints.

Thermoplastic acrylic resins also form the basis for strippable varnishes, which provide temporary protection for metal surfaces.

In addition, such acrylic resins find application in the coating of paper (e.g. overprint varnishes), plastic parts and leather.

Furthermore, it has also been recommended that acrylic resins containing methacrylic esters with long side-chains be added to alkyd paints to accelerate their initial drying velocity.

Currently, the most serious disadvantage of all the products described here is the low solids content of the resultant paints. Thermoplastic acrylic resins yield acceptable properties in terms of drying rate, adhesion, flexibility and relative resistance only if they have high molecular masses and high glass transition temperatures. The average molecular masses of the resins are between 30,000 and 100,000g/mol; the glass transition temperatures are between 60 and 95°C. Both properties lead to relatively high solution viscosities respectively low application solids. Clear coats containing such resins have application solids of 20 to 25 wt.%, and do no longer meet the requirements of VOC regulations. They must therefore be replaced by water-borne systems or by coatings based on resins which form films by crosslinking and which have much higher application solids.

Examples of commercial products: Plexisol (Röhm GmbH), Paraloid F (Rhom & Haas – Dow), Neocryl (DSM), Viacryl types (Cytec-Vianova), Synocryl types (Cray Valley)

Numerous products are made by suspension polymerization. They have the same properties and application fields as those made by solution polymerization. Acrylic resins made by suspension polymerization have broader molecular mass distributions (see Chapter 2.3.2).

Examples of commercial products: Neocryl B grades (DSM)

On account of their low glass transition temperatures, acrylic resins which contain just acrylic esters are termed "soft" resins. For example, poly-n-butyl acrylate is a liquid, although the polymer has a high molecular mass. Such resins are unsuitable as the sole binders for paints, but they do serve as additive resins, e.g. as polymer plasticizers.

Examples of commercial products: Acronal L types (BASF SE)

3.5.1.2 Acrylic resins with methylol acrylamides

As far back as the early 1940s, it was known that acrylic resins made from acrylamide or methacrylamide are water-soluble polymers that can be crosslinked by adding formaldehyde [15]. Back then, they were used for plastic parts. The 1950s saw the advent of coating resins which actually utilised this crosslinking mechanism. Initially, copolymers were made from acrylamide or methacrylamide (or blends thereof), which were then methylated with formaldehyde [16]. The resultant methylol ethers are unstable, and react with each other at ambient temperatures. Therefore, just like the amino resins, the process was modified so that the methylol ethers were be etherified with mono-alcohols [17]. These alcohols were primarily butanols, which already being used as process solvents. The resultant acrylic resin contains butoxymethyl (meth)acrylamide by way of monomer unit (see Chapter 2.2.3.4). Later, the complex monomer as such was made separately by reaction of acrylamide or methacrylamide with the semi-formal of butanol, and then copolymerized with other monomers in solution to yield acrylic resins. The first (meth)acrylamide polymers were water-soluble, whereas the newer resin class consists of combination monomers and comonomers which are relatively hydrophobic. The process solvents were mixtures of aromatic hydrocarbons (e.g. xylene) and butanols (e.g. n-butanol).

Acrylic resins containing alkoxymethyl acrylamides as building blocks can crosslink by themselves.

In the earlier literature, and down to this day, the first reaction is believed to involve a reaction between methylol ether and etherified methylol ether, with cleavage of water or alcohol. By analogy with the reaction of amino resins, here it is assumed that the crosslinking reaction starts at the imino groups, which then react with methylol ethers and etherified methylol ethers to form methylene bridges. However, the methylol ethers subsequently participate in crosslinking reactions among themselves and finally – depending on the crosslinking conditions – formaldehyde is cleaved (see Formula 3.7).

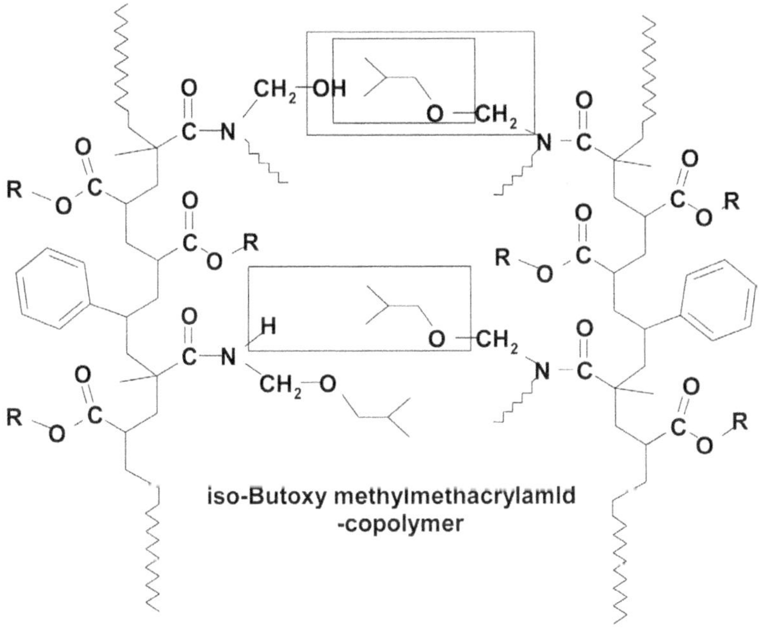

Formula 3.7: Crosslinking reactions of alkoxymethyl (meth)acrylamide copolymers

The crosslinking is carried out at temperatures of 150 to 180°C and can be accelerated with acid catalysts. However, there is no need to add acid catalysts if the acrylic resins already contain sufficient quantities of acrylic acid or methacrylic acid, generating acid values of 30 to 40mg KOH/g.

In a manner similar to the reaction of amino resins, the carboxyl groups do not really participate in the crosslinking process, despite what has been stated in the literature for a long time.

On the other hand, it is possible to copolymerize hydroxy-functional monomers into the acrylic resins. The hydroxyl groups react with alkoxymethyl (meth)acrylamides forming methylol ether bridges, with release of monoalcohols.

Alkoxymethyl (meth)acrylamides form films by self-crosslinking and are notable for their excellent adhesion on metal and outstanding chemical and solvent resistance. Their high chemical resistance – to acids, as well – is an indication that the molecular network of such coating films mainly contains methylene groups. Where resins additionally contain hydroxyl groups that form methylene ether bridges, the chemical resistance of such films is relatively moderate.

The attainable properties predestine such paint systems for the production of can coatings. These offer excellent resistance to all kinds of filling goods. For example, can coatings are tested by treatment with boiling lactic acid (sterilisation tests).

As the resistance to alkali and surfactants is excellent, coating systems based on alkoxymethyl (meth)acrylamide copolymers are ideal for domestic appliances, mainly for coatings for washing machines and refrigerators ("white goods"). A typical coating for washing machines is a white, one-layer coat based on an alkoxymethyl (meth)acrylamide copolymer. A disadvantage of such lacquers has been the relatively high solvent content. For example, a white-pigmented, one-layer coat would have had application solids of just 40 to 50 wt.%. This triggered the development and market launch of resins with lower molecular masses. Ultimately, though, this class of product was replaced. Currently, coating systems for so-called "white goods" are powder coatings based on aromatic epoxy resins and crosslinked by carboxyl groups (carboxylic acids and derivatives, carboxy-functional polyesters).

Examples of commercial products: Synthalat A 600, A 645 (Synthopol Chemie), Uracron CS types (DSM), Viacryl SC types (Cytec-Vianova), Synocryl 836, 839 (Cray Valley)

3.5.1.3 Hydroxy-functional acrylic resins crosslinked by amino resins

Surprisingly, it was not until the 1950s that combination resins for amino resins not only played a plasticizing role, but could also react by crosslinking (co-crosslinking) with free hydroxyl groups present on the functional groups of amino resins. First, the preferred resins were hydroxy-functional alkyd resins. The late 1950s saw the advent of the industrial production of hydroxy-functional acrylic monomers by making acrylic acid or methacrylic acid react with ethylene oxide and propylene oxide [18]. Numerous copolymers produced from such hydroxy-functional monomers and other acrylic or methacrylic esters, along with other comonomers, were developed in the 1960s and evolved into the most important class of resins, namely solution polymerization products.

Characterization of hydroxy-functional acrylic resins

First of all, hydroxy-functional acrylic resins are combined with amino resins as crosslinkers. The functional groups (methylol and methylol ether groups) on the amino resins react with the hydroxyl groups on the acrylic resins.

The reactions take place at elevated temperatures. Efficient crosslinking temperatures vary with the type of amino resin from 125 to 200°C, and the reactions may be catalysed by acids.

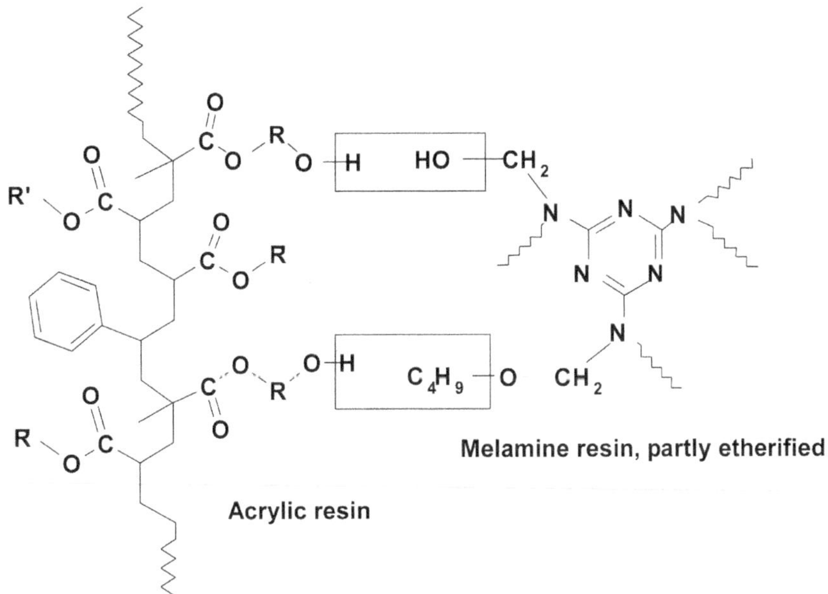

Formula 3.8: Crosslinking reactions of hydroxy-functional acrylic resins and melamine resins
Note: This is a model representation of an acrylic resin, which is repeated deliberately in the following formulas. Other compositions are conceivable.

The effective crosslinking conditions can be lowered to ambient temperatures by adding sufficient amounts strong acid (e.g. hydrochloric acid). Adding smaller quantities of organic acids (e.g. p-toluenesulphonic acid) lowers the effective crosslinking temperature for hydroxy-functional acrylic resin with a typical amino resin from 130°C to 80°C. This is important where heat-sensitive substrates (e.g. plastic parts) have to be coated. In addition, so-called internal acid catalysis exerts an influence. Hydroxy-functional acrylic resins always contain small quantities of carboxyl groups. These stem from small fractions of residual free acids from the industrial monomer-production process. However, hydroxy-functional acrylic resins are deliberately formulated with specific admixtures of free acid monomers whose purpose is to catalytically influence crosslinking with amino resins. Adequate storage stability of such paint systems is achieved by adding added mono-alcohols to act as solvents. The alcohols shift the equilibrium towards the reactants. For the same reason, salts of such acids with volatile amines are used instead of free acid catalysts.

Amino resins available for crosslinking are urea resins, melamine resins, benzoguanamine resins, and some of the carbamate resins. The most important crosslinkers for hydroxy-functional acrylic resins are the melamine resins on account of their weatherability, which dovetails well with the properties of acrylic resins. Formula 3.8 shows the possible crosslinking reactions between hydroxy-functional acrylic resins and melamine resins.

All amino resins – and hence melamine resins, too – undergo self-crosslinking. The relative extents of co-crosslinking and self-crosslinking influence the film properties significantly. Co-crosslinking confers flexibility, weatherability, and chemical resistance. By contrast, self-crosslinking boosts physical hardness, gloss, and solvent resistance. Principially, both types of reaction occur in parallel. However, it is possible to influence the extents of the different crosslinking reactions. Co-crosslinking is increased by having higher proportions of acrylic resin than melamine resin, higher proportions of hydroxyl groups, highly reactive hydroxyl groups, the use of less reactive melamine resins, less acid catalysis, lower crosslinking temperatures, and longer reaction times (effect of film-forming conditions).

Conversely, self-crosslinking reaction is increased by higher proportions of melamine resin, stronger acid catalysis, higher stoving temperatures, and short reaction times.

Typical acrylic resins for crosslinking with melamine resins have number-average molecular masses of 3000 to 8000g/mol; hydroxyl values of 60 to 130mg KOH/g, i.e. 5.3 to 11.6 hydroxyl groups per resin molecule (number-average molecular mass of 5000g/mol), and acid values of 10 to 25mg KOH/g. The preferred glass transition temperatures are between –10 and 50°C. The delivery form is 50 to 70%, dissolved in aromatic hydrocarbons (e.g. xylene, Aromatic 100), sometimes in combination with some more-polar solvents (e.g. butanols). Table 3.1 shows a typical example of hydroxy-functional acrylic resin for crosslinking with melamine resins in stoving enamels.

Table 3.1: Typical acrylic resin for crosslinking with melamine resins

Ingredients	wt.%
Xylene	37.3
Charge, heat to 140°C, hold temperature	
2-Hydroxypropyl methacrylate	15.4
Methacrylic acid	1.2
2-Ethylhexyl acrylate	40.0
Methyl methacrylate	23.5
N-Butyl methacrylate	19.9
Total monomers	**100.0**
Mix, add over 4h	
Di-tert.-butyl peroxide	1.0
Xylene	4.0
Mix, add in parallel, then hold for 0.5h	
Di-tert.-butyl peroxide	0.4
Xylene	2.0
Mix, add, hold for 2h, then cool down	
Total	**144.7**
Characteristics: NV (60‘ 130°C): 69.6% acid value (NV): 9.8mg KOH/g viscosity (50% in xylene) 500mPa s	

These types of acrylic resins are used for stoving enamels for industrial appliances (machines and equipment) and primarily for automotive OEM finishes, mainly topcoats, base coats and clear coats [20].

Examples of commercial products: Lioptal A types (Synthopol Chemie) Setalux (Nuplex), Synthacryl (Cytec Vianova), Uracron CR types (DSM), Synocryl types (Cray Valley)

Since the 1970s, various restrictions have been imposed on the emission of solvents during paint application. First, solvents were evaluated on the basis of their photolytic effects and their use was correspondingly restricted.

This spurred the development of so-called **NAD acrylic resins** (non-aqueous dispersions), initially in the U.S.A. Then followed measures and regulations aimed at reducing solvent quantities generally. There were various reasons for this: less use of products which are dependent on mineral oils, environmental protection, and health considerations. This led to the development of **high-solid resins**, later to **water-borne systems**, and finally to **powder coatings** based on acrylic resins.

NAD acrylic resins

At the end of the 1960s, analysis in the U.S.A. showed that photolytic compounds support smog formation and deplete the ozone layer. They led to regulations aimed at avoiding such compounds. The compounds in question were halogenated hydrocarbons and aromatic hydrocarbons. One part of the regulation was a ban on the use of chlorofluorohydrocarbons as cooling fluids in all kinds of cooling equipments. Coatings were affected on account of their content of aromatic hydrocarbons as solvents. At that time, there were no restrictions on the use of other solvents, e.g. esters, alcohols, aliphatic hydrocarbons, or ketones. Nowadays, but for different reasons, there are restrictions on the use of aromatic hydrocarbons as solvents in Europe, too. Some of the solvents are classified as toxic, are prohibited and are no longer used in paint systems: benzene and toluene. Other aromatic solvents are defined as harmful to health, and there are specific restrictions on the quantities which may be used in formulations: xylene, ethylbenzene, cumene, and 1,2,4-trimethylbenzene.

Until today, aliphatic hydrocarbons (white spirits) are considered to have only a minimum of healthy risks. However, aliphatic hydrocarbons are unsuitable for acrylic resin solutions, which require more-polar solvents to form stable solutions. Therefore acrylic resins were developed which form stable dispersions in aliphatic hydrocarbons. To differentiate them from the widely known aqueous dispersions, they are called non-aqueous dispersions (NAD). They are prepared in different ways, some of which are described below.

As in aqueous dispersions, stabilization of non-aqueous dispersions is warranted by surfactants and protective colloids. Various stabilizers are soluble in aliphatic hydrocarbons, as well as being compatible with hydroxy-functional acrylic resins.

There are polycaprolactones, which are prepared from ε-caprolactone, opened by water and formed in a polyaddition reaction. The carboxyl group on the one end of this polymer is made to react with glycidyl methacrylate, yielding macromonomers which are soluble in aliphatic hydrocarbons. Specific quantities of these macro-monomers are copolymerized in aliphatic hydrocarbons (white spirit) together with essentially more-polar monomers.

The macro-monomers become oriented on the surface of particles and carry and stabilize these particles in the non-polar solvent [22].

Other carriers are prepared by making hexahydrophthalic anhydride react with the glycidyl ester of neodecanoic acid. The resultant terminal carboxyl groups are made to react with glycidyl methacrylate as well. The ensuing non-polar soluble macro-monomer is also incorporated into a copolymer to form a stable dispersion in aliphatic hydrocarbons [23].

In other processes, non-soluble seed polymers are made in aliphatic hydrocarbons. Non-polar monomers are then added to the polymerization process, yielding more soluble shells and finally a stable, non-aqueous dispersion [24].

Another advantage accrues from using NADs, apart from the fact that a less-hazardous aliphatic hydrocarbon is involved. The viscosity behaviour of all polymer dispersions differs from that of organic solutions. To an extent depending on their solids content, they can have much steeper viscosity curves. Consequently, low viscosities are achieved at lower thinning rates. Paints based on non-aqueous dispersions therefore have relatively higher solids content at application viscosity than those based on comparable hydroxy-functional acrylic resins in organic solutions. Figure 3.5 shows such a comparison of viscosities.

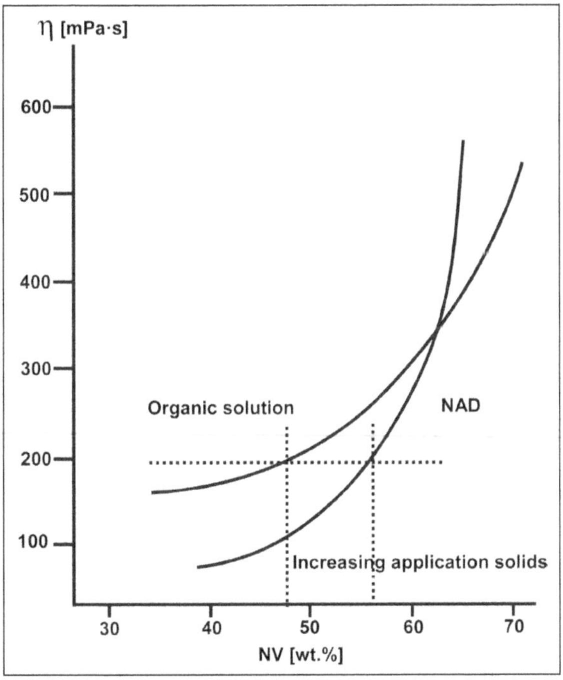

Figure 3.5: Comparison of viscosities of NAD and an organic solution of acrylic resin

Film formation of paints based on acrylic-based NADs takes place by evaporation of solvents and fusion of the dispersion particles, mostly at elevated temperatures (stoving enamels). The efficiency of film formation may be supported by adding high-boiling polar solvents. After the aliphatic hydrocarbons have evaporated, the high-boiling polar solvents form true solutions with the resins until they evaporate as well.

The preparation of an aqueous dispersion of crosslinked acrylic polymers (microgels) by emulsion polymerization constitutes a special process. The aqueous dispersion is transferred into a non-aqueous dispersion by adding solvents and extra resins; the water is separated and removed [25]. Because the particles are crosslinked, they are insoluble in organic solvents, and the added resins act as protective colloids for the polymer. As the particles are swellable in organic solvents, they confer pseudoplastic viscosity behaviour by virtue of particle-particle interaction. Where films undergo physical drying, the particles generate increasing levels of viscosity. This prevents the films from sagging on vertical surfaces. Due to this property, NAD microgels have a stabilizing effect on the orientation of flake-like pigments, e.g. aluminium pigments, and support formation of the metallic effect in solvent-borne automotive base coats, even when the base coats have relatively high application solids.

High-solid acrylic resins

Ongoing requirements aimed at reducing solvent emissions during paint application spawned the development of formulations with higher application solids. For example, high-solid contents of spray-paints are required at application viscosity of about 25" to 35", measured as flow-out viscosity (DIN 53211, 4 mm flow cup, at 23°C), corresponding to 85" to 110" (ISO 2431, 4 mm flow cup, at 23°C), or 120 to 180mPa s in a rotation viscometer at the same temperature.

Among others, the viscosity of paints depends on the viscosity of the polymer solutions (resins in the formulation). The latter varies with the polymer concentration (in this case, it should be as high as possible), the temperature, and the average molecular mass and molecular mass distribution of the polymer, the solubility of the polymer and the solvent power of the solvent. The first step to achieve these requirements was to reduce the average molecular masses of polymers. For acrylic resins, this means choosing those process conditions – the various influences are described above (see Chapter 3.3.3) – which are conducive to low molecular masses, and hence a low viscosity, thereby opening up the chance to increase the application solids. To achieve hydroxy-functional acrylic resins of lower molecular mass, a higher initiator concentration and higher polymerization temperatures are chosen. Prolonging the feed time has little effect. And, using large quantities of solvent for the polymerization contradicts the goal of achieving the highest-possible solids content.

However, lowering the molecular mass of acrylic resins does impair some coatings properties. The step from a resin molecule of low-molecular mass to crosslinked molecules of infinite molecular size by chemical film formation is much larger than it is when going from conventional resins. It is therefore necessary to increase the number of functional groups on the resin molecules at the same time as the molecular weight is reduced, in order that a comparable network structure may be formed in the films. That is why high-solid acrylic resins have consistently higher hydroxyl values than conventional resins. In addition, functional monomers are chosen which have high reactivity, e.g. 1,4-butanediol monoacrylate. However, high-solid acrylic resins containing highly reactive hydroxyl groups and having high hydroxyl values crosslink to molecular networks which are not as extensive, and such films have less flexibility. Furthermore, increasing the hydroxyl values of acrylic resins diminishes their solubility, and such solutions have higher viscosities than resins with lower hydroxyl values.

The second step in preparing suitable high-solid acrylic resins is therefore to optimize solubility. True, the viscosity of acrylic resins is lowered significantly and the flexibility is increased when monomers with long aliphatic side-chains are chosen. However, as such monomers introduce low glass transition temperatures, the films are not as hard and their resistance to chemicals and solvents is poor.

Therefore, the monomers are chosen for their ability to improve solubility – as it relates to coatings technology – while still having a high glass transition temperature that confers hardness and resistances. Examples of such monomers are tert.-butyl acrylate, cyclohexyl acrylate, 4-tert.-butylcyclohexyl acrylate, 3,3,5-trimethylcyclohexyl acrylate, iso-bornyl acrylate, and dihydrodicyclodipentadienyl acrylate [27].

These monomers are relatively expensive and some of them have a tendency to yellow at higher application temperatures and on weathering.

Table 3.2: Example of a hydroxy-functional acrylic resin for a high-solid paint

Ingredients	**wt.%**
Solvent naphtha (Aromatic 100)	33.3
N-Butanol	33.3
Charge, heat to 125°C, hold temperature	
2-Hydroxypropyl acrylate	20.0
N-Butyl acrylate	39.0
tert.-Butyl acrylate	39.5
Acrylic acid	1.5
Total monomers	100.0
Mix, add over 2h	
tert.-Butyl perbenzoate	1.5
Add in parallel, then hold for 0.5h	
tert.-Butyl perbenzoate	1.0
Add, hold for 2h, then cool down	
Total	**169.1**

Characteristics: NV (60‘ 130°C): 60.8%
acid value (NV): 16.2mg KOH/g
OH value: 86mg KOH/g
K-value (Fikentscher): 15.0
glass transition temperature (Fox): -4°C

Finally, the composition of acrylic resins not only is a compromise between the required application conditions and the required film properties of the intended application, but also must comply with VOC regulations.

Well-established hydroxy-functional acrylic resins for high-solid paints have number-average molecular masses of 1500 to 3000g/mol hydroxyl values of 120 to 160mg KOH/g, i.e. 4.3 to 5.7 hydroxyl groups per molecule (average molecular mass of 2000g/mol). For crosslinking with melamine resins, the acid values are 10 to 25mg KOH/g. Table 3.2 lists the composition of such a high-solid acrylic resin [28].

The example illustrates that there is not only the reduction of molecular mass of resin (the initiator concentration and polymerization temperature are rather moderate), but also the importance of influence on solubility of monomers. The acrylic resin in Table 3.2 can be combined with a partly etherified melamine resin to achieve an application solids of 50 wt.% [29].

Clear coats based on hydroxy-functional acrylic resins

The most important application area for hydroxy-functional acrylic resins is that of automotive OEM clear coats.

Since thermoplastic acrylic resins started replacing cellulose nitrate combination coatings, acrylic resins have been considered particular weatherable. When, in the late 1960s, clear coats for two-layer metallic coatings systems were introduced into the automotive OEM coatings market, clear coats in the U.S.A. consisted of hydroxy-functional acrylic resins combined with melamine resins. However, in Europe the clear coats contained alkyds that were modified with a combination of saturated fatty acids and melamine resins.

This combination was preferred on account of its application properties, which yielded particularly smooth and glossy surfaces with optimum hold-out properties (optical appearance and impression of optimum covering the subsurface). Nor did these clear coats undergo redissolving, which is the physical interaction during spray application that occurs between the clear coat and the only physically dried base coat layer (socalled wet-on-wet process); which may lead to disorientation of effect pigment flakes and impair the flip-flop effect (difference in brightness when the surface of effect coatings is viewed at different angles: the surface looks bright from above and dark from below). However, in the early 1970s, customers complained about poor weatherability. In Europe, therefore, alkyd-based automotive OEM clear coats were replaced by clear coats based on hydroxy-functional acrylic resins, following the example of the U.S.A. Nonetheless remembering the optimum properties of previous products the acrylic resins had to be developed and optimized for the OEM clear coat sector.

The main aspects were the improvements in application properties, prevent popping (which leads to defects, blisters and pin-holes during film formation due to uncontrolled evaporation of foam, solvents and volatile reaction products), prevent redissolving, and improve weatherability.

Thereafter, acrylic resins for high-solid clear coats had to be provided. As already described, acrylic resins for high-solid clear coats have lower molecular masses, higher quantities of functional groups per unit mass (not per molecule), and monomer building blocks which influence solubility as it relates to coating technology, without lowering hardness and resistances. In the formulation of high-solid, hydroxy-functional acrylic clear coats combined with melamine resins, the type and quantity of melamine resin contribute significantly to the value of the solids content at application viscosity. A conventional clear coat based on an acrylic resin with a molecular mass of about 5000g/mol (number-average molecular mass) and a partly etherified melamine resin (mixing ratio 65 : 35), yields an application viscosity of about 41 wt.%. The corresponding formulation for a clear coat based on a high-solid acrylic resin and a lower molecular melamine resin containing imino groups achieves about 49 wt.%. If such a high-solid acrylic resin is combined with an HMMM resin (hexamethoxymethylmelamine, the average number of melamine rings per molecule of resin is 1.3; mixing ratio: 80 : 20), the resultant application solid is about 57 wt.%. The mentioned mixing ratios are determined in practical tests, and are optimized in respect of clear coat film properties. It becomes apparent that, the type of melamine resin has an important influence on the application viscosity, and the very low molecular masses of HMMM resins depress the viscosity substantially. If higher amounts of HMMM resins are employed, application solids in excess of 60 wt.% are feasible. However, the resultant film properties are somewhat compromised. HMMM resins are fully etherified resins (in this case, with methanol). For effective crosslinking at the common OEM stoving temperatures of 125 to 150°C, these clear coats must be catalysed. However, adding acid catalyst has some disadvantages (e.g. sensitivity of aluminum pigments).

The 1980s brought acrylic resins for two-component clear coats (see Chapter 3.5.1.4), which were introduced mainly in Europe. This necessitated the development of automotive OEM clear coats optimized for application with water-borne base coats.

Since the late 1980s, water-borne automotive OEM clear coats were developed, followed by the development of powder clear coats in the 1990s.

Extensive trials are in progress to develop and introduce UV clear coats for automotive OEM coatings (see Chapter 5).

At present, solvent-borne clear coats based on hydroxy-functional acrylic resins still dominate the automotive OEM market.

The current focus of development work is on optimizing application properties (high values for "first-run OK rates") and scratch resistance, as well as compliance with VOC regulations.

Weatherability of clear coats

The complaints about damage to automotive OEM alkyd clear coats centred mainly on macroscopic cracks resembling crackle glaze. Acrylic clear coats at that time did not exhibit such effects or, if so, not until after much longer exposure. For the most part, the damage occurred following exposure of panels at test sites in Florida.

Weatherability tests in Florida reflect both the high levels of UV radiation in the sunlight there and the high humidity, which changes frequently. In some places, there is additional fall-out from industrial plant (e.g. Jacksonville, which has oil-fired power plants). Exposure of all new paint systems for two years in Florida is still a prerequisite particularly for clear coat approval. Although there are numerous other exposure sites around the world and although many accelerated weathering tests have been developed, the automotive OEM industry still insists on results of Florida tests before it will give its approval. The various accelerated artificial weathering tests yield important information that can help clarify physical interrelations. However, the chosen conditions lead to decomposition reactions which are different from those produced by natural weathering.

In the past, it was believed that the damage to clear coats containing alkyd and melamine resins originated from the poor saponification resistance of alkyds and polyesters. In addition, it was assumed that the UV radiation induced post-curing effects in the guise of additional crosslinking in the film matrix, which would lead to embrittlement.

As clear coats thereafter were exclusively based on hydroxy-functional acrylic resins, intensive studies of how to improve their weatherability were carried out. First of all the investigation focused on the analysis of physical processes behind the weathering phenomena. This entailed observing clear coat layers during and after exposure in Florida tests and in accelerated weathering tests under artificial conditions, and judging the level of gloss retention, measuring the time it took for the first cracks to occur, and the loss of film thickness. Comparative IR analyses and determinations of glass transition temperatures before and after exposure attempted to elucidate the molecular processes occurring during weathering.

Summarizing here we came to the following conclusions:

Neither the saponification reaction of polyester chains, nor post-curing play any role in weathering processes.

The critical factors leading to visual damage phenomena – cracking – in weathered clear coats are the quantities of aromatic molecular building blocks in the resins, closely followed by the influence exerted by the glass transition temperatures of the films, which are related to the glass transition temperatures of the resins (acrylic resins). High amounts of aromatic building blocks induce cracking at an early stage. Where the aromatics in two clear coat resins are present in the same proportion but have different glass transition temperatures, the clear coat containing the resin with the higher glass transition temperature will crack earlier than the other. Clear coat films containing resins of high glass transition temperatures but without aromatic building blocks crack much later, if at all.

The ivestigations showed, that clear coats based on styrene-rich acrylic resins exhibit the same damage symptoms over the same time period as those based on conventional alkyd resins, which – as is well-known – consist principally of aromatic building block, namely phthalic esters. Clear coats based on alkyd resins made without any aromatic building blocks, e.g. from hexahydrophthalic anhydride, trimethylol propane, and synthetic fatty acid, do not suffer cracking, even on long-term weathering (Florida tests, accelerated weathering tests). In fact, some results were actually even better (e.g. gloss retention) than the good results obtained for acrylic clear coats with no styrene content.

However, it is totally wrong to assume that aromatic building blocks in clear coat resins generate worse stability to UV light and other weathering impacts. It is well known that aromatic compounds absorb UV light, converting the energy into heat. This fact should benefit the resistance of the coating layer. And, in the physical sense, that is correct.

Clear coat layers containing acrylic resins free of styrene show early loss of gloss and film degradation, and lose layer thickness as weathering progresses. Clear coat layers containing styrene-rich acrylic resins remain glossy and do not degrade as early, but they exhibit cracking. Thus, styrene-rich acrylic resins are more stable to degradation in the physical sense, but exhibit unacceptable damage symptoms. Car owners will tolerate loss of gloss and loss of film thickness but not cracking of clear coat films, especially if cars are washed and polished relatively often.

For substantial optimization it is important to explain the intrinsic reasons of reasons and conditions for the behaviour in UV light exposure. This also means answering the question as to whether post-curing occurs during exposure to light. Unfortunately, there is still a widespread assumption that films of high crosslinking density are inherently brittle. There are numerous examples to prove that such coating films are definitely elastic.

The reason of the brittleness is not the high crosslinking density itself, but the fact that the crosslinked molecular networks are not extensive enough. Less extensively crosslinked molecular networks result very often – but not always – when highly reactive and highly functional compounds participate in crosslinking reactions. Positive examples of the combination of high crosslinking density and optimum flexibility are wire enamels based on polyester imides, as well as reactive acrylic resins, cured by UV light (see Chapter 5). If post-curing in clear coats would to take place during exposure to UV light, the film properties would improve. However, here was found that all weathering led to degradation reactions. The effective principle here is the combination of UV light and humidity, which yields free-radicals and ions. The greatest weak-point in the molecular network of clear coat films is the melamine resins, mainly the methylol ether groups, and not the ester groups of the partner resins. IR analysis reveals that carbonyl, carboxyl and primary amide groups are generated first. Increasing the level of such polar groups in the molecular network causes the glass transition temperature to rise. However, the level of elasticity modulus is lower in the elastic state. Among others, that is an indication that network structures have decomposed. An increase in glass transition temperature combined with a simultaneous decrease in network density can lead to brittleness. Therefore, the damage in question – cracking – can stem from a collapse in tension within the molecular network and be triggered by degradation reactions. We assume that molecular gaps form first, which then combine with each other to eventually cause macroscopic cracking.

As already mentioned, higher amounts of co-crosslinking is beneficial to weatherability as it generates molecular networks which are much more weatherable than clusters of self-crosslinked melamine resins. The more extended the molecular network is, the less pronounced is the effect of degradation on film properties.

Optimum acrylic resins for weatherable clear coats contain monomer combinations which generate moderate glass transition temperatures and contain sufficient hydroxyl groups for effective co-crosslinking. Incidentally, a limited quantity of styrene in the monomer combinations has a thoroughly beneficial effect.

Although the various acrylic resins react differently in clear coat formulations, the aforementioned study results require additional measurements aimed at protecting clear coat films against degradation during weathering. For this reason, since the late 1970s, light stabilizers have been added to automotive clear coats. Commonly, these consist of a combination of UV absorbers and free-radical quenchers. Chemically, the UV absorbers are oxalanilides, hydroxyphenylbenzotriazoles, and hydroxyphenyltriazines [30]. The available products mainly contain longer aliphatic side-chains for better compatibility with the film matrix containing acrylic resin. Such products are less sensitive to humidity. In addition, they have relatively high molecular masses for preventing migration in films. UV absorbers act like colored dyes, the difference being that the absorption maxima are between 250 and 400nm. In other words, they are colourless, but absorb UV light.

Free-radical quenchers are hindered amines (HALS, hindered amine light stabilizers), and also their derivatives, such as amides and amine oxides. These chemical compounds are variously substituted – including N-substitution – 2,2,6,6-tetramethylpiperidines [31]. The sterically influenced nitrogen atom can absorb and chemically bind oxygen and hydroxyl radicals. Like UV absorbers, hindered-amine light stabilizers are substituted with aliphatic compounds and have relatively high molecular masses to prevent evaporation and migration and to optimize solubility and compatibility with the other ingredients of clear coat formulations. Currently, all automotive clear coats contain a combination of UV absorber and free-radical quencher. These are admixed in amounts of about 0.5 to 2.0 wt.%, expressed in terms of the film matrix.

It should be mentioned that for pigmented topcoats other conditions apply in respect of achieving adequate weatherability. In pigmented topcoats, the UV light is absorbed by pigments, albeit to different extents. Not only that, but the entire spectrum of electromagnetic radiation – including UV light – is dispersed by pigment particles. The molecular composition of acrylic resins is of minor importance where the weatherability of pigmented topcoats is concerned. Satisfactory results have been yielded by topcoats containing acrylic resins with relatively high quantities of styrene. The most important factors in these cases are the properties of the pigments and more particularly the reaction at interfaces.

SCA modification of clear coats

A key development goal for reliable application is avoidance of sagging. This is the tendency of clear coats to run down vertical surfaces, and mainly occurs if they are applied in thick layers by automatic spray equipment in a short time. Sagging manifests itself as tear-like **runners** or **sliding** of sections of coating material on the surface. Corners and folds are prone to it because film thicknesses tend to be high there. It is particularly pronounced in high-solid paints, which have to pass through relatively low viscosities during film formation. Such films have a low viscosity when they start coalescing in the initial stages of stoving because the solvents have not yet evaporated; however, the temperature and, with it the viscosity, starts rising before crosslinking has a chance to occur. The problem can be solved by adding additives which introduce pseudoplastic viscosity into the coatings. These additives must not increase the application viscosity, as otherwise the solids content has to be reduced. Nor must they be too effective before the solvent evaporates during the onset of physical film formation. Furthermore, they must not impair the other coating properties.

Effective rheological additives for preventing sagging do not impair flow, levelling or gloss. Of course, clear coats have to be absolutely transparent.

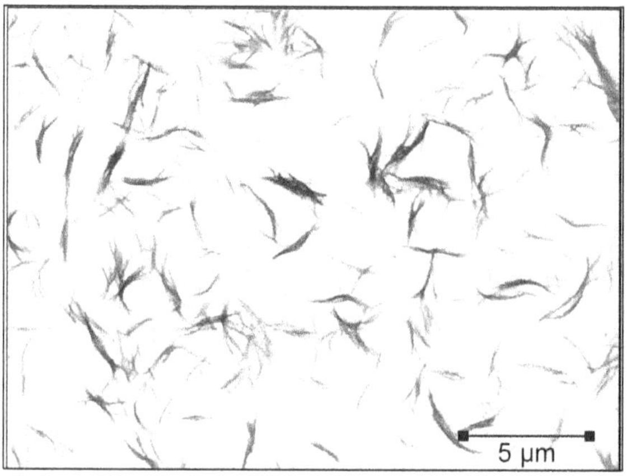

Figure 3.6: Micrograph of SCA particles

Substantial, the rheological additives employed are colloidal silica, bentonites of very fine particle size, micro-crystalline polyethylene waxes, polyamide dispersions and crystalline ureas. Crystalline ureas make the best sag control agents (SCAs) for clear coats. They are colourless and their refractive indices are close to those of acrylic resins. In addition, they can melt at elevated temperatures, whereby the urea groups react with the melamine resins, losing their particle character and ultimately becoming an integral part of the film matrix. The best-established SCA is the product of the reaction of two moles of benzylamine and one mole of hexamethylene diisocyanate (HDI, [32]). Optimal acting SCAs of this type have uniform, small needle-like particles. Such urea particles result only if the urea is precipitated at a low concentration in a resin solution under stirring at high shear rates (dissolver). Figure 3.6 [33] shows a micrograph of such particles prepared in acrylic solution.

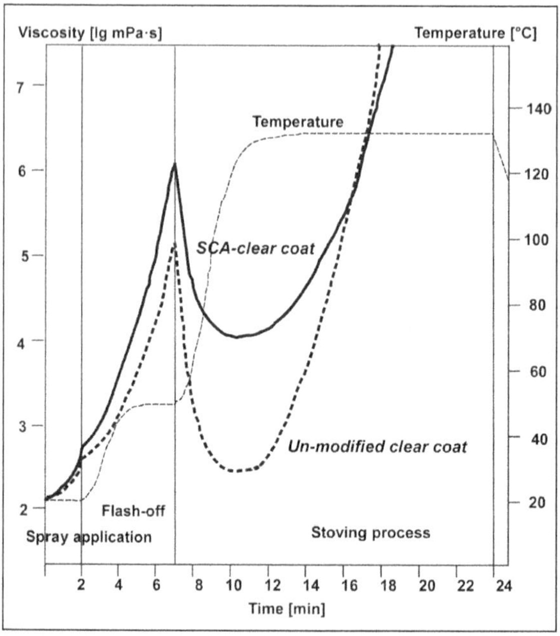

Figure 3.7: Viscosity behaviour of clear coats, with and without SCA modification, during film formation

Figure 3.7 schematically illustrates the viscosity behaviour of clear coats with and without SCA modification during application, flash-off, and stoving.

By way of summary, Table 3.3 provides an example of a clear coat [34] containing acrylic resin, melamine resin, a light stabilizer combination and SCA modification.

Table 3.3: Clear coat based on acrylic resin, melamine resin, light stabilizer combination, SCA modification

Ingredients	Product	NV	wt.%
Hydroxy-functional acrylic resin (65% in solvent naphtha/butanol 97/3)	Setalux 1761 [Nuplex]	27.87	42.88
Acrylic resin, SCA-modified (60% in solvent naphtha/ butanol)	Setalux 11761 [Nuplex]	8.73 0.62	15.59
Melamine resin (70% in n-butanol)	Setamin US 138 [Nuplex]	15.61	22.30
Polyether-modified silicone oil (2% in xylene)	Baysilon OL 17 [Borchers]	0.05	2.60
Radical quencher (n-methyl-2,2,6,6-tetramethyl piperidine derivative)	Tinuvin 292 [BASF Ciba Spec.]	0.52	0.52
UV absorber (95% in methoxypropyl acetate) (substituted hydroxy benzotriazole)	Tinuvin 384 [BASF Ciba Spec.]	0.49	0.52
Butyl diglycol	[BASF]		0.52
Aromatic hydrocarbon (Aromatic 150) (b.p. 180 to 210°C)	Solvesso 150 [Exxon]		9.36
Xylene	[Exxon]		5.71
Total		**53.89**	**100.00**

Scratch resistance of clear coats

When a car with, e.g., a dark metallic base coat and clear coat, emerges from a car-wash into sunlight, extensive, circular, capillary scratches are noticeable on the surface. This already applies to relatively new cars. Naturally, suppliers of clear coats need to eliminate such quality deficiencies. Comprehensive analysis has shown that scratches occur mostly when the brushes in the car-wash move dirt particles across the surface of the paintwork, that damage by the brushes themselves is a secondary effect and that erosion by fine sand particles as a car is being driven is a minor consideration. Scratch damage itself falls into two categories, namely abrasive and plastic. The former occurs when material is removed from relatively brittle film layers while

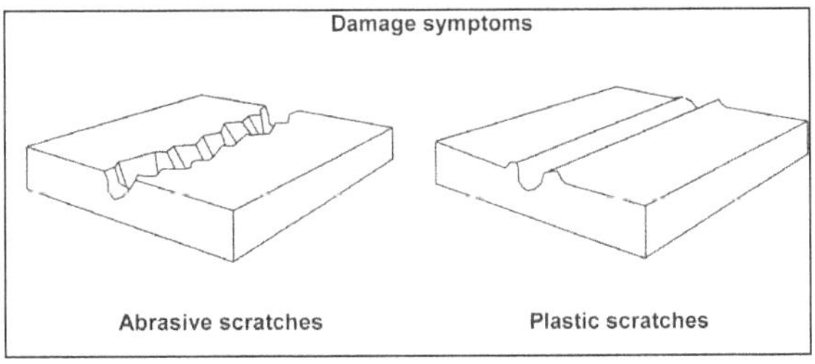

Figure 3.8: Models illustrating the structure of different scratches

the latter stems from deformation of flexible films. The structure of these two types of scratch is shown in Figure 3.8.

Plastic-deformation scratches may possibly disappear after a certain time, and if heated (self-healing effect). Abrasive scratches, by contrast, are permanent. It would seem then that the best scratch-resistant clear coats should contain enough flexible material to prevent scratches or render them temporary. They can in fact be prepared from hydroxy-functional acrylic resins which have low glass transition temperatures and which generate a low crosslinking density. However, such clear coats unfortunately fail to meet the requirements of chemical resistance, due to their low diffusion density. Ultimately, the search to improve scratch resistance must cover the additional need for optimum chemical resistance. To balance results of both requirements there are different approaches.

One is to use nanoparticles in clear coats. These very fine particles are embedded effectively into the clear coat matrix. They counteract the stress of mechanical scratching by dissipation (evasion and re-forming), with the scratching energy being consumed as heat. The nanoparticles can also increase the film hardness and significantly improve the diffusion density, leading to better chemical resistance. An example [35] of such nanoparticles is colloidal silica doped with functional siloxanes. The siloxane ensures that the particles are optimally incorporated into the clear coat matrix of acrylic resin and crosslinker.

3.5.1.4 Hydroxy-functional acrylic resins for crosslinking with isocyanates

Reactions and properties

The hydroxyl groups of acrylic resins can react with free isocyanates to form urethane groups via transfer of the hydrogen atom of the hydroxyl group (addition reaction). This reaction occurs already at ambient temperatures. Reaction rates vary largely, and are mainly influenced by the nature of the isocyanate and hydroxyl groups. Unlike the reaction of melamine resins, this crosslinking reaction is well defined. Side-reactions occur only if the isocyanates react with atmospheric humidity to form carbamic acid, which decomposes spontaneously into carbon dioxide and a primary amine. The primary amine reacts very rapidly and efficiently with residual isocyanate to form urea groups. These become integral molecular constituents of the film matrix. Consequently, the film properties are not normally impaired. Formula 3.9 illustrates the possible reactions of isocyanates.

For paint systems, hydroxy-functional acrylic resins are mixed with isocyanates, preferentially in stoichiometric amounts, to achieve optimum film properties. It is therefore possible – in contrast to the case for melamine resins where the best

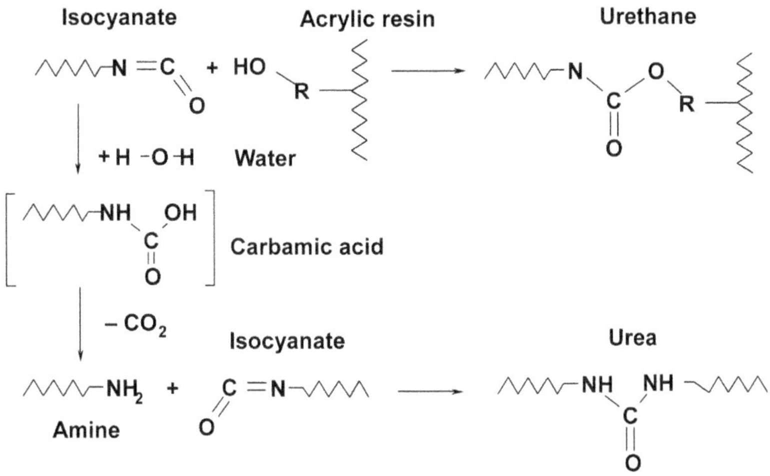

Formula 3.9: Possible reactions of isocyanates

mixing ratios need to be established by practical tests – to calculate the mixing ratios with the aid of stoichiometric formulas. The OH values of the acrylic resin and the NCO content (in wt.%) of the isocyanate crosslinker yield the mixing ratio, as shown in Formula 3.10:

$$\frac{m_{OH}}{m_{NCO}} = \frac{M_{KOH} \cdot 10\ \text{NCO-\%}}{M_{NCO} \cdot \text{OHZ}} = 13.35 \cdot \frac{\text{NCO-\%}}{\text{OHZ}}$$

m_{OH} = mass amount of acrylic resin, containing hydroxyl groups
m_{NCO} = mass amount of polyisocyanate
M_{KOH} = molecular mass of kaliumhydroxide (reference to OH-value: 56.11g/mol)
M_{NCO} = molecular mass of isocyanate group (42.02g/mol)
NCO-% = isocyanate content of crosslinker in %
OHZ = OH-value of acrylic resin in mg KOH/g

Formula 3.10: Calculation of stoichiometric mixtures of acrylic resins and isocyanate crosslinkers

Here it transpires that the values of film properties change only marginally if the stoichiometric proportions of the OH groups (n_{OH}) and isocyanate groups (n_{NCO}) of the crosslinker vary between 0.9 : 1.0 and 1.2 : 1.0.

As mixtures of hydroxy-functional acrylic resins and isocyanate crosslinker react at already ambient temperatures, the two ingredients must be delivered separately. Such systems are therefore called two-component coatings.

On the market, these take the form of a base paint component which contains the acrylic resin and – when indicated – pigments, additives and solvents, and of a hardener consisting of solutions of polyisocyanate adducts. Although most other paint systems also contain a minimum of two binder compounds – such as those containing acrylic resins and melamine resins – they are defined as one-component coatings because they are supplied ready for use.

Low-molecular polyisocyanates are toxic, have relatively high vapour pressures at ambient temperatures and the vapours are corrosive to mucous membranes, and are otherwise harmful. For this reason, the polyisocyanates are transformed into high-molecular adducts which still contain free isocyanate groups for the crosslinking reactions. These adducts are resin-like and have virtually no vapour pressure at ambient temperatures. They are licensed for sale, but direct contact with skin and mucous membranes must be prevented. Furthermore, special protective measures are necessary during application (e.g. spraying).

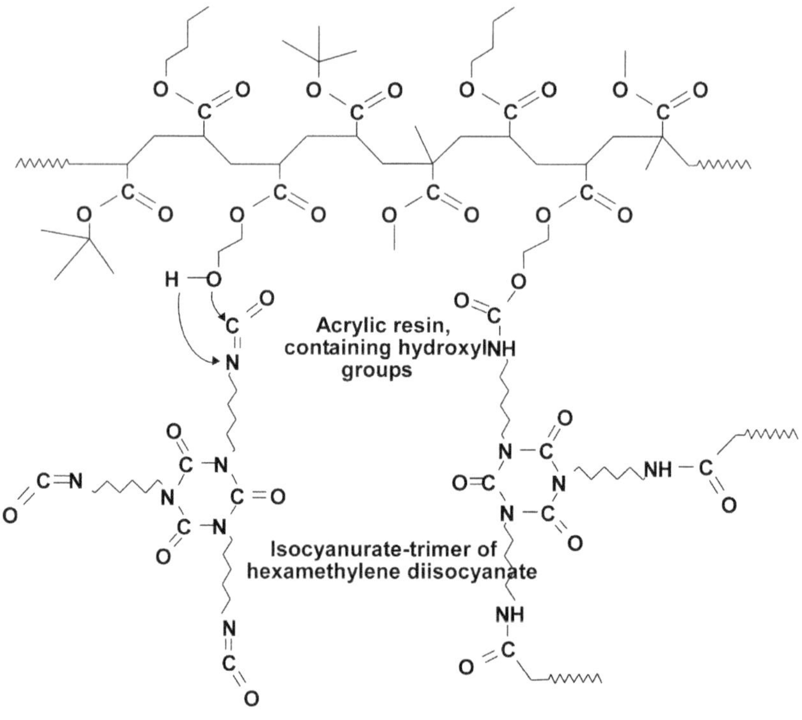

Formula 3.11: Reaction between hydroxy-functional acrylic resin and polyisocyanate

The reaction rate of isocyanates decreases significantly in the following order:

aromatic > aliphatic at primary C atoms > aliphatic at secondary C atoms > aliphatic at cycloaliphatic rings > aliphatic at tertiary C atoms.

Crosslinkers based on highly reactive aromatic isocyanates do not confer weatherability. The aromatic urethanes are degraded by UV light and the coating films yellow, embrittle and crack. Such polyisocyanates are solely used for primers and indoor coatings (e.g. wood furniture).

As hydroxy-functional acrylic resins have outstanding weatherability they are mostly combined with aliphatic and cycloaliphatic polyisocyanate adducts, to generate weatherable films. The best-known products are the biuret trimer of hexamethylene diisocyanate and the isocyanurate trimers of hexamethylene diisocyanate and isophorone diisocyanate (3-isocyanatomethyl-3,5,5-trimethylcyclohexyl isocyanate). The reaction sequence is shown in Formula 3.11.

Since aliphatic, and especially cycloaliphatic isocyanates, react relatively slowly, the reaction rate is accelerated by adding catalysts. Suitable catalysts are organic salts of heavy metals, tertiary amines, and heterocyclic compounds with nitrogen atoms. A particularly effective compound is a dialkyl tin salt, the di-n-butyl tin dilaurate (DBTL). As organo-tin compounds are classified as dangerous for the environment, substitutes have to be found, even though the fraction of such catalysts in coating systems is very low. Currently, organo-vanadium salts or combinations of organo-zinc salts with heterocyclic N compounds are recommended [36].

As soon as the base coat and hardener are mixed, the crosslinking reaction starts. The viscosity of the mixture increases slowly at first, and then exponentially. After a specific period of time, the viscosity is so high that the coating cannot be applied properly. That period of time is called the pot-life. In most cases, the pot-life is the time it takes for the viscosity to double.

The value of the pot-life depends on the reactivity of the mixed compounds. Certainly, coatings which contain catalysts have much shorter pot-lives than comparable products which do not. Hardeners containing aliphatic polyisocyanates bearing primary isocyanate groups (e.g. the isocyanurate trimer of hexamethylene diisocyanate) generate shorter pot-lives than hardeners containing cycloaliphatic isocyanate groups (e.g. the isocyanurate trimer of isophorone diisocyanate). However, the different reactivities of the hydroxyl groups have to be taken into consideration. Given comparable viscosities and the same hardener, the pot-lives of the following hydroxy-acrylic resins increase in the order given:

4-hydroxybutyl acrylate < 2-hydroxyethyl acrylate < 2-hydroxyethyl methacrylate < 2-hydroxypropyl acrylate < 2-hydroxypropyl methacrylate.

The high reactivity of 4-hydroxybutyl acrylate stems from the exposed primary hydroxyl group. Although high reactivity is an advantage, short pot-lives are disadvantageous to application properties, not only because of the time restriction, but also in terms of optimum film formation (flow and levelling).

For industrial application, two-component spray guns have been developed in which the base coat and hardener are metered by volume and delivered separately to a static mixer inside the gun. That is the best way to circumvent the pot-life time restriction. Only the spray guns need to be cleaned after use.

On account of the high reaction rate and the uniform crosslinking reaction – essentially only co-crosslinking reactions occur – paints consisting of hydroxy-functional acrylic resins and hardeners containing aliphatic or cycloaliphatic polyisocyanate adducts yield films which are much more weatherable than films containing melamine resin crosslinker. In addition, there is a much better balance between film hardness and flexibility. Urethane bridges are more chemically resistant than the methylol ether groups of crosslinked melamine resins. Urethane networks tend to have poorer solvent resistance, though, because the wide-meshed molecular structure confers a lower diffusion density on them. A further disadvantage is that polyisocyanate crosslinkers are more expensive than melamine resins.

In theory, the same hydroxy-functional acrylic resins used for crosslinking with melamine resins lend themselves to crosslinking with polyisocyanates. However, there are some special points to bear in mind. The hydroxyl values of acrylic resins for polyisocyanate crosslinking cover two large areas. There is the group of acrylic resins of relatively low hydroxyl values of 50 to 90mg KOH/g. They are mainly used for two-component industrial coatings and wood finishes. Their low hydroxyl values necessitate low quantities of polyisocyanate hardeners. These paint systems are therefore less expensive. Acrylic resins of higher hydroxyl values of 80 to 120g are used wherever higher crosslinking densities and thus better resistance are required. Two-component systems that meet even higher resistance requirements and are suitable for high-solid paints contain acrylic resins of hydroxyl values ranging from 120 to 160mg KOH/g.

The acid values of hydroxy-functional acrylic resins for crosslinking with polyisocyanate are generally low, mainly below 10mg KOH/g, as carboxyl groups do not act catalyse the crosslinking reaction. High acid values tend to be disadvantageous. The glass transition temperatures of the chosen acrylic resins depend on the application area and the type of polyisocyanate crosslinker. They tend to be lower for hydroxy-functional acrylic resins for plastic coatings, but higher for metallic surfaces. Polyisocyanate adducts based on hexamethylene diisocyanates depress the glass transition temperatures of the overall coating system. This can be compensated by using acrylic resins of higher glass transition temperatures.

Table 3.4: Composition of hydroxy-functional acrylic resins

Ingredients	wt.%
Aromatic solvent (b.p.: 158 to 170°C)	55.7
Charge, heat to 140°C, hold temperature	
Styrene	10.0
N-Butyl methacrylate	62.3
4-Hydroxybutyl acrylate	25.6
Acrylic acid	2.1
Total monomers	**100.0**
Mix, add over 4h	
tert.-Butylperethyl hexanoate	4.3
Aromatic solvent (b.p.: 158 to 170°c)	4.3
Mix, add in parallel over 4.75h, hold for 2.0h, then cool down	
Total	**146.3**
Characteristics: NV (60' 130°C): 59.1%	
acid value (expressed in terms of solids): 9.8mg KOH/g	
OH value(expressed in terms of solids): 100mg KOH/g	
viscosity (original, 23°C) 660mPa s	
molecular weight (number average, GPC): 2241g/mol	
molecular weight (weight average, GPC): 7211g/mol	

The converse tends to be the case for polyisocyanate adducts based on isophorone diisocyanate, as the cycloaliphatic ring structure confers higher glass transition temperatures.

Given the relatively low molecular masses of the polyisocyanate adducts and – for coatings purposes – the optimum solubility primarily of products based on hexamethylene diisocyanate, high application solids can be achieved with suitable hydroxy-functional acrylic resins (high-solids). If, for example, an acrylic resin with hydroxyl values of 120 to 160mg KOH/g is combined stoichiometrically with the isocyanurate trimer of hexamethylene diisocyanate, solids contents of 55 to 60 wt.% at application viscosity are achieved. As crosslinking reactions with isocyanates are so effective, quantities of low-molecular compounds bearing hydroxyl groups can be added to further increase the solids content. Such compounds are usually highly branched aliphatic or cycloaliphatic diols [37].

The aforementioned properties of coating systems based on hydroxy-functional acrylic resins and polyisocyanate adducts open up numerous applications for such systems. These included two-component wood finishes (for high-quality furniture coatings), metal paints (for machines and appliances), and coatings for plastics and leather. However, the most important application fields are automotive OEM coatings and automotive repair finishes. Topcoats and clear coats for use in automotive OEM coatings contain hydroxy-functional acrylic resins crosslinked by polyisocyanate adducts. These are chosen for their weatherability and chemical resistance, and even outperform coatings based on acrylic resins and melamine resins.

Examples of commercial products: Desmophen A (Bayer); Joncryl 587, 902, 909, 942 (BASF-Johnson Polymers); Setalux (Nuplex); Synthalat A (Synthopol Chemie); Worléecryl A (Worlée), Synocryl types (Cray Valley)

Table 3.4 lists the composition of an hydroxy-functional acrylic resin, while Table 3.5 shows the resultant two-component clear coat.

Table 3.5: Composition of a two-component clear coat

Ingredients	Product	wt.%
Base coat		
OH-Acrylic resin (60% in aromatic solvent (b.p.: 158 to 170°C)	patent example V1 in US 6,013,739	90.0
Radical quencher (n-methyl-2,2,6,6-tetramethylpiperidine derivative	Tinuvin 292 [BASF Ciba-Spec.]	0.9
UV absorber (100%) (substituted hydroxybenzotriazole)	Tinuvin 1130 [BASF Ciba-Spec.]	0.9
Dibutyl tin dilaurate (1% in xylene)		2.0
Polyether-modified silicone oil (5% in xylene)	Baysilon OL 44 [Borchers]	3.8
Butylglycol acetate	[BASF]	2.4
Total		**100.0**
Thinner		
Xylene		20.0
Aromatic solvent (b.p.: 158 to 170°C)	Solvesso 100 [Exxon]	15.0
White spirit (b.p.: 135 to 180°C)	[BP]	10.0
Butylglycol acetate		5.0
N-butyl acetate		50.0
Total		**100.0**
Hardener		
HDI-uretdione adduct	Desmodur N 3400 [Bayer]	19.5
HDI-isocyanurate adduct (90%)	Desmodur N 3390 [Bayer]	24.0
N-Butyl acetate		49.0
Butyl glycol acetate		6.0
Dibutyl tin dilaurate (1% in xylene)		1.5
Total		**100.0**
mixing ratio: base coat : thinner : hardener = 100 : 30 : 50 solid content: 42 wt.% viscosity: flow cup DIN 53211, 4 mm, 23°C 15"		

Of course – these two-component clear coats are applied by two-component spray guns, and frequently by robots. The reaction rate of the isocyanates for OEM finishes is accelerated not by means of a catalyst, but rather by elevated temperatures. Automotive OEM two-component clear coats are essentially crosslinked under the same conditions as clear coats based on melamine resins. A logarithmic plot of the reaction velocity of such two-component systems against temperature (reciprocal values) is significantly less steep than that of clear coats based on melamine resins (Arrhenius plots). After stoving, residual isocyanate groups remain. It is assumed that some sort of post-curing also occurs. In addition, physical formation processes may conceivably take place.

The particular requirements imposed on automotive repair coatings necessitated the development of special hydroxy-functional acrylic resins.

Modification of acrylic resins crosslinked with isocyanates

As already mentioned, the first automotive topcoats consisted of cellulose nitrate combinations which formed films at ambient temperatures. These were partially replaced by thermoplastic acrylic resins, mainly in the U.S.A., because they promised better weatherability. For a period, automotive repair topcoats in Europe consisted of oxidatively crosslinkable resins, mainly of alkyd resins of medium oil length, which have greater solvent resistance than cellulose nitrate combinations. Since the 1970s, automotive repair coatings have consisted of two-component coating based on hydroxy-functional acrylic resins and polyisocyanate adducts, and are distinguished by their excellent weatherability, solvent resistance, and chemical resistance. Positive experience gained previously with alkyd resins, which are notable for their good wetting properties, optimum flow and levelling, high gloss and excellent topcoat hold-out, led to a search for acrylic resins offering corresponding properties. Unfortunately, these alkyd properties are not the emerging properties of conventional hydroxy-functional acrylic resins. Consequently, specific acrylic resins were developed which better meet the required application properties.

The first such group of resins were combinations of alkyd and acrylic resins. The alkyd resins chosen were modified with saturated fatty acids, mostly synthetic fatty acids. It emerged during development work that mixtures of such alkyd resins and hydroxy-functional acrylic resins tend to be incompatible. One response was to carry out solution polymerization in alkyd solution. There have been different approaches advanced to explain why one alkyd resin is incompatible in a mixture, yet compatible after **in situ** polymerization. The first postulates a chain-transfer reaction between molecular chains of alkyd resins and acrylic resin during chain propagation.

The second proposes that molecular weights become smaller if the concentration of free monomer declines over time relative to the mass fraction of alkyd resin solution. Finally, there is the possibility that the presence of alkyd resin gives rise to special secondary molecule structures composed of growing acrylic resin chains (tacticity). All these mechanisms conspire to produce adequate compatibility. The outcome is products with a high-build effect.

As long as alkyd resins are mostly made from phthalic anhydride, the weatherability of the resultant clear coats is not that good, due to the aromatic building block. However, combinations of acrylic resins with alkyd resins based only on aliphatic and cycloaliphatic building blocks (e.g. hexahydrophthalic anhydride) are available which meet both requirements, namely optimum application properties with hold-out and high build on one hand and adequate weatherability on the other.

In the second modification method, the acrylic resins themselves are imbued with alkyd resin-like character through modification with saturated fatty acid. As an esterification process is much too expensive, the modification is carried out by an epoxy addition reaction. In this case, free carboxyl groups of acrylic resin are made to react primarily with the glycidyl ester of neodecanoic acid [40]. The addition reaction is shown in Formula 3.12.

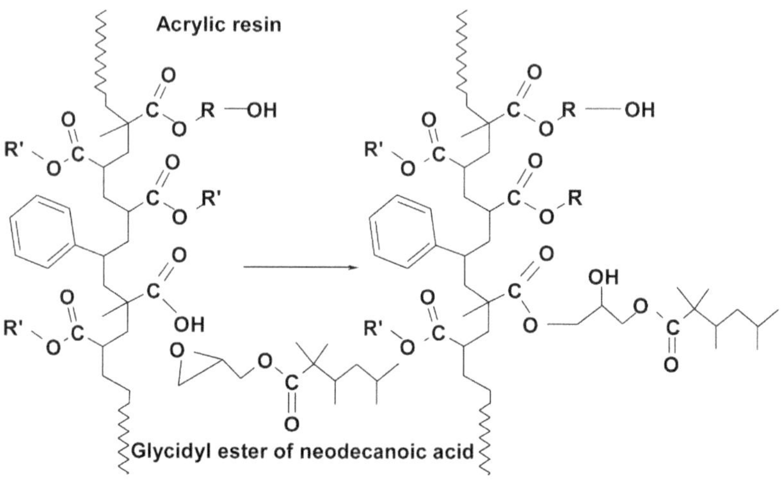

Formula 3.12: Addition reaction by the glycidyl ester of neodecanoic acid

The reaction introduces larger, branched side-chains into molecules of acrylic resins, which act as spacers and prevent the formation of dense molecule coils. The solubility of such acrylic resins is shifted in the direction of non-polar solvents. Both these effects are the reasons that the acrylic resins have alkyd-resin-like properties. The branched side-chain depresses the glass transition temperature only marginally.

Therefore, hardness and initial drying rate of acrylic resins modified in this way are still high. However, as shown in Formula 3.12, the addition reaction of glycidyl ester additionally generates secondary hydroxyl groups, which are moreover sterically hindered by the bulky side-chain. These hydroxyl groups react relatively slowly. Stepwise reaction of hydroxyl groups of different reaction rate may in fact be an advantage. It is believed that wide-meshed and extended molecular networks are formed first by reaction of highly reactive hydroxyl groups. Thereafter the network is densified by the reaction of less-reactive hydroxyl groups. The overall outcome is the formation of molecular network structures which confer optimum film properties (weatherability, chemical resistance, and the like).

The modification can be accomplished by different processes. First, the acrylic acid or methacrylic acid can react with the glycidyl ester to yield a special monomer that can then be polymerized with other monomers by standard solution polymerization [14].

It is also possible to prepare a carboxy-functional acrylic resin by solution polymerization and then to introduce the modification with glycidyl ester by a polymer-analogue addition reaction [42]. This yields polymers with relatively high viscosities.

The third process offers some specific advantages. The glycidyl ester and the process solvent are combined at the start of the process and heated to reaction temperature. As the polymerization process is started by feeding monomer mixture containing specific quantities of acrylic acid or methacrylic acid, the addition reaction takes place simultane-

Table 3.6: Composition of hydroxy-functional acrylic resin modified with glycidyl ester of neodecanoic acid

Ingredients	wt.%
Xylene	45.33
Ethyl glycol acetate*	21.62
Glycidyl ester of neodecanoic acid	29.85
Charge, heat to 147°C, hold temperature	
Methyl methacrylate	20.22
2-Hydroxyethyl methacrylate	18.82
Styrene	22.32
Acrylic acid	8.79
tert.-Dodecyl mercaptane	0.28
Total of polymer building blocks	100.00
Mix, add over 2h	
di-tert.-butyl peroxide	0.98
Add in parallel, hold for 6h at 140 to 145°C	
Total	**167.93**

Characteristics: NV (60‘ 130°C): 60.0%
acid value (expressed in terms of solids): 7 to 8mg KOH/g
OH value (expressed in terms of solids): 4.33 to 4.73%
viscosity (50% thinned with xylene, 23°C) 300 to 400"
[flow cup DIN 53211, 4mm, 20°C]
* due to hazard classification of ethyl glycol acetate, it must currently be replaced by methoxy propyl acetate, which has somewhat different solvent power and also a different regulating effect

ously with the polymerization reaction. Of course, the polymerization reaction is much faster than the addition reaction. One of the advantages is that – due to the fraction of glycidyl ester present – the concentration of free monomer during the process is relatively low, a fact which leads to smaller polymer molecules. This process is ideal for preparing high-solid acrylic resins, or for economising on initiator. Table 3.6 shows the typical composition of an acrylic resin modified in this way [43].

Examples of commercial products: Macrynal SM 510 (Cytec-Vianova), Synthalat A 150 (Synthopol Chemie)

There are also acrylic resins for automotive repair finishes which contain acrylic or methacrylic esters bearing long aliphatic side-chains. Cycloaliphatic ester monomers also bestow the aforementioned properties on automotive repair coatings.

Acrylic resins for two-component plastics coatings have low glass transition temperatures which result from soft monomers such as acrylic or methacrylic esters bearing long aliphatic side-chains. Such monomers suffer a marginal decline in resistance properties. These different tendencies need to be compensated. One approach is to modify hydroxy-functional acrylic resins bearing polyester side-chains which have been made from cyclic esters. Hydroxyl groups of acrylic resins can in particular react with ε-caprolactone (2-oxocyclo-1-heptanone). The ring-opening reaction yields an ester and an exposed hydroxyl group, which is the preferential site of further addition reactions with residual ε-caprolactone molecules. The overall outcome is acrylic resins which have a segmented structure, an acrylic backbone and more or less long polyester side-chains. The hydroxyl groups are located on the monomer chain and at the ends of the side-chains. The modification reaction is shown in Formula 3.13:

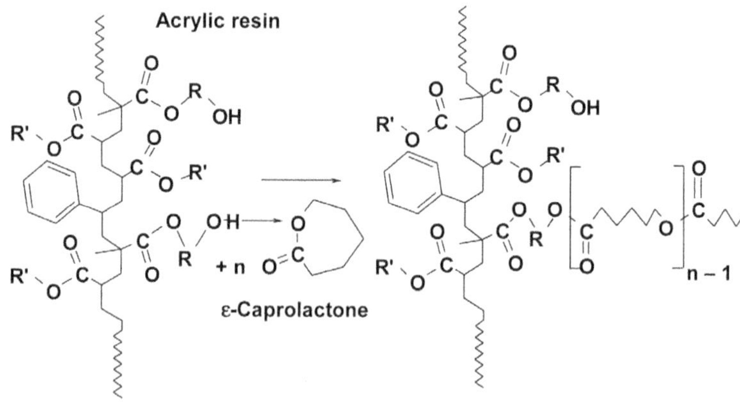

Formula 3.13: Addition reaction of ε-caprolactone on hydroxy-functional acrylic resins

There are monomers on the market which already contain such side-chains [44].

It is also possible to modify hydroxy-functional acrylic resins in a polymer-analogue reaction with ε-caprolactone.

Just as with the modification of hydroxy-functional acrylic resins with glycidyl esters, modification with ε-caprolactone is carried out by combining process solvent and ε-caprolactone in the reactor first. The ε-caprolactone is unequally distributed along the hydroxyl groups on the polymer chain. The outcome is polyester side-chains bearing several units of reacted ε-caprolactone and residual free hydroxyl groups on the resin backbone. That is why such resins have a segmented structure. Using resins of this structure in two-component coatings creates an optimum balance of hardness and flexibility, thereby meeting the property requirements for plastic coatings. In particular, these resins yield films that are sufficiently flexible at low temperatures (-40°C).

Acrylic resins modified with ε-caprolactone are also crosslinked with melamine resins for use in stoving enamels.

Crosslinking of hydroxy-functional acrylic resins with blocked polyisocyanates

Although two-component paints can be applied by means of two-component spray guns, their pot-life is a hindrance for several other application fields. To exploit the advantage of urethane crosslinking in one-component stoving paints, so-called blocked polyisocyanates were developed.

The addition reaction of hydroxyl groups and isocyanate groups at higher temperatures is an equilibrium reaction. Urethanes revert to isocyanates at higher temperatures. For urethanes formed from isocyanates and primary alcohols, the corresponding temperatures are well above 200°C. Special partner compounds exist for isocyanates that have significantly lower decomposition temperatures. Of course, such adducts must be stable at ambient temperatures. If such adducts are mixed with hydroxy-functional acrylic resins, the resin mixture can react at elevated temperatures, high the first addition compound and forming a stable urethane network. Compounds that form thermally unstable adducts with isocyanates are called blocking agents. The various blocking agents have different minimum temperatures for their reaction with resins containing hydroxyl groups. These temperatures also vary with the type of polyisocyanate. For example, as aromatic isocyanates are much more reactive than aliphatic or cycloaliphatic polyisocyanates, the reaction temperatures are about 20°C lower, given the same blocking agent.

As already mentioned, most acrylic resins confer excellent weatherability, and so they are combined with blocked aliphatic or cycloaliphatic isocyanates for optimum weatherability.

Also phenol, which is a blocking agent, is not used for blocked isocyanates for acrylic resins due to its tendency to yellow and its low weatherability. Table 3.7 lists the preferred blocking agents for aliphatic and cycloaliphatic polyisocyanates. They are arranged in order of decreasing minimum (effective) reaction temperature.

Table 3.7: Blocking agents and their effective reaction temperatures

Blocking agent	Effective reaction temperature
ε-Caprolactam	~160°C
1,2,4-triazole	~155°C
Methyl ethyl ketoxime	~150°C
3,5-Dimethyl-1,2-pyrazole	~145°C
Ethyl acetoacetate	~140°C
Diethyl malonate	~130°C

However, there are some restrictions on the use of these blocking agents. First of all, there are health concerns and environmental risks associated with methyl ethyl ketoxime. The solubility of the adducts of ε-caprolactam and 1,2,4-triazole renders them problematic. The effective reaction temperatures of adducts with ε-caprolactam, 1,2,4-triazole, and methyl ethyl ketoxime are too high for some application fields (e.g. for the conditions in which automotive OEM topcoats and clear coats are applied).

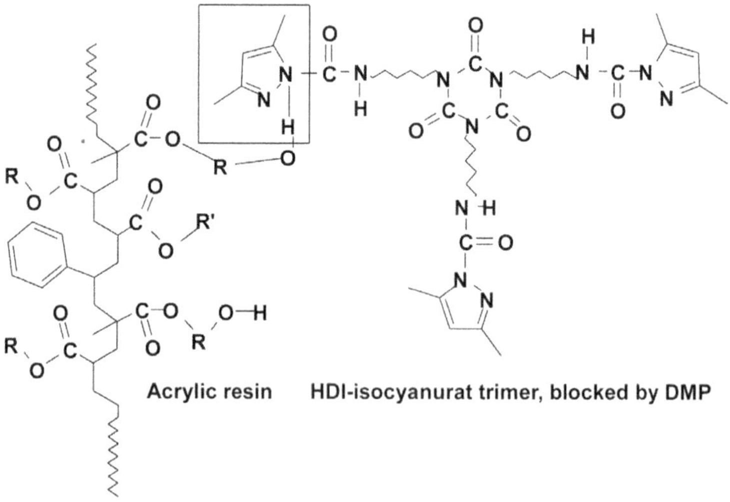

Formula 3.14: Reaction of a polyisocyanate blocked with 3,5-dimethyl pyrazole.

Ethyl acetoacetate and diethyl malonate react at least partly by transesterification, but that is not automatically detrimental to the resultant film properties. Adducts with ethyl acetoacetate tend to yellow during stoving, though. Adducts with diethyl malonate exhibit troublesome solvent properties, forming physical gels in some solvents; this was once interpreted as instability and chemical gelation. However, it is possible to circumvent this phenomenon [45]. Diethyl malonate is the most reactive blocking agent. Blocked polyisocyanates with diethyl malonate react almost exclusively by transesterification reaction, cleaving ethanol but not the blocking agent itself. 3,5-Dimethyl pyrazole seems to be the blocking agent which has the best properties, but it is relatively expensive.

In the past, it was assumed that the reaction of blocked polyisocyanates is a two-step reaction, in which the free isocyanate groups are re-formed and then react with the hydroxyl groups of acrylic resin. Kinetics studies and extensive analyses have shown that the reaction akin to a trans-urethanisation, definetly it is a exchange reaction.

Formula 3.14 shows the reaction of a polyisocyanate blocked with 3,5-dimethyl pyrazole.

The reaction rate of blocked polyisocyanates with hydroxy-functional acrylic resins can be accelerated by adding catalysts. In most cases, the catalysts employed are the same as those used in the case of free polyisocyanates. However, those catalysts are unsuitable for polyisocyanates blocked with ethyl acetoacetate or diethyl malonate. Acid catalysts tend to be used for the transesterification in that case.

Table 3.8: Typical composition of an acrylic resin for a clear coat, crosslinked with melamine resin and blocked polyisocyanate adduct

Ingredients	wt.%
Aromatic solvent (b.p.: 158 to 170°C)	93.9
Charge, heat to 150°C, hold temperature	
tert.-Butyl acrylate	46.3
N-Butyl methacrylate	15.0
2-Hydroxypropyl methacrylate	30.0
4-Hydroxybutyl acrylate	6.0
Acrylic acid	2.7
Total monomers	**100.0**
Mix, add over 4h	
tert.-Butyl perbenzoate	6.0
Aromatic solvent (b.p.: 158 to 170°C)	6.0
Mix, add in parallel over 4.5h, then hold for 2h, then	
Aromatic solvent (b.p.: 158 to 170°C)	-43.3
Distill off, then add:	
Methoxypropyl acetate	8.3
Total	**170.9**

Delivered 60% in aromatic solvent (b.p: 158 to 170°C) + methoxypropyl acetate (35 : 5)
Characteristics: NV (60‘ 130°C): 59.5%
acid value (expressed in terms of solids): 23.6mg KOH/g
OH value (expressed in terms of solids): 139mg KOH/g
viscosity (as delivered, at 23°C) 390mPa s

The film properties of coatings prepared from hydroxy-functional acrylic resins and blocked polyisocyanates resemble those of films prepared with free polyisocyanates: excellent chemical resistance, very good weatherability if aliphatic or cycloaliphatic polyisocyanates are used, and optimum flexibility. Some of the blocking agents cause yellowing, sometimes only temporarily. However, the coatings are not noted for their high physical hardness. To achieve higher hardness and, not least, to lower the material costs of crosslinking components, combinations of blocked polyisocyanates and melamine resins were chosen. These combinations generated the best overall properties. Accordingly, they serve as the basis for automotive OEM clear coats which are distinguished by excellent weatherability, optimum resistance to chemicals and solvents, and an optimum balance of hardness and flexibility [46]. Table 3.8 shows an example of an acrylic resin for a clear coat, which is described in Table 3.9.

Here the interpretation is that the overall positive properties are achieved by the step-wise reaction of the two crosslinkers. The change in reaction velocity over temperature (Arrhenius plot) is significantly steeper for melamine resins than for blocked polyisocyanates. The smaller quantity of faster-reacting melamine resin tends to favour co-crosslinking. The corresponding fractions of the blocked isocyanates are responsible for expansion of the molecular network. The overall outcome is an aggregation of positive properties and not a compromise.

Table 3.9: Clear coat, with acrylic resin crosslinked with melamine resin and blocked polyisocyanate adduct

Ingredient	Product	NV	wt.%
OH-acrylic resin (60% in aromatic solvent, b.p.: 158 to 170°C)	Example 1 patent WO 93/15849	31.6	52.7
Melamine resin, etherified with n-butanol, containing imino groups (80% in n-butanol)	"Cymel" 1158 [Cytec]	9.0	11.3
HDI-isocyanurate trimer, blocked with diethyl malonate (60% solution)	example 5 patent WO 93/15849	7.9	13.2
UV-absorber (95% in methoxypropyl acetate) hydroxy benzo triazole	"Tinuvin" 384 [BASF Ciba-Spec.]	1.6	1.7
Radical quencher N-methyl-2,2,6,6-tetramethyl piperidine derivative	"Tinuvin" 292 [BASF Ciba-Spec]	1.6	1.6
Xylene	[Exxon]		13.0
Polyether-modified silicone oil (5% in xylene)	Baysilon OL 44 [Borchers]	0.1	2.0
n-Butanol	[BASF]		4.9
Total		**51.8**	**100.00**
Thinned with xylene/butanol to viscosity: (flow cup DIN 53211, 4mm, 20°C) 23"			

There is a particular compound, which may belong to the class of blocked polyisocyanates, namely trisalkyl carbamatotriazine (TACT) [48]. Formally, it is the product of the reaction between 2,4,6-triisocyanato-1,3,5-triazine and a mixture of n-butanol and methanol. However, it is prepared by the reaction between melamine and methyl/butyl carbamate. While urethanes of polyisocyanates and primary mono-alcohols, e.g. adducts of hexamethylene diisocyanate or isophorone diisocyanate, react only at temperatures much higher than 200°C, alkyl carbamates react with hydroxyl groups of acrylic resins at 130°C, for example. The reason for this is the directive influence of the triazine ring on the urethane-like carbamate groups of TACT. The reaction is presented in Formula 3.15.

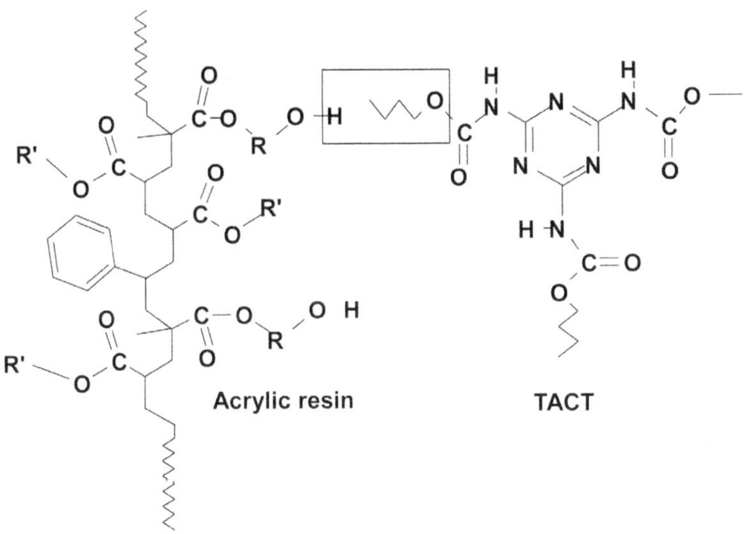

Formula 3.15: Crosslinking of hydroxy-functional acrylic resins with TACT

Crosslinking of hydroxy-functional acrylic resins with TACT in stoving enamels leads to excellent weatherability, optimum chemical resistance, and high flexibility combined with high hardness. Not only is the film hardness positively influenced by the triazine ring, but so is the gloss, as the triazine ring introduces a relatively high refractive index.

3.5.1.5 Comparison of hydroxy-functional acrylic resins with other resins

Other resin classes containing free hydroxyl groups can be crosslinked with melamine resins and polyisocyanate adducts and compete against hydroxy-functional acrylic resins.

They are saturated polyesters and alkyd resins, mainly those which have been modified with saturated fatty acids. Application properties do not result exclusively on material composition (primary structure), the size of the molecular masses and a possibly branched structure (secondary structure), but also of the colloidal properties of their solutions or melt (tertiary structure). An important indication of differences in colloidal structures is provided by the viscosity of the solutions of the aforementioned coating resins.

Hydroxy-functional acrylic resins have much lower viscosity values in solution than solutions of polyesters or alkyd resins of comparable molecular masses and comparable solubility. The reason is that molecules of acrylic resins are almost exclusively linear, but form dense coils in organic solutions. Saturated hydroxy-functional polyesters and hydroxy-functional alkyd resins contain branched molecules which are de-coiled by solvents. This leads to larger particles or the formation of more-colloidal particles, resulting in higher solution viscosity. The tendency of acrylic resins to form denser molecule coils confers some advantages. The lower solution viscosity supports the preparation of high-solid paints and helps to meet the VOC-requiremets. In addition, initial drying tends to be faster, a property which is important for repair coatings. Furthermore, the colloidal behaviour leads to acrylic resin films of higher diffusion density followed by better chemical and solvent resistance. However, there are also disadvantages. Due to the formation of dense molecule coils in colloidal solutions, the functional groups are less accessible and the diffusion ability of crosslinkers is restricted. Consequently, the crosslinking efficiency is lower than in the case of polyester or alkyd resin solutions, a fact which has a deleterious effect on the balance of hardness and flexibility. Also, the wetting properties of acrylic resin solutions for pigments and surfaces are inferior to those for solutions of polyesters or alkyd resins.

As explained above, the weatherability of coatings depends on the composition (primary structure) of the resins, which is why there are big differences in all resin classes. Furthermore, the weatherability varies with the type of crosslinker, which may be the same for the three resin types; this will now be discussed here for reasons of comparison. The view that acrylic resins are generally more weatherable than the other resins needs to be put in perspective. That said, though, when all aspects of application and properties are taken into account, the overall weatherability of acrylic resins is positive.

Although the three resin systems under comparison exhibit significant differences in paint properties, an attempt will now be made to evaluate the trends in the properties of the various resins. Table 3.10 ranks the three classes by their application behaviour and film properties.

When the rankings in each column are totalled, there are surprisingly few differences between the three classes. The upshot is that all resins have their strengths and weaknesses. Of course, the preferred application fields of the different resin types reflect the corresponding property profile.

Paints based on alkyd resins are distinguished particularly by efficient wetting properties and optimum flow and levelling. They are preferred for highly pigmented paints, for topcoats and for one-coat systems.

Saturated polyesters offer the best crosslinking behaviour and thus a good balance of hardness and flexibility. They are preferred for can-coating and coil-coating systems, for primers, primer-surfacers, and base coats.

Table 3.10: Rankings for the application and film properties of different hydroxy-functional resins

Property	OH-acrylic resins	OH-polyesters	OH-alkyd resins
Wetting	3	2	1
Adhesion	3	2	1
Physical drying	1	2	3
Flow and levelling	3	2	1
Hold out	3	2	1
Gloss	1	3	2
Crosslinking efficiency	3	1	2
Hardness	1	2	3
Flexibility	3	1	2
Chemical resistance	1	2	3
Solvent resistance	1	2	3
Weatherability	1	2	3

Finally, hydroxy-functional acrylic resins are distinguished by good initial drying and excellent resistance properties, compared with the other resins. Consequently, paints based on such acrylic resins dominate the formulation of topcoats, and – primarily – clear coats for general industrial application, automotive OEM coatings and repair coatings. From an application point of view, acrylic resins are ideal for wet-on-wet, two-coat systems comprising clear coat and effect base coat, due to their slight redissolving action.

3.5.1.6 Acrylic resins and alternative crosslinking reactions

Hydroxy-functional acrylic resins are the most important class of solution polymerization products. However, alternative crosslinking reactions are being researched. This research is aimed at:

- Improvements in film properties
- Avoidance of hazardous ingredients (primarily isocyanates)
- Avoidance of two-component effects (e.g. pot-life)
- Economies in solvents (compliance with VOC regulations)
- Broadening the use of acrylic resins

Carbamate-functional acrylic resins

One way of introducing urethane groups into coating systems is to dope acrylic resins with carbamate groups. To this end, hydroxy-functional acrylic resins are made to react with methyl carbamate to form acrylic resins bearing carbamates as functional groups. The conversion is shown in Formula 3.16.

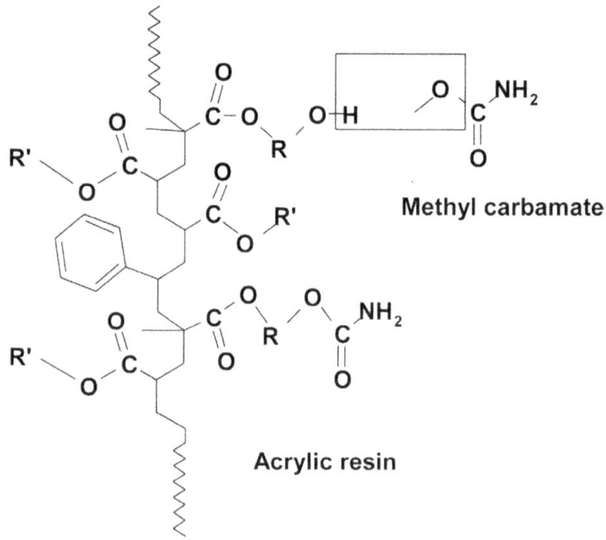

Formula 3.16: Preparation of carbamate-functional acrylic resins

Carbamate-functional acrylic resins are combined with low-molecular, fully-etherified melamine resins (HMMM resins) for crosslinking in stoving enamels. The resultant coating films are distinguished by excellent weatherability, optimum chemical resistance, and mechanical resistance. Urethane groups are formed during the crosslinking reactions. The products were initially compared with systems based on hydroxy-functional acrylic resins and blocked polyisocyanates. It was assumed that the urethane structure itself is responsible for the particular properties. However, it was found that the carbamate NH_2 group is much more reactive than the hydroxyl groups of conventional acrylic resins. Thus, the film-forming reaction involves a much greater degree of co-crosslinking, resulting in more extended molecular networks and explaining their good film-forming properties. The crosslinking reaction of carbamate-functional acrylic resins is shown in Formula 3.17.

Oxazolidine-functional acrylic resins

Oxazolidines hydrolyse to yield a secondary amine group and a primary hydroxyl group per molecule. Both functional groups formed in this way can be cross-

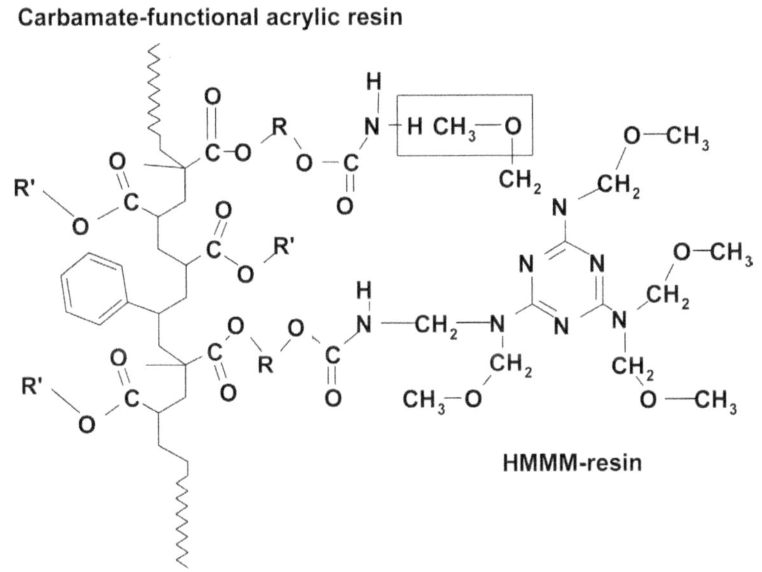

Formula 3.17: Crosslinking reaction of carbamate-functional acrylic resins

linked. This is usually done with epoxy resins or polyisocyanates. Although the groups react very quickly with these crosslinkers – primarily the NH group – it is possible to circumvent the two-component effect.

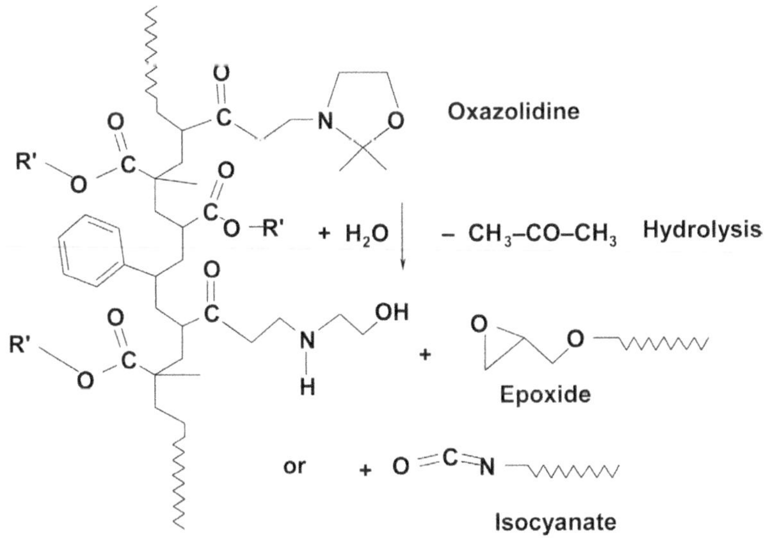

Formula 3.18: Reaction of an oxazolidine-functional acrylic resin

Suitable coatings formulations based on acrylic resins doped with oxazolidine groups and combined with epoxy resins or polyisocyanate adducts are stable under the proper storage conditions. The oxazolidines hydrolyse after application and only then form the reactive functional groups. Formula 3.18 shows the reaction of an oxazolidine-functional acrylic resin.

Besides avoiding the two-component effect, such paint systems are distinguished by fast and effective crosslinking and good adhesion of the films. Thus they can be re-coated after a short time. This is essential for primers and primer-surfacers used in automotive repair coatings.

Acetoacetate-functional acrylic resins

Acetoacetates are esters of acetoacetic acid and are classed as β-diketo compounds. On account of keto-enol tautomerism, which is particularly pronounced in this product class, the hydrogen atoms on the CH_2 group are very reactive. They can enter into an addition reaction with olefinic double bonds (Michael addition) to yield C-C bridges. Acetoacetates can also react with primary amines to form ketimines or enamines. The possible reactions of acetoacetate-functional acrylic resins are presented in Formula 3.19.

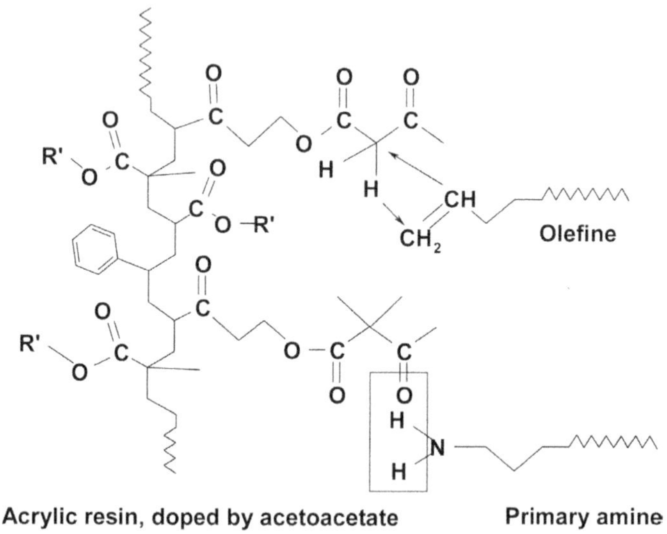

Formula 3.19: Possible reactions of acetoacetate-functional acrylic resins

The reactions take place already at ambient temperatures.

Acrylic resins containing carbonyl groups can also react with dihydrazides.

Crosslinking of carboxy-functional acrylic resins

Copolymerization of monomer mixtures containing sufficient amounts of acrylic acid or methacrylic acid leads to the formation of carboxy-functional acrylic resins. Expressed in terms of solid resin, the resins have acid values mainly between 40 and 100mg KOH/g. Carboxy-functional acrylic resins can be crosslinked in different ways.

Carboxyl groups react with epoxy groups at higher temperatures, faster with aromatic epoxies than with aliphatic epoxies. Coatings based on this chemistry are notable for their good adhesion and chemical resistance, and are used for industrial stoving enamels, primers, topcoats and one-coat systems. They are also suitable for water-borne coatings (see Chapter 3.5.2). Formula 3.20 shows the reaction between a carboxy-functional acrylic resin and an epoxy resin.

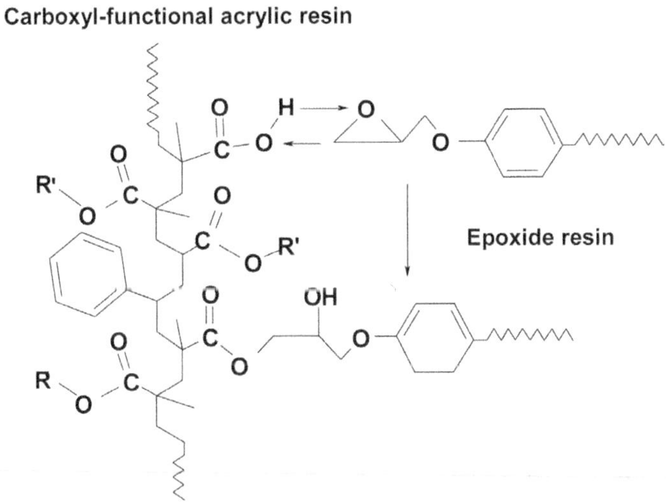

Formula 3.20: Reaction between a carboxy-functional acrylic resin and an epoxy resin

Example of a commercial product: Synocure 886 (Cray Valley)

The carboxyl groups of acrylic resins react with dihydrazides already at ambient temperatures. This is the method used for crosslinking primary acrylic dispersions (see Chapter 4).

Acrylic resins with glycolacid etheramides

Acrylic resins containing methylacrylamido-glycolacid methylester (MAGME) can react with compounds containing active hydrogen atoms to form crosslinked molecular networks [49]. The methyl ester releases methanol. Possible reaction partners are amides, amines, primary hydroxyl groups, and carboxyl groups. One reaction is shown in Formula 3.21.

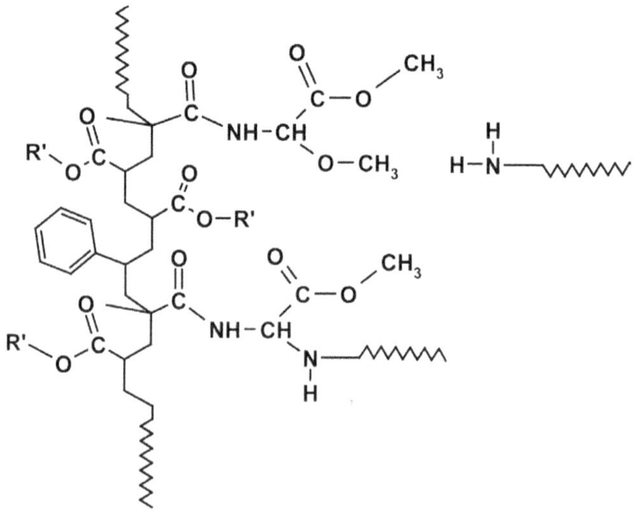

Formula 3.21: Reaction of an MAGME-functional acrylic resin

Attention is drawn to these products as they do not cleave formaldehyde during crosslinking reactions. It is claimed that the resultant coatings have at least the same properties as those of acrylic resins crosslinked with melamine resins.

Acrylic resins with siloxanes

Alkoxy silanes can react with compounds containing active hydrogen atoms. Copolymers of acrylic esters and monomers with alkoxy-siloxane groups react with compounds containing amino groups, hydroxyl groups, or carboxyl groups. However, as alkoxy siloxanes hydrolyse to yield silanol groups, they can crosslink with themselves. The reason is the amphoteric behaviour of silanol groups, which have mobile hydrogen atoms, but can also cleave off hydroxyl groups. The, in this case, resultant Si–O–Si bridge is much more stable than the Si–O–C bridges formed by the reaction of alkoxy siloxanes with hydroxyl groups. Trimethoxy silanes are preferred on account of their high reactivity. Consequently, the potential crosslinking density is quite high. It is believed that siloxane ring clusters are formed due to the characteristic properties of silicon atoms. The crosslinking reactions confer durability and high hardness on the coating films. The self-crosslinking of acrylic resins containing alkoxy siloxane groups is presented in

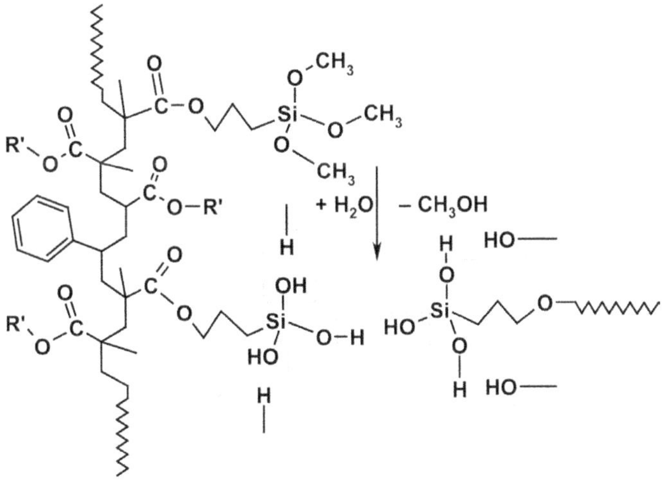

Formula 3.22.

Formula 3.22: Self-crosslinking of alkoxy siloxane functional acrylic resins

Crosslinking of hydroxy-functional acrylic resins by Michael addition

The relatively high reactivity of exposed, primary hydroxyl groups of hydroxyalkyl acrylics can be utilised for an addition reaction at olefinic double bonds. Influenced by acid catalysts (Lewis acids), such hydroxy-functional acrylic resins can crosslink easily at common stoving temperatures [50]. The preferred monomer for this is 4-hydroxybutyl acrylate. Suitable crosslinkers are compounds containing more than one double bond, which are also used for radiation-curing systems (see Chapter 5). The crosslinking process is shown in Formula 3.23.

In addition, such acrylic resins bearing exposed hydroxyl groups can crosslink by themselves under the influence of catalysts by a transesterification reaction. Acrylic resins modified with ε-caprolactone can react in this manner, too.

Thus far, alternative crosslinking processes have played only a minor role in the coatings market and are only used in special cases. The market for solvent-borne coatings is still dominated by hydroxy-functional acrylic resins crosslinked with either melamine resins or polyisocyanates (bearing free or blocked isocyanate groups).

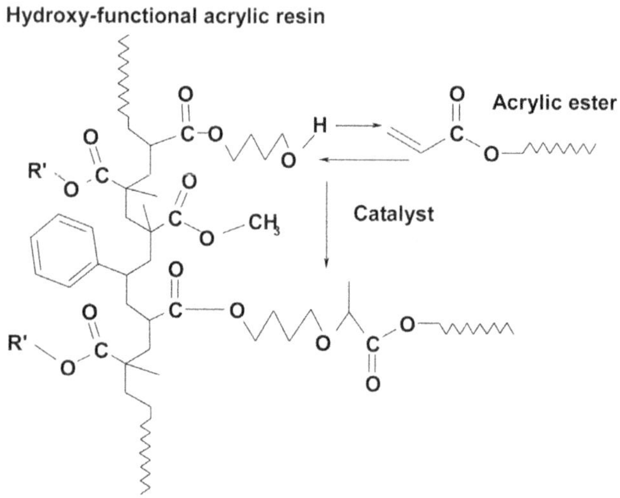

Formula 3.23: Crosslinking of hydroxy-functional acrylic resins by Michael addition

3.5.2 Acrylic resins prepared by solution polymerization for water-borne coatings

3.5.2.1 Water as solvent and dispersing agent

One of the important ways in which emissions of organic solvents during application of paints can be reduced in compliance with VOC regulations is to choose water as solvent or dispersing agent. Only a few coating resins are truly water-soluble. Nonetheless, nearly all resins can be converted into an aqueous processing form via special modifications. Water has some special properties that set it apart from other common solvents and that must be reflected in the coatings formulation.

On account of their high polarity, water molecules tend to form clusters. That is one reason for the comparatively high boiling temperature of water (despite its low molecular mass), its high density, and its high surface tension, which hampers wetting of non-polar surfaces and may promote formation of foam. In addition, water has a high evaporation enthalpy and so more energy is needed for optimum initial drying of films. The low specific electrical resistance of water requires special measures to be adopted in electrostatic spray application.

Low viscosities at application solids and the lower evaporation rates of water-borne systems necessitate the use of thickeners and other rheological additives, amongst others, to avoid sagging on vertical objects. Retention of water in coating films during initial drying can cause popping (forming of blisters and pin-holes).

These disadvantages of water as solvent or dispersing agent are counterbalanced by its advantages. Water is available in almost infinite amounts, it is totally harmless, it is not flammable and, not least, it is relatively cheap – relatively, because the water used for coatings systems is deionized or distilled.

There are numerous ways in which coatings formulations can be optimized to allow for the specific properties of water as solvent or dispersing agent. Most of all, there are many different additives available for water-borne paints.

3.5.2.2 Production of secondary dispersions of acrylic resins

Dispersions of acrylic resins produced by emulsion polymerization are defined as primary dispersions, because the monomers are directly transformed into a polymer in the aqueous phase in one step (see Chapter 4). If finished acrylic polymers are transferred into an aqueous phase, the resultant products are defined as secondary dispersions, because the entire process involves two steps. The first step is the polymerization and the second step is the formation of an aqueous dispersion. The acrylic polymer for secondary dispersions can be produced by bulk, suspension, or – most commonly – solution polymerization.

Only a few acrylic resins are inherently water-soluble, e.g. polyacrylamides. Most polymers are markedly hydrophobic. Therefore, particular measures need to be adopted in order that the acrylic polymers may be stabilized in aqueous phase.

In principle, there are two ways to prepare stable secondary dispersions:

- Use of surfactants (emulsifiers)
- Doping of polymers with ionic carrier groups

There are a number of products which utilise a combination of both.

Surfactant-stabilized acrylic dispersions can be made with two groups of emulsifier: ionic and non-ionic. The former consist of alkali or amine salts of carboxylic acid or sulphonic acids containing long aliphatic side-chains. The latter include alcohols and alkylphenols with long, ethoxylated side-chains.

Commonly used emulsifier consist of long-chained aliphatic alcohols, which are ethoxylated, and endgroups on the polyethylene part of sulfonic acid or sulfonic anions.

The way in which emulsifiers work is that the hydrophobic part of the molecule associates with the acrylic polymer molecules (akin to the process of dissolution), whereas the hydrophilic part of the emulsifier molecule aligns itself with the aqueous phase. The type and quantity of emulsifier determine the size of the polymer particles formed, which are covered with a layer of surfactant molecules in the described arrangement; the outcome is a stable dispersion. The larger the amount of emulsifier the smaller are the dispersion particles.

For reason of energy minimisation, the particles are ideal spheres and perform well-defined interfaces. Structurally, the particles of secondary dispersion are similar to particles of primary dispersions which have been made by emulsion polymerization (see Chapter 4).

There are also two ways to dope acrylic polymers with ionic carrier groups. The predominant way utilises anionic carrier groups, which are generated by specific amounts of acrylic or methacrylic acid in polymers, which are neutralized by alkali or amines, to yield carboxylate anions. These are very hydrophilic and have the ability to carry the otherwise hydrophobic polymer molecules into the aqueous phase. The polar carboxylate anion surrounds itself with water molecules – much in the manner that solvates are formed in organic polymer solutions. The effect is to stabilize the distribution of the polymer particles in aqueous phase.

Formula 3.24 shows the principle behind the neutralization of the carboxyl groups of an acrylic resin.

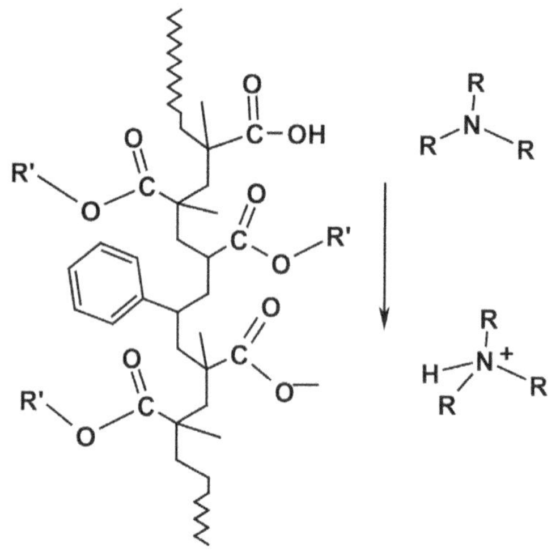

Formula 3.24: Principle behind the neutralization of carboxyl groups in an acrylic resin

There are also polymers that contain sulfon groups (R–SO_2–OH).

The second way is to dope acrylic polymers with cationic carrier groups. Such copolymers contain significant quantities of tertiary aminoalkyl acrylates or methacrylates. The amino groups are neutralized by volatile acids (e.g. acetic acid). The resultant ammonium ions are hydrophilic, too; they also surround themselves with water molecules, and can act as carrier groups in aqueous phase. Formula 3.25 shows the principle behind the neutralization of tertiary amine groups in acrylic resins.

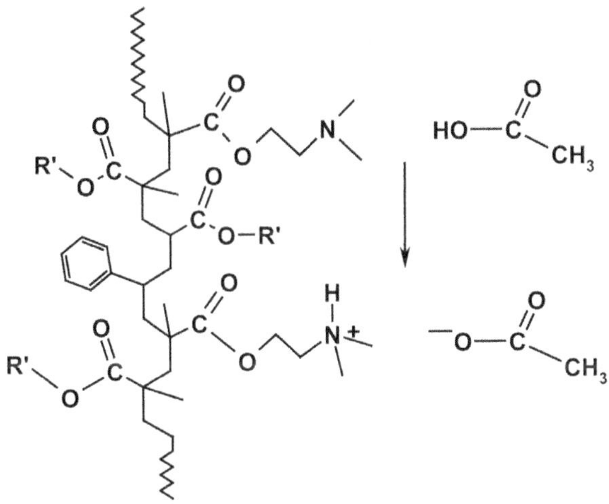

Formula 3.25: Principle behind the neutralization of tertiary amine groups in acrylic resins

As was the case with surfactants, the type and quantity of ionic carrier groups determine the structure, size, and number of particles. As the ionic groups are distributed randomly along the polymer chains, the number of carrier groups that become oriented on particle surfaces is important. Positive for those orientation are polymer molecules which are mobile. Mobility of polymer molecules is achieved by lower glass transition temperatures or lower average molecular masses. Orientation of the carrier groups is supported by co-solvents. The latter are solvents which have adequate solubility in water, but also act as solvent for the corresponding polymers. They diffuse to some extent into the particles of polymer coils, lowering the local viscosity, de coiling the polymers and finally supporting the surface orientation of ionic carrier groups. Unlike particles of primary and secondary dispersions stabilized by surfactants, these particles have no well-defined particle interfaces. If they are small enough, they tend to resemble particles of colloidal organic solutions more closely. That confers a number of benefits on film-forming properties.

The first step in the production of secondary dispersions is to prepare the acrylic polymer by solution polymerization as described above. The choice of process solvent depends on the intended use of the dispersion. On one hand, it might be intended to remove the process solvent by distillation. In that event, the solvents are low-molecular ketones (e.g. methyl ethyl ketone) or low-molecular alcohols (e.g. ethanol, iso-propyl alcohol). The preferred process solvents acting simultaneously as co-solvents – when the polymer solution is transferred into aqueous phase – are glycol ethers (e.g. butyl glycol, butyl diglycol), glycol ether acetates (e.g. methoxypropyl acetate).

On the other hand, if it is intended to transform the solution polymerization product into a secondary dispersion stabilized by surfactants, the water and emulsifier are added to a vessel and the polymer solution is admixed regarding defined shear conditions. Alternatively, the emulsifier may be added to the acrylic resin solution, or to distribute it in both phases. The outcome is a dispersion composed of more or less fine particles. Optional, the process solvent may then be removed by distillation.

If it is intended to transform the acrylic polymer into a secondary dispersion with ionic carrier groups, these first are partially neutralized with amine or volatile acid to generate the ionic groups. Then the solution is admixed to water. The secondary dispersion is formed spontaneously, and no specific shearing conditions are needed. The number and size of the particles depends on the quantity of potential carrier groups and the degree of neutralization, i.e. on the available number of carboxylate or ammonium groups.

3.5.2.3 Properties and use of aqueous, secondary acrylic dispersions

Aqueous, secondary acrylic dispersions for crosslinking with amino resins

In aqueous paints containing melamine resins as crosslinkers, the preferred reaction partners are secondary dispersions of acrylic resins containing hydroxyl groups as well as carboxyl groups. The dispersions are anionically stabilized. In contrast to acrylic resins of primary dispersions, acrylic resins prepared by solution polymerization and used for secondary dispersions have significantly lower molecular masses. This is an advantage for coating systems which form films by crosslinking, including water-borne paints. The common number-average molecular masses are 2500 to 5000g/mol. The hydroxyl values lie mainly between 55 and 125mg KOH/g and the acid values usually between 30 and 60mg KOH/g. In most cases, the products are delivered in aqueous dispersions which have already been neutralized. The neutralization agents are amines: N,N-dimethylethanolamine, triethylamine, di-iso-propanolamine, 2-amino-2-methyl-1-propanol (AMP) (in order of decreasing importance). In special cases, triethanolamine, morpholine, and ammonia are used. The degree of neutralization is between 0.7 and 1.0 (moles amino groups: moles carboxyl groups). In most of these cases, the pH is 7.5 to 8.5.

Dispersions which contain surfactants for stabilization in aqueous phase may have much lower pH-values.

The solids content of the ready-to-use secondary dispersions is between 35 and 50 wt.%. The viscosities are usually relatively low. Besides the influence of the solids content, the viscosity is determined by the degree of neutralization or the resultant pH-value. If the dispersion has the same solids content but higher pH-values, the viscosity will be significantly higher. According to the manufacturers, the particle size of such secondary dispersions is about 200nm.

Acrylic resins produced by solution polymerization and intended for making secondary dispersions may also be shipped in the form of organic solutions. The solvent is the process solvent, which also makes a suitable co-solvent. Such solutions must be at least partially neutralized in order that they may form dispersions. Thinning with water gives rise to anomalous viscosity behaviour. As the water is added, the viscosity initially rises to a maximum, whereupon it rapidly decreases. The effect is called the "water hill" of viscosity. The reason is believed to lie in the variations in particle sizes arising from the different amounts of water and co-solvent in the particles and in the equilibrium established with the external phase. Initially, the particles absorb comparatively large amounts of water and co-solvent, become larger, and the viscosity rises. After a maximum has been reached, the particles release mainly water and less co-solvent, and the viscosity decreases.

Co-solvents not only help stabilize aqueous dispersions, they also benefit the application properties by smoothing out anomalous viscosity behaviour, improving wetting of pigments and surfaces, ensuring uniform evaporation of water, and assisting the flow, levelling, gloss and smoothness of coating films. The most important co-solvents are: butyl glycol, butyl diglycol, methoxybutanol, butoxypropanol, and sec-butanol. Such hydroxy-functional solvents groups also help to avoid premature reactions with melamine resins while paints are in storage.

Beside of co-solvents, which are already contents of delivery form of acrylic dispersions, they are added preparing the paint formulation. Secondary aqueous acrylic dispersions tolerate small quantities of solvents that are not water-compatible, e.g. aromatic hydrocarbons.

Some of the melamine resins which are intended here as crosslinkers are among the few resins which are fairly soluble in water, namely melamine resins etherified with methanol. Such melamine resins – mainly the low-molecular, fully etherified HMMM resins – are chosen for combination with acrylic resins in aqueous secondary dispersions. On account of the broad solubility and compatibility of HMMM resins and their low molecular masses, these crosslinkers can diffuse efficiently into the dispersion particles of acrylic resins – during film-forming, at the latest. This is a good pre-condition for optimum crosslinking. After evaporation of the amines, the released carboxyl groups can act as catalysts for crosslinking reactions. Amine salts of stronger acids (e.g. sulfon acids) may also be added as catalysts.

A combination of acrylic resin dispersions and melamine resins forms the basis for stoving enamels usually stoved at temperatures of between 130 and 180°C.

Such combinations are suitable for one-coat systems, topcoats and clear coats for general industrial application as well as for automotive OEM coatings.

Examples of commercial products: Bayhydrol A types (Bayer), Luhydran (BASF), Worléecryl A-W types (Worlée), Macrynal types (Cytec-Vinanova)

Special variants of such acrylic resin dispersions are suitable for electrodeposition coatings. In this process, the anionically stabilized acrylic resins are precipitated at the anode (the object to be coated). In this case, the crosslinker resins are melamine resins that have low solubility in water or also contain anionic carrier groups, which are deposited together with the acrylic resins.

The most common electrodeposition coatings – mainly for automotive OEM primers – contain aromatic epoxy resins, functionalised with amines, in combination with blocked polyisocyanates. These systems offer excellent adhesion on metals and outstanding corrosion resistance. However, they have a tendency to yellow and are definitely not weatherable.

By contrast, electrodeposition coatings based on acrylic resins do not yellow and, to an extent depending on the type of crosslinker, are highly weatherable. For this reason, electrodeposition coatings containing anionically stabilized acrylic resins and melamine resins are preferred for one-coat metal systems for general industrial application, e.g. for radiators or metal implements that have complex surfaces (e.g. fences, grids, and metal furniture).

Aqueous acrylic dispersions crosslinked by polyisocyanates

As already discussed, isocyanates can react with water to form carbamic acid, which spontaneously decomposes into primary amines and carbon dioxide. The primary amine can react very rapidly with further isocyanates to form ureas (see Formula 3.9).

Nevertheless, it is possible to crosslink hydroxy-functional acrylic resins with isocyanates in aqueous phase, too. The reason is that aliphatic and cycloaliphatic isocyanates are markedly hydrophobic. They are therefore much more compatible with the acrylic resins in the dispersion particles, which are hydrophobic as well. As high acid values of acrylic resins tend to interfere the crosslinking reaction with polyisocyanates negatively, secondary acrylic dispersions stabilized mainly by surfactant are preferred. The acrylic resins themselves contain only small quantities of acid groups (acid values of about 10mg KOH/g), which are neutralized with small amounts of amines to support dispersion stability. The hydroxyl values of such dispersions are commonly between 60 and 140mg KOH/g.

Compared to corresponding solvent-borne paints, the aqueous combinations of acrylic resins and polyisocyanates also have a two-component effect, i.e. a pot-life. However, the pot-life in this case cannot be measured via the rise in viscosity of ready-mixed paints over time. The reason is that the viscosity of aqueous dispersions, in contrast to that of organic solutions, is dependent not on molecular masses, but on the number of particles and the interaction of these particles via surface tension. In some cases, the particles of the combinations discussed here increase in density during crosslinking, and so the viscosity may even decrease. Of course, crosslinked particles cannot flow or form homogeneous, smooth films. Thus, the pot-lives of water-borne two-component paints are best determined by application tests. Such tests must take the temperature and atmospheric humidity into consideration.

Although the hydrophobic properties of aliphatic and cycloaliphatic polyisocyanates protect the compounds from reacting prematurely with water, they also harbour a disadvantage. Paint users are accustomed to manually mixing the components of solvent-borne two-component paint just before use. And a static mixer is more than adequate for use with a combination of base coat and hardener in a two-component spray gun. Naturally, the hardeners for water-borne two-component paints are not stored and delivered in aqueous form. If such non-aqueous hardeners are mixed together with base coats containing acrylic dispersions without any big effort, the polyisocyanates will form only large emulsion particles. However, for effective crosslinking to take place, the particles of both components must mix intimately by diffusing into each other. This is made all the more difficult by the fact that the mass fractions of the acrylic resin are much higher than those of the polyisocyanate. Figure 3.9 tries to convey an impression of the situation for a mixture of acrylic resin dispersion and polyisocyanate in aqueous milieu.

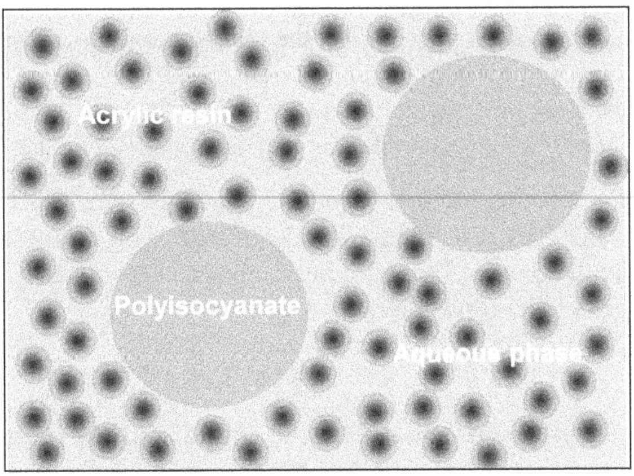

Figure 3.9: Acrylic resin dispersion and polyisocyanate in aqueous milieu

An improvement is achieved if the mixing effort is significantly increased. Various types of equipment for mixing water-borne two-component paints have been proposed.

Particularly effective is a so-called jet stream disperser [51], in which the aqueous base coats and hardener is sprayed into a mixing chamber under pressure.

As few customers are prepared to purchase expensive mixing equipment, other materials-focused approaches have been developed.

First, polyisocyanate adducts with relatively low viscosities have been developed. These are distinguished by narrow molecular mass distributions, or are prepared by special oligomerisation processes, or by reaction with compounds that lower the solution viscosities (e.g. formation of allophanates from diisocyanates and 2-ethylhexanol). Lower-viscosity polyisocyanates are much easier to emulsify into aqueous phase, and are much more amenable to diffusion processes. In addition, it is possible to lower the viscosity of hardeners by adding a co-solvent which is at least partially water-tolerant. Of course, the solvents must not have active hydrogen atoms, which would otherwise react with the polyisocyanate. A suitable solvent is methoxypropyl acetate.

Second, polyisocyanate adducts modified with hydrophilic compounds have been developed. An example of this is the reaction between a small amount of isocyanate groups and mono-functional polyethylene oxide (e.g. the conversion product of methanol with a number of ethylene oxide molecules). The polyether structures impart compatibility with water, making it much easier to introduce hardener into the aqueous phase. As a result of the compatibility with water, in turn, less mechanical effort is needed to generate relatively smaller particles of hardener emulsions. However, the modification reaction also means that some of the isocyanate groups are not available for the crosslinking reaction. Furthermore, the superior water compatibility of isocyanates leads to more side-reactions between the isocyanate groups and water molecules. These possible side-reactions are the reason that it is virtually impossible to calculate precise stoichiometric ratios for acrylic resin and isocyanate crosslinker in the manner commonly employed for two-component, solvent-borne coating formulations. It is assumed that the isocyanate quantity must be 120 to 150 wt.%, compared to conventional systems. Of course, the material costs are correspondingly much higher.

Co-solvents for such two-component systems need to have a certain degree of water tolerance, but they may not react with isocyanates. Suitable co-solvents are methoxypropyl acetate, butyl glycol acetate, butyl diglycol acetate, diethylene glycol dimethyl ether.

Otherwise, aqueous two-component paints containing secondary acrylic dispersions consist of pigments (optional) and different additives, such as reaction catalysts, antifoam agents, rheological additives and levelling agents.

Water-borne two-component coatings containing secondary acrylic resin dispersions form films and crosslink at already ambient temperatures, but they form films more efficiently at elevated temperatures. Application fields include wood coatings, plastic coatings, general industrial coatings, automotive repair coatings, coatings for trucks, buses, machines, agricultural implements and vehicles, and automotive OEM coatings. The preferred products are water-borne primers, one-layer coatings, topcoats and clear coats made from the aforementioned combinations. Table 3.11 presents the composition of a water-borne, two-component primer for general industrial application [52].

Table 3.11: Composition of a water-borne two-component primer for general industrial application

ingredients	Product	NV	wt.%
Base coat			
Acrylic resin (45%, OH: 3,2%)	Setalux 6520 AQ-45 [Nuplex]	9.8	21.8
Water, deionized			7.8
Antifoam additive	Foammaster [Cognis]		0.6
Dispersion agent	Disperse aid W 22 [Daniel Products]		0.5
Titanium dioxide	Kronos 2059 [Kronos]	9.9	9.9
Talcum	Talcum AT [Norw. Talc]	2.3	2.3
Barium sulphate	Blanc fixe micro [Sachtleben]	10.4	10.4
Mica	Plastorit micro [Naintsch]	6.9	6.9
Zinc phosphate	Delaphos 2 [ISC Alloys]	11.6	11.6
Yellow iron oxide	Bayferrox 3910 [Bayer]	3.8	3.8
Anticorrosion additive	Forbest 600 [Lucas Meyer]	0.4	0.4
Grind, then add			
Acrylic resin (45%, OH: 3,2%)	Setalux 6520 AQ-45 [Nuplex]	10.0	22.3
Wetting agent	Byk 348 [Altana-Byk]		0.4
Rheology additive	Viscalex HV 30 [Ciba Specialties]	0.1	1.3
Total		**65.2**	**100.0**
Hardener			
HDI trimer (NCO: 23%)	Desmodur N 3600 [Bayer]	10.2	10.2
Butyl acetate			5.4
Butyl glycol acetate			1.7
Total		**10.2**	**17.3**
PVC (film): 33Vol.%, nOH : nNCO = 1.0 : 1.5			

Examples of commercial products: Bayhydrol types (Bayer), Macrynal types (Cytec-Vianova), Worléecryl types (Worlée), Synocryl types (Cray Valley)

Aqueous acrylic dispersions crosslinked by blocked polyisocyanates

It is possible to dope the aforementioned blocked polyisocyanates (see Chapter 3.5.1.4) with anionic groups to achieve a certain degree of water compatibility. This is the case, for example, if dimethylol propionic acid is made to react with two moles of diisocyanate to form an adduct which still has two free isocyanate groups and a tertiary carboxyl group. The isocyanate groups may be blocked with different blocking agents. The tertiary carboxyl group can be neutralized with amines to form a carrier group for conferring water solubility. This same principle is also used to generate anionic carrier groups for polyurethane dispersions.

As carboxyl groups more or less hinder the isocyanate reactions – also those of blocked isocyanates – a more convenient method for blocked polyisocyanates in the aqueous phase is to use acrylic dispersions which have been stabilized with surfactants. To this end, hydroxy-functional acrylic resins are prepared by solution polymerization using water-tolerant solvents. The blocked polyisocyanate, too, is diluted in such a solvent. The two organic solutions are mixed in accordance with the above-mentioned mixing rules, and are transferred into the aqueous phase by adding emulsifiers, protective colloids and – of course – water. Where necessary small quantities of acid groups are neutralized to support the stabi-

Table 3.12: Hydroxy-functional acrylic resin for aqueous dispersions

Ingredients	wt.%
Methyl ethyl ketone	93.54
charge, heat on 80°C, hold temperature	
tert.-Butyl acrylate	38.50
n-Butyl methacrylate	15.00
Cyclohexyl methacrylate	20.00
2-Hydroxy ethyl methacrylate	25.50
Acrylic acid	1.00
Total monomers	**100.00**
mix, add in 5h	
tert.-Butyl perbenzoate	10.00
Methyl ethyl ketone	6.53
add parallel, hold until solid content is achieved	
Methyl ethyl ketone	-55.00
distill off	
Total	**155.07**

Characteristics: NV (60‘ 130°C): 71.4%
acid value (NV): 10.1mg KOH/g
OH-value: 110mg KOH/g
viscosity (50% in MEK, 23°C): 800mPa•s

lity of the dispersion. The process solvents are removed by distillation. The result is a practically solvent-free, aqueous dispersion, which consists of homogeneous, uniform particles containing acrylic resin and crosslinker.

Table 3.12 lists the composition of a typical acrylic resin for the crosslinking method described here [53].

The acrylic resin in the example is mixed with a solution of a blocked polyisocyanate and the mixture is transferred into an aqueous dispersion [54]. Table 3.13 shows the composition of the dispersion, which is converted to a clear coat.

Such products are used in automotive OEM clear coat applications.

Furthermore, acrylic resins which consist of tertiary aminoalkyl acrylates (e.g. N,N-dimethylaminoethyl methacrylate, see Chapter 2.2.3.3) are suitable for the production of cationically stabilized secondary dispersions.

Table 3.13: Clear coat based on an aqueous dispersion of hydroxy-functional acrylic resins, and DMP-blocked polyisocyanate

	Product	**wt.%**
OH-acrylic resin (70% in MEK)	Preparation example 1 of US 2003,022,985	37.36
Blocked polyisocyanate adduct* (70% in MEK)	Preparation example 2 of US 2003,022,985	21.75
UV absorber hydroxy benzotriazine	"Cyaguard" 1164 [Cytec]	0.42
Radical quencher N-methyl-2,2,6,6-tetramethyl piperidine derivative	"Tinuvin" 1130 [BASF Ciba-Spec.]	0.42
N,N-Dimethylethanolamine	[BASF]	0.36
Water, deionized		54.64
Total		**114.95**
Methyl ethyl ketone, distilled off	**Ingredient**	-17.73
Thickener 1	"Acrysol" RM 8 [Rohm & Haas]	2.20
Thickener 2	"Viscalex" HV 30 [Allied Colloids]	0.58
Total		**100.00**
* adduct of TMP (1mol), IPDI (3mol) and 3,5-dimethyl pyrazole (3mol)		
NV (nonvolatiles): 43.2% application: three-layer system (wet-on-wet-on wet, stoving conditions 30' at 145°C		

Such dispersions can be used in electrodeposition processes. The resins are deposited at the cathode (object to be coated). The neutralization agents employed are volatile organic acids, e.g. acetic acid or formic acid. If such acrylic resins additionally contain hydroxyl groups, these can react with blocked polyisocyanates to form crosslinked films. In that case, the amine groups present act positively as catalysts for the isocyanate reaction.

3.5.2.4 Comparison of aqueous acrylic resins in secondary dispersion with other resins

The differences between primary dispersions and secondary dispersions of acrylic resins are not just restricted to the way in which they are made.

As already mentioned, acrylic resins prepared by solution polymerization and converted into aqueous dispersions have significantly lower molecular masses than resins made by emulsion polymerization. That fact is an advantage for film-forming reactions that involve crosslinking.

True, ways of crosslinking acrylic resins prepared by emulsion polymerization have been described, but such crosslinking reactions can only be incomplete (see Chapter 4).

In particular, aqueous secondary acrylic resin dispersions form homogeneous films that exhibit good levelling and gloss. Film formation is supported by co-solvents. The reason is that ionically stabilized dispersion particles have totally different particle structures. There are no pronounced particle interfaces. During initial drying, the particles can inter-diffuse readily, which is a pre-condition for forming homogeneous, dense films. Diffusion of the crosslinkers added to the aqueous paint can also promote optimum film formation. However, there are other resins which are dispersed in aqueous milieu and which are much better at diffusion. The colloidal particles of saturated polyester are less densely coiled than the particles of acrylic resins. This has benefits for application and film formation, including better crosslinking efficiency. However, such resins suffer from a major disadvantage, namely that they are less resistant to saponification. For this reason, secondary aqueous acrylic resin dispersions are preferred to other resins, not only on account of their superior storage stability, but also because they meet the demands for optimum resistance to chemicals, solvents, mechanical impact and weather exposure.

3.5.3 Acrylic resins for powder coatings

Powder coatings are 100% systems, and so they fully meet the requirements of avoiding emissions of volatile organic compounds during the production and application of paint materials.

Currently, the most important class of powder coatings in volume terms is a combination of aromatic epoxy resins and carboxy-functional saturated polyesters. These powder coatings are distinguished by excellent adhesion, high hardness, optimum chemical resistance and outstanding corrosion resistance. They are employed in a wide range of application areas. However, they are not weatherable.

In contrast, weatherable powder coatings contain acrylic resins. There are two groups of acrylic-based powder coatings, which differ in their crosslinking mechanism. The first are the hydroxy-functional acrylic resins, which are crosslinked by blocked polyisocyanates. The other are epoxy-functional acrylic resins, which are crosslinked by carboxylic acids or carboxylic anhydrides.

3.5.3.1 Powder coatings based on acrylic resins and blocked polyisocyanates

Powder coatings are made from solid resins. The crushed or already powder-like ingredients are mixed and then homogenized in a heated extruder (melt kneading). The extruder product is crushed and milled, sieved and classified. As the milling process generates heat energy, the ingredients, mainly the resins, must have minimum softening temperatures. The limit on the softening temperature is 70 to 90°C. Suitable resins here have glass transition temperatures of more than 55°C. Furthermore, adequately high glass transition temperatures are also needed to ensure the storage stability of the powder coatings. The choice of monomer mixtures must reflect these constraints.

Extrusion is carried out at temperatures of 90 to 110°C. The mixtures of resins and crosslinkers must remain stable at those temperatures. Consequently, the effective crosslinking temperatures cannot be lower than 150°C.

Hydroxy-functional acrylic resins for powder coatings are prepared by standard solution polymerization. The monomer composition of the acrylic resins must be chosen such that the resultant glass transition temperatures are high enough. After polymerization, the process solvent is removed by distillation. The polymer melt is discharged onto cooling conveyor belts and then crushed.

The blocked polyisocyanates, too, must have higher glass transition temperatures. The best solution is to choose isocyanurate trimers of isophorone diisocyanate (IPDI) blocked with ε-caprolactam, 1,2,4-triazoles, or ketoximes [55]. During crosslinking, the isocyanates form urethanes with the acrylic resins, and the blocking agents are set free, partly remaining in film where they act as a kind of plasticizer.

The powder coatings contain the usual additives (degassing agents, flow agents, levelling agents and, where necessary, light stabilizers).

Resins, crosslinkers and any pigments, along with the aforementioned additives are mixed, fed into the extruder and homogenized efficiently. The mix is extruded, discharged onto a cooling conveyor belt and crushed. Then the product is milled, preferably on impact mills, and classified in cyclones. Both coarse quantities and fine fractions are separated. The coarse fractions are returned to the milling process and the fines, to the extruder. The average particle size of powder coatings is 20 to 40µm. At average particle sizes of about 20µm, the particle size distribution ranges from somewhat below 10µm to somewhere above 50µm. Although powder coatings melt and flow during stoving, it is not possible to achieve smooth and homogeneous films in thin layers. Conventional powder coatings do not form smooth, homogeneous films unless the thickness is at least 65µm.

For application, the powder coats can be fluidised with compressed air, which means they are converted by an air steam into an aerosol which behaves like a liquid.

In an older process, fluid-bed sintering, heated objects are dipped into the aerosol. The powder particles melt on the surface of the hot objects and form films. The resultant film can be cured by stoving. This process is still employed for mainly small objects.

Table 3.14: Composition of hydroxy-functional acrylic resins for crosslinking with blocked polyisocyanates in powder coatings

Ingredients	**wt.%**
Toluene	93.54
Charge, heat to 125°C, hold temperature	
Methyl methacrylate	36.99
N-Butyl methacrylate	23.99
Styrene	18.88
2-Hydroxy ethyl methacrylate	19.12
Acrylic acid	1.01
Total monomers	**100.00**
Mix, add over 4h	
tert.-Butyl perbenzoate	4.43
Toluene	6.75
Add in parallel over 5h, hold, then cool down	
Total	**202.14**
Characteristics: NV (60‘ 130°C): 50.7% acid value (NV): 7.4mg KOH/g OH value: 80mg KOH/g viscosity (23°C): 1880mPa s glass transition temperature (DSC, solid resin): 67°C	

The predominant application method for powder coatings is that of electrostatic spraying. In this, particles of a coating powder are charged electrostatically in a spray gun. The charging causes them to be transported in an electrical field to an earthed object, where they adhere. The transfer efficiency is relatively high. The coated objects are then stoved and the powder melts to form a film. At the same time, crosslinking takes place. The effective transfer efficiency can be increased to 95 wt.% by collecting those powder particles which did not adhere to the object and returning them to the process. However, this recovery harbours the risk of contamination.

Extrusion ensures that the ingredients are sufficiently homogenized, e.g. white coating powders pigmented with titanium dioxide (pigment volume concentration of 18 vol.%) yield high-gloss films. There are also processes which offer even better homogenization. These mix together the resin, crosslinker and additives in organic solution. The solvent is then removed by distillation and the resultant solid is crushed, milled, sieved and classified. This process is akin to the production of aqueous secondary dispersions with a homogeneous particle composition (see Chapter 3.5.2.3 and Tables 3.12 and 3.13). This procedure is illustrated in the following examples for hydroxy-functional acrylic resins [56], whose composition is presented in Table 3.14, and the resultant powder coat [57] (Table 3.15).

Coating powders based on hydroxy-functional acrylic resins and blocked polyisocyanates are preferred for high-grade, general industrial coatings, for automotive OEM coatings, one-layer coatings, topcoats and clear coats. They are distinguished by excellent weatherability, chemical resistance, and high flexibility. They are relatively expensive.

Table 3.15: Coating powder based on hydroxy-functional acrylic resin and blocked polyisocyanates

Ingredients	Product	wt.%
OH-acrylic resin (50% in toluene)	Example A 1 From patent US 5,508,337	130.36
Polyester (for modification)	Example B 1 from patent US 5,508,337	3.41
Blocked polyisocyanate adduct* (75% in toluene)	Example from patent C 2 US 5,508,337	39.35
Toluene		34.13
Dibutyl tin dilaurate		1.42
Flow additive	"Perenol" F 45 [Cognis]	0.47
Total		**209.15**
Toluene, lost by drying		109.15
Total		**100.00**

*Adduct of isocyanurate trimer of hexamethylene diisocyanate and methyl ethyl ketoxime (75% in toluenel)

Mix and spray-dry

3.5.3.2 Powder coatings based on epoxy-functional acrylic resins

Copolymerization of epoxy acrylate monomers with other monomers in solution polymerization leads to epoxy-functional acrylic resins. The preferred monomer for such resins is glycidyl methacrylate. Epoxy-functional acrylic resins can react with carboxyl groups of polycarboxylic acids or their derivatives (see

Table 3.16: Formulation and production of an epoxy-functional acrylic resin

Ingredients	wt.%
Xylene	44.42
Charge, heat to 120°C, hold temperature	
Methyl methacrylate	55.67
Glycidyl methacrylate	21.32
N-Butyl acrylate	13.33
Styrene	9.68
Total monomers	**100.00**
Mix, add over 4h	
tert.-Butyl perethyl hexanoate	4.44
Add in parallel over 4.5h, hold 1h	
Total	**148.86**
Then heat to 180°C, vacuum distillation, discharge the melt onto a cooling belt	

Chapter 3.5.1.6, where the reverse reaction of carboxy-functional acrylic resins with epoxy resins is described). Carboxylic anhydrides and polyanhydrides also serve here as crosslinker. In these cases, only small quantities of hydroxyl compounds are required to open the anhydride groups; after that, the reaction proceeds with the resultant β-hydroxyl esters, which leads to a higher crosslinking density.

These reactions have the advantage that crosslinking is an addition reaction, without any cleavage products. The risk of generating defects in films, even thick ones, is therefore avoided. In addition, there is no

Table 3.17: Powder clear coat prepared with epoxy-functional acrylic resin

Ingredients	Product	wt.%
GMA-acrylic resin (100%)	Example A from patent DE 42 27 580	73.37
Polyanhydride of dodecanoic diacid	Example B from patent DE 42 27 580	20.30
Trimethylol propane		1.59
Degassing additive	Benzoin	0.40
Flow additive	Perenol F 40 [Cognis]	0.40
Radical quencher (2,2,6,6-Tetramethyl piperidine derivative)	Tinuvin 144 [BASF Ciba-Spec.]	2.37
UV absorber (hydroxy benzotriazole)	Tinuvin 900 [BASF Ciba-Spec.]	1.57
Total		**100.00**
Buss co-kneader, 90 – 110°C, mechanical impact air-classifier mill ACM 2L Hosokawa Mikro Pul, cyclone, average particle size 30µm.		

retention of by-products as in the case of the blocking agents ε-caprolactam or 1,2,4-triazole for polyisocyanate adducts.

Powder coating films prepared with this resin combination have excellent weatherability. They are therefore recommend for use in automotive OEM clear coats. Table 3.16 presents the formulation and production of an epoxy-functional acrylic resin [58] while Table 3.17 shows the resultant powder clear coat [59].

Example of a commercial product: Fineplus AC (DIC)

3.5.3.3 Powder slurries based on acrylic resin

For many application fields, flawless application of thick films by means of powder coatings is definitely an advantage. That is why powder coatings based on epoxy resins and carboxy-functional saturated polyesters dominate the field of single-coat systems for household appliances ("white goods"). These coatings, in thicknesses of about 80μm, are distinguished by excellent corrosion resistance and chemical resistance.

However, for other application fields, such high film thicknesses are a disadvantage on account of the high material consumption, associated costs and sometimes the larger heaviness of coated objects.

There are numerous approaches aimed at preparing powder coatings which yield homogeneous, smooth, glossy films at significantly lower film thickness. As particle size is the essential factor for such film formation, these approaches concentrate on preparing smaller particles and generating the narrowest-possible particle size distributions. The production of smaller particles requires exponentially greater energy input (due to the large surface areas which must be formed) and special cooling (in the extreme case with liquid nitrogen). However, the main problem is the fact that very fine powders like this cannot be fluidised on account of particle-particle interaction, which prevents the formation of suitable aerosols.

Nevertheless, this problem can be circumvented. If the powder coating, made by the usual production process, is converted into an aqueous dispersion, the dispersion can be milled by a standard process to much smaller particle sizes. The resultant powder slurry exhibits optimum free-flow properties. Interestingly, milling can be performed with common grinding mills with fitted stirrers (sand mills, pearl mills). These types of mill are normally used for pigment dispersion, which definitely is not a milling process, but rather is a measure for generating surface wetting under high shear. Nonetheless, using the mills for powder slurries generates much smaller particles. Figure 3.10 shows the particle size distribution of a powder clear coat compared with that of the resultant powder slurry.

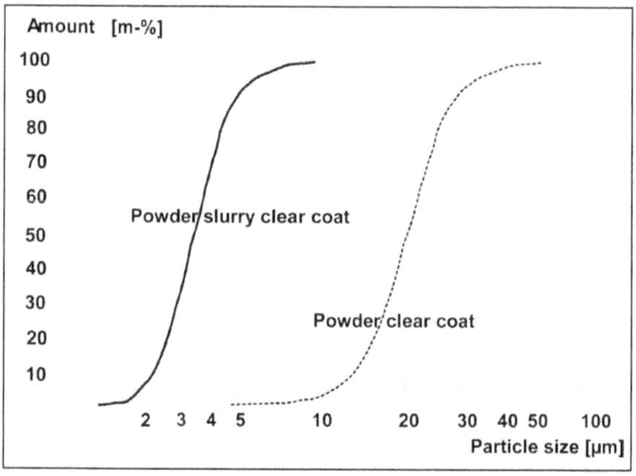

Figure 3.10: Particle size distribution of powder coat and powder slurry

Powder slurries are solvent-free; they can be applied with standard equipment for water-borne coatings (electrostatic spray guns with separate charging). Powder slurries based on acrylic resins are mainly used in automotive OEM clear coats [60]. The optimum film thickness is 40µm, which is the same as that of solvent-borne clear coats. The coating films are highly weatherable and offer adequate chemical resistance. Of course, the powder slurry is relatively expensive due to the relatively high raw material costs and the complex production process.

The composition and preparation process for powder slurry are illustrated in the following tables. Table 3.18 describes the composition of the acrylic resin [61], Table 3.19 presents the coating powder while Table 3.20 shows the resultant powder slurry [62].

Table 3.18: Acrylic resin for a powder slurry

Ingredients	wt.%
Xylene	24.39
Charge, heat to 130°C, hold temperature	
Methyl methacrylate	12.46
n-Butyl methacrylate	29.47
Styrene	19.98
Glycidyl methacrylate	38.08
Total monomers	**100.00**
Mix, add over 4h	
tert.-Butyl perethyl hexanoate	5.20
Xylene	5.62
Add in parallel over 4.5h, hold 1h	
Total	**135.21**
Then heat to 180°C, vacuum distill, discharge the melt onto a cooling belt	

3.6 Outlook

Since the late 1970s, much effort has been expended on lowering emissions of volatile organic compounds (VOCs) in the production and application of coating materials. This has led to the development of high-solid paints and water-borne coating systems. Progress has also been made in the development of 100% systems, i.e. powder coatings as well as reactive acrylic resins for radiation

Table 3.19: Coating powder for a powder slurry

Ingredients	Product	wt.%
Acrylic resin	Example 1 from patent DE 19744561	73.50
Dodecanoic dicarboxylic acid		17.80
TACT	[BASF SE]	5.00
UV absorber (100%) (hydroxy benzotriazole)	"Tinuvin" 1130 [BASF Ciba-Specialties]	2.00
Radical quencher (2,2,6,6-tetramethyl piperidine derivative)	"Tinuvin" 144 [BASF Ciba-Specialties]	0.90
Flow additive	"Additol" XL 490 [Cytec]	0.40
Degassing additive	Benzoïn	0.40
Total		**100.00**
Buss PLK extruder, mechanical impact air-classifier mill Hosokawa ACM 2, cyclone average particle size 22µm		

curing (see Chapter 5). In Europe, a VOC regulation came into force on 21 August 2004 which requires a two-step reduction in the level of emissions of volatile organic compounds during application of coatings. The deadlines for the two steps were 1 January 2007 and 1 January 2010. Different regulations apply to industrial applications on one hand and workshop and do-it-yourself applications on the other.

Against this background, it was assumed ten years ago that the use of solvent-borne coatings would accordingly fall substantially. However, it is still the case that solvent-borne coatings play an important role in several application fields. Naturally, VOC regulations must be complied with. New regulations with further restrictions are certain to follow.

Conservatism on the part of coatings users is not believed to be the reason for the continued popularity of solvent-borne on the market.

Table 3.20: Composition of a powder slurry clear coat

Ingredients	Product	wt.%
Water, deionized		64.04
Charge, continuously stirring:		
Antifoam additive	"Troykyd" D 777 [Troy]	0.19
Dispersing additive	"Orotan" 731 [Dow]	0.19
Emulsifier	"Tergitol" TMN 6 [Dow]	0.02
Thickener	"Acrysol" RM 8 [Dow]	5.44
Powder coating	Example 2.3 from patent DE 19744561	30.10
Add slowly in two equal portions		
Sand mill 3.5h, under cooling, then add		
Wetting agent	"Byk" 345 [Altana-Byk]	0.01
Total		**100.00**
Average particle size 4µm		

There are definitely technical reasons. On one hand, a solution is the optimum physical state for the various application processes, promoting optimum compatibility of coating ingredients, optimum film-formation during physical drying and homogeneous crosslinking. All these benefit the appearance of coating films (smoothness, gloss, hold-out, brilliance) and resistance to weathering, chemicals, solvents, and mechanical impacts. For industrial application, a second aspect is more important, namely application reliability and the associated "first-run OK rate". This is even truer now that industrial applications are increasingly making use of robots. Solvent-borne systems are much easier to handle here than other systems. What is more, optimum application reliability is an important cost-saving aspect.

To be sure, the market share of solvent-free coatings keeps on growing, but is reaching its limits. Powder coatings are conquering more and more application fields, and there are now even coating powders for wooden objects. The general problem with coating powders is achieving optimum flow and smoothness in thin layers. UV coatings yield particularly outstanding films, but are limited for the most part to planar surfaces. In addition, different concepts have been advanced for circumventing inhibition of UV curing by atmospheric oxygen, but the measures involved are expensive. The trend towards high-solid paints necessitates compromises on film properties. The difference is whether to build up optimally crosslinked film from components of lower molecular weights or to use resins of high molecular masses.

We assume that water-borne coatings will likely take on an important role in the near future, in spite of water's specific properties as solvent and dispersing agent. The best examples of this are aqueous secondary acrylic dispersions, whose particles already contain homogeneously dispersed polymer resin and crosslinker.

3.7 Literature

General Literature

a) Wagner Sarx: Lackkunstharze (Coating Resins) Carl Hanser Verlag, München 1971
b) Kittel: Lehrbuch der Lacke und Beschichtungen (Textbook of Paint and Coatings) Vol. 2: Bindemittel für lösemittelhaltige und lösemittelfreie Systeme (Resins for Solvent-borne and Solvent-free Systems), S. Hirzel Verlag, Stuttgart, 1998
c) Kittel: Lehrbuch der Lacke und Beschichtungen (Textbook of Paint and Coatings) Vol. 3: Bindemittel für wasserverdünnbare Systeme (Resins for Water-borne Systems); S. Hirzel Verlag, Stuttgart, 2001
d) Stoye, Freitag: Lackharze, Chemie Eigenschaften und Anwendungen; (Coating Resins, Chemistry, Properties and Applications) Carl Hanser Verlag, München 1996
e) Wileys Resin Technology; P. Oldring, P. Lam: Vol. I: Water-borne & Solvent Based Acrylics and their End User Application; John Wiley & Sons, New York 1996

References

[1] Rule 66 in California
[2] SGO Process of Johnson Polymers
[3] The production process discussed here is described in examples V1, V2 E1 to E4 of patent USA 6,013,739 held by BASF Coatings
[4] Gel effect, defined by Norrish and Trommsdorff
[5] Reinhard Volmer, Beate Köster: Statistical experimental design of influence parameters on acrylic resins prepared by solution polymerisation process (Research at BASF Coatings, not published)
[6] Model resin, not used industrially, containing n-butyl acrylate, styrene, methyl methacrylate, 2-hydroxyethyl acrylate, and acrylic acid
[7] Described for example in patent USA 6,552,144 held by Johnson Polymers
[8] Mark-Houwink equation defines the relation between molecular mass (mass average) and viscosity: $[\eta] = k \cdot M^{\alpha}$
[9] Half-life temperatures of initiators in a brochure issued by Interox Chemicals
[10] Technical data sheets on initiators, issued by AKZO-Nobel, DuPont, Interox Chemicals, and Wako Chemicals
[11] For example, tert.-amyl peroxide and tert.-amyl peresters from Arkema
[12] Hansen solubility parameters, defined in 1971, in Allan F. M. Barton: Handbook of solubility Parameters, CRC Press 1983
[13] R. F. Patella: Vinyl Acrylic Copolymers, Modern Plastic & Coatings 1978/07
[14] Products of Lilly Corp.
[15] Described for example in patent DE 947338 held by BASF
[16] Described for example in patent DE 1005730 held by BASF
[17] Described for example in patent DE 1035363 held by Bayer
[18] Described for example in patent USA 3,314,998 held by Rohm & Haas
[19] Example C of patent DE 2938304 held by BASF Coatings
[20] Ulrich Poth: Automotive Coatings Formulation – Chemistry, Physics, and Practices; Vincentz Network 2008
[21] NAD development in the U.S.A., initiated by Regulation in California (Rule 66) prevention of emissions of photolytically active compounds.
[22] Described for example in patent WO 95/21895 held by BASF Coatings
[23] Description by Nippon Paint
[24] Described for example in patent USA 4,845,147 held by BASF
[25] Described for example in patent USA 4,377,661 held by Cook Paint & Varnish
[26] Acrylic Resins with exposed hydroxyl groups, in patent DE 2938304 held by BASF Coatings
[27] Described for example in patent USA 2003/059,547 held by BASF Corporation
[28] Example 1 in patent DE 3148051 held by BASF SE
[29] Acrylic resin of example 1 in patent DE 3148051, combined with a partly etherified melamine resin (Luwipal 015 from BASF SE) mass fraction ratios are 65 : 35
[30] UV absorbers hydroxyphenyl benzotriazoles from BASF Swiss (Ciba Specialties, Tinuvin) and hydroxyphenyltriazines from Cytec (Cyaguard) and BASF Swiss (Ciba Specialties, Tinuvin)
[31] Free-radical quencher 2,2,6,6-tetramethylpiperidine (HALS = hindered amine light stabiliser) from BASF Swiss (Ciba Specialties, Tinuvin)
[32] Setalux products from Nuplex Resins

[33] Micrograph of product of the reaction of benzylamine and hexamethylene diisocyanate
[34] Recommended formulation from AKZO Resins, now Nuplex Resins
[35] Described for example in patent USA 2004/176,529 held by PPG
[36] Catalysts from King Industries
[37] Described for example in patent DE 2500309 held by BASF SE
[38] Comparative example 1 in patent USA 6013739 held by BASF Coatings
[39] Example V 1 (Table III) in patent USA 6013739 held by BASF Coatings
[40] The commercial product "Cardura" E 10 of Hexion consists mainly of the glycidyl ester of 2,2,3,5-tetramethyl hexanoic accid
[41] Pre-production of the monomer with glycidyl ester is described for example in patent DE 2709784 held by Hoechst AG
[42] Polymer-analogous reaction of glycidyl ester, described for example in DE 1.520.688 held by Reichold AG
[43] In situ reaction of the glycidyl ester, described for example in patent DE 2065770 held by Hoechst AG
[44] Building blocks: Tone monomers from Dow; and Placcel products from Daicel
[45] Described in patent DE 19748584 held by BASF Coatings
[46] U. Röckrath, K. Brockkötter, Th. Frey, U. Poth, G. Wigger: Investigations of the Crosslinking Mechanism of Etch Resistant Clear coats (Int. Conf. of High-Solid & Water-borne Paints, Athens 1996)
[47] Described in patent WO 93/15849 held by BASF Coatings
[48] TACT was developed by Cytec, currently produced and delivered by BASF SE as Laro-TACT
[49] Described for example in patent USA 5,369,204
[50] Described for example in patent WO 9201025 held by BASF Coatings
[51] Comparable dispersing equipment is described as suitable in patent EP 0101007 held by Bayer
[52] Recommended formulation from AKZO-Resins, now Nuplex Resins
[53] Preparation example 1 (acrylic resin) in patent USA 2003/0,022,985 held by BASF Coatings
[54] Example 1 (clear coat dispersion) in patent USA 2003/0,022,985 held by BASF Coatings
[55] Numerous patents regarding blocking agents for powder coatings
[56] Example A1 (hydroxy-functional acrylic resins for powder coatings) in patent USA 5,508,337 held by Bayer
[57] Example 1 (powder coating) in patent USA 5,508,337 held by Bayer
[58] Example A (acrylic resin containing glycidyl methacrylate, for powder coatings) in patent DE 4227580 held by BASF Coatings
[59] Example 3 (powder coating) in patent DE 4227580 held by BASF Coatings
[60] Application of powder-slurry based on acrylic resin as clear coat in a three-layer system, containing functional layer, water-borne base coat, and the aforementioned clear coat, ("integrated coatings application concept" of BASF Coatings) for automotive OEM coatings
[61] Example 1 (acrylic resin containing glycidyl methacrylate) in patent DE 19744561 held by BASF Coatings
[62] Example 2.3 (powder coating) in patent DE 19744561 held by BASF Coatings
[63] Example 3 (powder slurry) in patent DE 19744561 held by BASF Coatings

Note

For better comparability, the quantities quoted in the patents have been expressed in terms of 100 parts monomer of the acrylic resins, and as percentages in the paint formulations.

4 Primary dispersions of acrylic resins

Manfred Schwartz

Definitions and introduction

Generally, a dispersion is a multi-phase system [1 – 4] in which at least one phase in a state of microscopically fine distribution (the disperse phase: liquid or solid, for example) is present within a continuous phase (e.g. liquid or gas). In polymer dispersions, the disperse phase consists of spherical polymer particles with a diameter which is usually less than 1µm; the continuous phase is water.

Aqueous polymer dispersions are usually milky liquids whose viscosity varies from low, like that of water, to high, like that of whipped cream. By analogy with the milky sap of the plants which provide natural rubber, they are also often referred to as latex, and the polymer particles as latex particles.

A single millilitre of polymer dispersion contains on average approx. 10^{15} particles. In turn, from 1 to 10,000 macromolecules are present per particle, each of these macromolecules being composed of approx. 100 to 10^6 building blocks (monomers).

Polymer dispersion systems are not thermodynamically stable per se. The polymer particles have a tendency to try to minimize the large internal surface area of the system by agglomeration, coagulation, or settling. Attaching charge carriers (charge or Coulombic stabilization) or uncharged spacers of medium to high molecular weight (steric or entropic stabilization) to the surface of the polymer particles, however, can stabilize the disperse state [5, 6]. Nevertheless, external influences, such as shearing (as a result of shaking, pumping or stirring), freezing, pressure or salt exposure, can in adverse circumstances cause the stabilization to fail; the dispersion then coagulates.

Polymer dispersions are classified as either primary dispersions, which are made by polymerizing the monomers directly in the liquid phase (via emulsion polymerization (EP) in water, for example), or secondary dispersions, which are

Poth/Schwalm/Schwartz: Acrylic Resins

ISBN: 978-3-86630-857-2

prepared by distributing or dispersing a polymer, such as a solution polymer or film-forming resin, in the medium in a second step that usually requires input of mechanical energy [7]. Most important are the primary dispersions, which are readily obtainable industrially by EP and are relatively inexpensive to make. Among the secondary dispersions, the most important class is that of polyurethane dispersions [8], which arc used primarily in the industrial coatings sector.

4.1 Binder classes, polymerization and polyacrylates

The chemistry of the polymers in the field of acrylic resins (primary dispersions) is very diverse. The most important classes of binder in the coatings sector are:

- Acrylate copolymers (pure or straight acrylics)
- Acrylate-styrene copolymers (or styrene-acrylates)

The present work concentrates on acrylic resins; a comparison with other binder classes is presented in Chapter 4.7. Thus, in the following, we will concentrate on primary dispersions of acrylic polymers.

4.1.1 Polyacrylates by polymerization

4.1.1.1 Free-radical polymerization

The acrylic resins which are used as binders are exclusively copolymers, whose fundamental properties are achieved by free-radical polymerization of a specific combination of different α,β-unsaturated organic building blocks (monomers) [9–16]. The free-radical polymerization is characterized by its rapid, exothermic course (described in Chapter 2).

The monomers most frequently employed for acrylic dispersions are given in Table 4.1; the basic structures of the principal classes of monomer are as follows:

$H_2C=CH\text{-}CO\text{-}OR$	**$H_2C=C(CH_3)\text{-}CO\text{-}OR$**
Acrylates	**Methacrylates**

4.1.1.2 Emulsion polymerization

Unlike solution polymerization (Chapter 3.3) in which the reaction partners can be found in one and the same phase, there is a further, crucial factor governing polymer performance in EP [6, 17–30]: the influence of the multiple phase system.

Mechanism of emulsion polymerization

The preparation of the aqueous dispersions takes place via a specific form of free-radical polymerization known as emulsion polymerization (EP). In this process, the monomers are made to react in water in the presence of surface-active compounds of low-molecular weight (emulsifiers) or polymers (protective colloids) by adding a water-soluble, free-radical initiator and heating. The emulsified monomers polymerize to form the dispersed macromolecules.

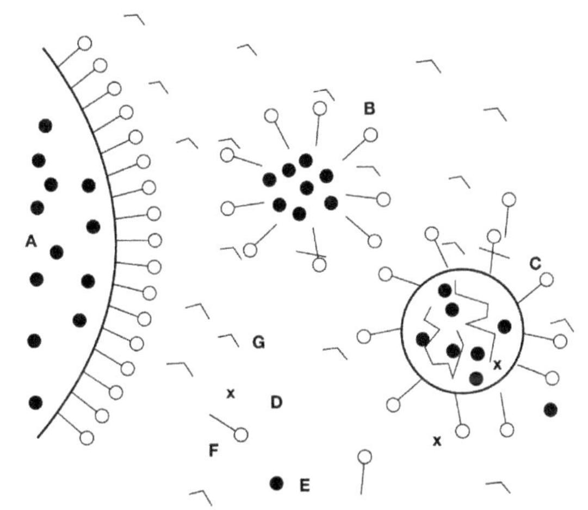

A: Monomer droplet with monomers (E) • and emulsifier molecules (F) -O
B: Micelle with monomers
C: Polymer particle, stabilized by emulsifier molecules, containing several macromolecules, one of which has a reactive, radical-chain end (x) and monomer (A)
D: Water-soluble initiator free-radical (x)
E: Monomer in the water phase
F: Dissolved emulsifier molecule, molecularly dispersed
G: Water molecule

Figure 4.1: Mechanism of emulsion polymerization

The micellar mechanism of the EP of styrene was first described by *Harkins* [31], and by *Smith* and *Ewart* [32] (Figure 4.1). According to those authors, the monomers in the polymerization reactor prior to addition of initiator are distributed between emulsifier-stabilized monomer droplets (having a diameter of 1 to 10μm) and so-called micelles, i.e. aggregates of 20 to 100 emulsifier molecules (with a diameter of 5 to 15nm). The fraction of monomer present in molecular solution in the water is very small. On heating, the initiator breaks down in the aqueous phase to form free-radicals which initially grow by attachment to the small water-dissolved monomer fraction to form oligomer free-radicals. Since the number of micelles in the reactor per unit volume (approx. 10^{18} per cm^3) is significantly higher than that of the monomer droplets (approx. 10^{10} per cm^3) and since the total surface area of the micelles is much greater than that of the monomer droplets, the oligomer free-radicals enter almost exclusively into the micelles.

There, the chains continue to grow, as a result of which the micelles should in fact become increasingly depleted of monomer. This does not occur, however, as sufficient monomer molecules are transported from the monomer droplets to the micelles via the aqueous phase. Consequently, the monomer concentration in the aqueous phase remains constant while there are still monomer droplets present in the reactor. Meanwhile, in the micelles, the polymer chains grow, building up latex particles until all of the monomer droplets have disappeared. In the course of the polymerization, therefore, the growing, polymer-filled micelles turn into emulsifier-stabilized latex particles.

The theory of micellar particle formation as described above was supplemented by *Fitch* and *Tsai* [33] who developed the principle of "homogeneous nucleation", which was later further developed by *Ugelstad* and *Hansen* [34, 4b]. According to this principle, initiation by a water-soluble charged peroxide free-radical which adds to monomer units in the aqueous phase is the trigger for the growth of oligomeric macro-radicals. Beyond a chain length specific to each monomer (2 to 100 units), the limit of solubility is exceeded and primary particles are formed. These primary particles are usually unstable and undergo agglomeration until they reach a state of colloidal stability: secondary particles. The diameter of the secondary particles is limited by the quantity of emulsifier and the polarity of the polymer.

The aforementioned mechanisms of particle formation represent idealized cases [35]. If polymerization is initiated with peroxodisulphates and performed using polar monomers (such as VAc, MMA, etc.), however, a considerable proportion of homogeneous nucleation can always be assumed. Irrespective of the mechanism discussed, a prerequisite for EP is that the monomers used be at least slightly soluble in water. Consequently, while monomers such as styrene or 2-EHA still readily undergo EP, polymer dispersions of very hydrophobic, long-chain and hence water-insoluble (meth)acrylates, such as lauryl (meth)acrylate or stearyl acrylate, are not obtainable by conventional EP.

At industrial scale, the monomers nowadays are usually pre-emulsified in water. The emulsion thus prepared and the initiator solution is then metered separately into the polymerization reactor over a defined period of time. This semi-continuous or feed process has a very high rate of instantaneous conversion of the monomers (usually > 90%), yielding a randomly constructed copolymer, largely irrespective of differences in reactivity and copolymerization parameters. Moreover, an advantage of the semi-continuous process over the formerly used batch or one-pot process is that the heat of polymerization produced can be regulated via the metering time and be dissipated in a controlled fashion by external cooling. In comparison to solution polymerization (see Chapter 3), EP, in which the resultant polymer particles are finely distributed in water, affords the additional advantage of permitting high molecular weights (more than 1 million Daltons) combined with a low system viscosity. Industrial polymer dispersions therefore usually have high polymer contents of 40 to 60 wt.%.

4.1.2 Polyacrylates; straight acrylics and styrene-acrylate copolymers

The acrylate copolymer dispersions [1, 36 - 45] to which this book is devoted can be divided into two classes:

- straight acrylics and
- styrene-acrylates.

Straight acrylics are polymer dispersions composed exclusively of acrylate and/or methacrylate monomers. Acrylate-styrene copolymers contain styrene as well. Chapter 4.7 provides a comparison of these two classes.

For both types of copolymer, numerous monomers are available (Table 4.1) which differ greatly in glass transition temperature (T_g) and the polarity of the homopolymers prepared from them [39, 46].

Table 4.1: Water solubilities and glass transition temperatures of principal monomers for acrylic dispersions [36]

Monomer building blocks	Water solubility at 25°C in g/100 cm³	Glass transition temperature (T_g) of the homopolymer [°C]
Acrylates		
Methyl acrylate (MA)	5.2	+22
Ethyl acrylate (EA)	1.6	-8 (or -17 [12])
n-Butyl acrylate (n-BA)	0.15	-43
iso-Butyl acrylate (i-BA)	0.18	-17
t-Butyl acrylate (t-BA)	0.15	+55
2-Ethylhexyl acrylate (EHA)	0.04	-58
Lauryl acrylate (LA)	<0.001	-17
Methacrylates		
Methyl methacrylate (MMA)	1.5	+105
n-Butyl methacrylate (n-BMA)	0.08	+32
iso-Butyl methacrylate (i-BMA)	0.13	+64
Styrene (S)	0.02	+107
Acrylonitrile (AN)	8.3	+105
Vinyl acetate (VAc)	2.4 to 2.5	+42 (or +28 [12])

The special features of the poly(meth)acrylates which justify their relatively high price are the generally very good weatherability and UV stability, high transparency (to UV radiation as well), good water resistance and yellowing resistance, and great ease of variation in toughness, flexibility and hardness. The high level of film gloss combined with good gloss retention under weathering, in conjunction with the chemical resistance to alkali, acid and water (resistance to hydrolysis), likewise contributes to the excellent suitability of this class of polymer for the most diverse applications.

Structure and properties

The main polymer properties, such as glass transition temperature, film mechanics and polarity, are influenced by the structure of the main chain and side-chains. The water solubility of the monomers (Table 4.1) can be regarded as a measure of the polarity of the homopolymers. As the solubility in water increases, there is an increase in the polarity of the resultant polymers. The free acids (M)AA are of unlimited water solubility, especially in the neutralized state. In the case of the esters, water solubility falls with increase in side-chain length.

The C-C-linked main chain is largely chemically inert and provides the poly(meth) acrylates with chemical resistance and weatherability.

4.1.3 Film formation by polymer dispersions

Film formation by a dispersion [47–57] is significantly more complex than for a solution polymer, in which evaporation of the solvent is followed by the production of a continuous film (see Chapter 3.4) via entanglement of the polymer chains. In contrast, film formation in polymer dispersions is a multistage process (Figure 4.2).

Drying phases

During drying, the dispersion polymer particles move closer and closer together (see a in Figure 4.2) as the water evaporates, until they make contact with one another (first phase of film formation, see b). As the water continues to evaporate, the capillary and surface tension forces press the particles against one another and, above a temperature which is characteristic of every polymer, namely the minimum film-forming temperature (MFFT), cause extensive deformation of them. Up to this point, however, the former particle boundaries are still visible (second phase of film formation, see c). The last step consists in inter-diffusion of polymer chains across the former particle boundaries.

As a result, the polymer particles coalesce with one another at the points of contact to form a continuous film (third phase of film formation, see d).

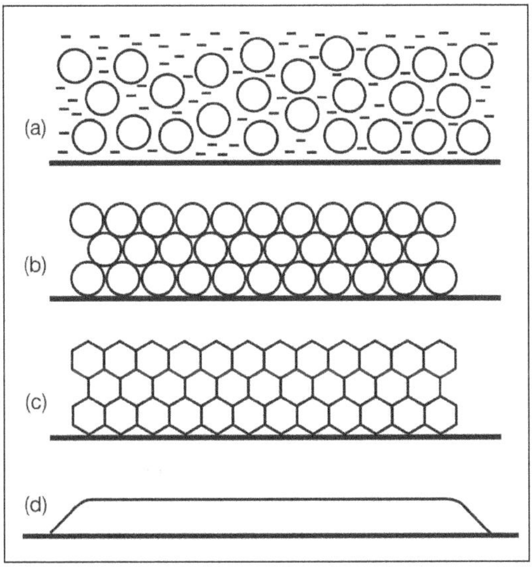

Figure 4.2: Scheme of film formation

On account of the fact that the water-soluble components, such as emulsifiers and salts, are deposited at the former particle boundaries in the form of a so-called interstitial phase during film formation, the dispersion film never completely "forgets" its particulate past, even at temperatures far above the MFFT. Consequently, in electron micrographs, it is usually still possible to discern a hexagonal or honeycomb network, which is formed from the former particles (examples in [57], Figure 4.2). The hydrophilic components in the interstitial phase are the reason that dispersion films and emulsion paints are more sensitive to water than films or coatings based on solution polymers.

Fine dispersions are better at forming films than coarse dispersions of the same composition [58], a fact which is usually also reflected in a somewhat lower MFFT [59], improved film quality and water resistance, and a higher level of film gloss. That explains why acrylate-based dispersion binders are usually fine dispersions with particle diameters below 300nm, and mainly from 100 to 200nm.

The MFFT is usually situated just below the T_g of the polymer. The difference between MFFT and T_g is particularly marked in the case of polar dispersions. In that case, the phenomenon of hydroplasticization occurs, i.e. plasticizing of the polymer due to swelling of the particle surface by water. A consequence of this is that polar dispersions have an MFFT which is up to 15°C lower than that of non-polar dispersions of the same T_g.

4.1.4 Parameters and properties of coatings binders

The key parameters of a polymer dispersion for coatings [60] are set out below:

- Solids content/polymer content
- Coagulum and microcoagulum (specks) content

- Particle size (particle diameter)
- Viscosity
- pH
- Colloidal stability (to shearing, electrolyte exposure, freeze/thaw cycling)
- Minimum film-forming temperature/glass transition temperature;
- Molecular weight/crosslinking density;
- Surface tension; and
- Residual monomer content/VOC content/odour

The significance of T_g and MFFT is discussed in Chapter 4.3. The remaining items are addressed in more detail below.

Solid content

The solid content is a measure of the quantity of active substance in the dispersion. It is the ratio of the dry mass of the dispersion (following evaporation of all volatile fractions) to the overall mass. The solid content is composed of the polymer, the stabilizers and the inorganic salts (initiator decomposition products or buffer substances, for example). The solid content is normally determined in accordance with DIN ISO 1625 or DIN EN ISO 3251.

Coagulum

Coagulum or micro-coagulum, is the term used to refer to more or less coarse polymer material or precipitate in the dispersion. It can be quantified by collecting it in a filter of defined mesh size (e.g. 100µm) and weighing it (e.g. in accordance with DIN ISO 4576). It comprises fragments of skin, dried-on foam or coarse aggregates of polymer particles, all of which may have a disruptive effect on the quality of the dispersion and coating film. To the eye, the fine coagulum fractions appear as specks when the dispersion has been applied with a doctor blade onto a glass plate and is viewed in transmitted light. These may lead to defects in the coating film, especially on spray application of emulsion paints, for example.

For this reason, manufacturers nowadays pass polymer dispersions through fine-mesh filters (usually from 20 to 200µm in mesh size). Since it is usually impossible to rule out foaming and skinning completely during transport and handling of the dispersions, it is advisable – especially for high-quality paint systems with little or no pigmentation – to filter the dispersion again before using it to prepare paints.

Particle size

The particle size of a dispersion has an influence on many important binder properties, such as film-forming tendency (see Chapter 4.3), film gloss (see Chapter

4.6), binding power (Chapter 4.6) and penetration capacity into porous substrates (Chapter 4.6). Moreover, the particle size strongly determines the internal surface area of the system and thus the quantity of stabilizer required. The particle size is therefore an important parameter of the binder. A simple method of determining it is to measure the haze or light transmission (LT) of a highly diluted sample (e.g. 0.01 wt.% solution). As the particle size decreases, there is an increase in the LT of the dilute dispersion. Similarly, the appearance of the undiluted dispersion changes from opaque white (for particle sizes greater than 200nm) to a bluish shimmer (for less than 100nm) down to virtually transparent for very fine dispersions having particle sizes below 30 to 40nm. The haze of 0.01% dispersion relative to water is referred to as the light transmission value. Certain dispersion manufacturers specify it as a measure of particle size. However, the LT is determined not only by the particle size but also by the difference in refractive index between the polymer and water.

Owing to the higher refractive index of polystyrene compared with poly(meth) acrylates, styrene-acrylate dispersions have lower LT values than straight acrylics of comparable average particle diameter. Nowadays, precise determination of the particle diameter is carried out with the aid of laser-assisted light scattering methods. To obtain precise information about the breadth of the particle size distribution, more complex methods are employed, such as analytical ultracentrifuges, electron microscopy with image analysis, and capillary hydrodynamic fractionation (CHDF) [60].

Surface tension

For a certain particle size and solid content, the surface tension of a dispersion is dependent essentially on the polarity of the base polymer and on the nature and quantity of the hydrophilic comonomers (e.g. AA) and the surface-active components (e.g. emulsifiers). This parameter is of considerable importance to the application of paints and coatings. For instance, effective wetting of a substrate and defect-free film formation are guaranteed only if the dispersion, or paint or other coating, has a sufficiently low surface tension. The majority of polymer dispersions have surface tensions of between 30 and 50mN m^{-1} (for comparison: that of pure water is 73mN m^{-1}). In practice, the surface tension of a dispersion is usually measured by the *Du Nouiy* ring method (see DIN ISO 1409). This involves removing a platinum ring oriented parallel to the surface from the dispersion, and measuring the force required to remove the adhering liquid lamella until it breaks off. Dispersions of high viscosity must be diluted beforehand in this method.

Viscosity and rheology

The flow properties, i.e. the viscosity of the polymer dispersion in the liquid phase, are of considerable significance for its preparation, handling and processing.

Besides the particle diameter, the polymer content of the dispersion and its surface functionalization – arising from the use of water-soluble comonomers, such as (M)AA or (M)AM – regulate the viscosity and flow behaviour. The overall viscosity is a function of the viscosity of the aqueous phase, and of the volume fraction and the packing of the particles. The viscosity of a dispersion may be described mathematically by the *Mooney* equation:

$$\ln \eta = \ln \eta_K + \frac{K_F \cdot V_D}{1 - (V_D / K_P)} \qquad \textit{Mooney} \text{ equation}$$

where

η = viscosity, η_K = viscosity of the aqueous phase, K_F = form factor (2.5 for spherical particles), V_D = volume fraction of the disperse phase, and K_p = packing factor (0.637 for monodisperse, close-packed spheres).

Accordingly, given a low volume fraction, the viscosity is determined essentially by the water-soluble polymer fractions and increases slowly as the volume fraction of the polymer rises. However, if the volume fraction is close to the packing factor, then the viscosity rises very rapidly [61]. As the particle diameter decreases, the increase in viscosity takes place at lower and lower volume fractions of polymer, due to increasing packing density and increasing particle interaction.

This increase occurs more readily for mono-modal dispersions (i.e. dispersions with particles of uniform size) than for dispersions with a broad or even multimodal particle size distribution [61 - 63].

Influence of solid content and viscosity

Very fine, mono-modal dispersions (particle size 30 to 80nm) are fluid only up to solid contents of 35 or 45 wt.%, whereas standard binders with particle sizes of 100 to 200nm retain their fluidity for solid contents of 50 up to a maximum of 55 wt.%. For dispersions, unlike polymer solutions, the effect of molecular weight on the viscosity is relatively minor.

A solid content of more than 60 wt.% – at an acceptable viscosity – can only be obtained with dispersions that have a bimodal or multimodal particle size distribution, i.e. for mixes of particles differing in size [61 - 63].

In such cases, the fine particles fill the gaps between the coarse particles. This is clear from the atomic force microscopy (AFM) surface micrograph of a film of an essentially bimodal dispersion (Figure 4.4). Multimodal, highly concentrated dispersions of this kind can be produced in a controlled fashion by means of special EP techniques. This is normally done by generating one or more new particle generations in the course of the polymerization, by means of targeted addition of emulsifier or seed latex (see Chapter 4.4).

Viscosity control

Aqueous polymer dispersions and paints without added thickener generally exhibit a more or less pronounced pseudo-plastic flow behaviour, i.e. their viscosity falls as the shear rate increases. This is in contrast to solution polymers (Chapter 3) and conventional solvent-borne paint systems (e.g., alkyd coatings), which tend to display Newtonian behaviour, i.e. virtually no shear thinning.

A simple method of determining the flow behaviour is to work with flow cups (Ford cup, DIN cup, or as per ISO 2431, EN 535). These are funnel-shaped cups which are filled with a defined volume of the dispersion. The efflux time as a function of the outflow geometry (length and diameter) is then a measure of the viscosity of the dispersion.

Alternatively, the viscosity can be determined with a rotational viscometer (e.g. as per DIN 53214, DIN EN ISO 3219 or Brookfield measurements as per ISO 2555, ISO 1652). In this case, a metal cylinder turns concentrically in a dispersion-filled cup. By measuring the torque on this cylinder, it is possible to measure the shear stress when the shear rate is varied. From the relationship that the viscosity is equal to the ratio of shear stress and shear rate, the viscosity at different shear rates can then be calculated (flow curve).

Normally, manufacturers will specify the viscosity of the dispersion at a certain shear rate (e.g. $100s^{-1}$ or $250s^{-1}$). For example, architectural coating binders will normally be located within the range 50 to 1500mPa s (at $100s^{-1}$ as per DIN EN ISO 3219).

Stability

The colloidal stability of the dispersion in preparation and use, as already described, is a function of the quantity of stabilizer (e.g. emulsifiers) and of the particle size and solid content. Owing to the rapid increase in internal total surface area, the stability of the dispersion falls sharply for a given particle volume and quantity of stabilizer as the particle size decreases.

With rise in solid content, colloidal stability diminishes owing to growing interaction between the particles due to a higher packing density. In general, the stability can be adversely affected by a number of factors, such as mechanical stress, temperature exposure, exposure to freezing or an electrolyte, or addition of solvent.

If, for example, the temperature falls below the freezing point of the aqueous phase, then in many cases the polymer dispersions coagulate or aggregate irreversibly. As far as possible, therefore, the dispersions should be stored under frost-free conditions. With modern-day dispersions, the low-temperature stability can be improved to such an extent by adding sterically stabilizing nonionic emulsifiers that in some cases they can withstand a number of freeze/thaw cycles while still allowing complete redispersion.

An overly high temperature likewise impairs the stability of polymer dispersions. As the temperature rises, the number and energy of the particle collisions grows, and thus the tendency of the system to coagulate also increases. In general, however, the effect of an increase in temperature is fairly small, provided that there is no severe loss of water due to evaporation and, where reactive monomers are used, that no chemical changes occur.

pH value

The pH of coating binders is usually within the neutral to weakly basic range (pH 6 to 9, measured in accordance with DIN 53 785 or ISO 976). This is a consequence of the fact that the stability of the binder, which is usually formed by copolymerization of (M)AA and which accordingly is functionalised with carboxyl groups, increases greatly as the pH goes up (especially above a pH of 5 to 7). The reason for this is increasing deprotonation and thus improved charge or Coulombic stabilization. Moreover, standard dispersant agents based on polycarboxylic acids likewise exert their agglomeration-preventing action on fillers and pigments only above a pH of 6.5 in the finished paint, which must therefore be neutral to slightly alkaline.

The pH also has an influence on the viscosity of dispersion and paint. As the pH goes up, the viscosity rises as a function of the quantity of copolymerized carboxylic acid (depending on type and quantity).

Residual monomers and residual volatiles

In the case of EP, the residual unreacted monomers (residual monomers) in the dispersion is up to 1% (10,000ppm) in the case of styrene-acrylate copolymers and up to 0.1% (1000ppm) in the case of straight acrylics. The total resultant residue is determined by the monomers used, the quantity of initiator and the polymerization conditions. Controlled addition of a redox initiator system, such as hydrogen peroxide/ascorbic acid, after the end of the polymerization proper allows the quantity of residual monomers to be reduced significantly. In this way, it is usually possible to meet requirements such as that of the German "Blue Angel" for gloss emulsion paints [64], i.e. residual monomer content in the binder of less than 500ppm. Stripping, i.e. passing steam through the dispersion under reduced pressure in the manner of water vapour distillation, can substantially eliminate residual monomer fractions (down to < 50ppm), provided the colloidal stability of the dispersion is sufficient. This method can also be used to eliminate the other volatile fractions (volatile organic compounds, VOCs), such as by-products or decomposition products, and also impurities from the starting materials used for the dispersion [65]. A total VOC of less than 500ppm is possible. Residual monomer contents and VOCs are usually determined by means of GC [66].

4.2 History

4.2.1 Chronological development

Polymer dispersions can be seen as having evolved historically in three main classes:

a) Natural latices
b) Synthetic polymer dispersions, and
c) Secondary dispersions

a) Natural latices

The most important naturally occurring polymer dispersion is latex, the milky juice of the Hevea brasiliensis tree. This produces a dispersion which is a polyisoprene dispersion composed of > 97% cis-1,4-polyisoprene and having a particle size of 100 to 1500nm. The actual rubber which retains its consistency is obtained by curing the latex with sulphur.

A further and, for some time, also quite important natural source was the gutta percha tree (Palaquium gutta), whose milky juice contains practically pure trans-1,4-polyisoprene.

b) Synthetic polymer dispersions

As already mentioned, in the early 20th century, research into synthetic polymer dispersions made by aqueous-phase polymerization advanced rapidly and was highly successful. At that time, the polymerization process used was EP; it was only later that suspension polymerization and mini-emulsion polymerization, a special kind of suspension polymerization, were developed.

c) Secondary dispersions

At the end of the 20th century, more and more research was done on creating secondary dispersions by dispersing a polymer in the aqueous phase. This can be accomplished by dispersing a solution polymer solution in water and then removing the solvent by distillation. Other methods are precipitation of a solution polymer in a water-miscible solvent by adding water, and the progressive dispersion of water in a polymer until phase inversion occurs. Polyurethane dispersions are typical secondary dispersions that are available commercially and are made by dispersing a solution polymer.

4.2.2 Technological development

The historical development of dispersions can also be described in terms of technological advances, especially in respect of coatings.

In the days before emulsion paints, indoor and outdoor walls were whitened with chalk-based paints based on CaO. Silicate paints were already in use in the 19th century. In some regions, glossy drying-oil and alkyd-resin paints were used for façades and walls. The binder in classical oil paint was linseed oil plus wood oil, with white lead or white zinc serving as pigment. For interior use, there were also wall paints based on glues and casein.

Polymer dispersions, generation by generation

A look at developments in water-borne coatings over the last eight decades reveals that the various changes occurred in a number of steps or generations [67].

1st Generation – "water, not solvent"

The first generation featured the change from combustible solvents to water. The first polyvinyl acetate dispersion came onto the market in 1934. It was used for coatings, adhesives, leather finishes and for textiles [68]. As PVAc is a hard polymer, a plasticizer needs to be added. This plasticizer either evaporates from the substrates as they progressively age or migrates into the wall. A further disadvantage of PVAc is its limited saponification resistance, which on alkaline substrates, such as concrete and mineral plasters, leads to coating damage. Better solutions to the priming/sealing problem took the form of new sealants based on finely divided polymer dispersions and solution polymers.

2nd Generation – "plasticizer-free"

The weaknesses of the first-generation polymers soon led to new developments. These took off in two directions. First was the development of a homopolymer with an MFFT near 0°C, which in 1955 led to the production of a protective-colloid-stabilized dispersion based on vinyl propionate. Second was the copolymerization of the "hard" VAc with a "soft" comonomer. These new "internally plasticized" dispersions marked a huge step forward in the formulated products.

3rd Generation – "more alkali resistance"

The biggest advances in polymer dispersions came with the 3rd generation. The use of new monomers in VAc and acrylic dispersions led to coatings which met requirements for high saponification resistance, low water absorption and low dirt pick-up. This new generation was the outcome of specific breakthroughs. One was undoubtedly the development of finely divided acrylate-styrene copolymer dispersions, which were launched for architectural paints in 1965. Emulsion paint formulators soon discovered that the styrene-acrylate dispersions had high pigment binding power (PBP), i.e. that the binder content of matt indoor paints could be reduced. Popular quality requirements at the time, such as "scrub resistance" and "wash resistance", which were later standardized by DIN were achieved with less binder. The success with indoor paints was followed by more widespread use in architectural paints and synthetic-resin-bound plasters, where their specific advantages of high saponification resistance and low dirt pick-up came to bear (Chapter 4.6).

New developments occurred in the field of VAc copolymers, too. Among these were dispersions produced by pressure polymerization of VAc and ethylene. The use of the monomers ethylene and vinyl chloride, which cannot be saponified, dramatically improved the alkali resistance of the VAc copolymers and the PBP.

Finally, there came advances in straight acrylics, which contain MMA by way of "hard monomer" instead of styrene. In some countries, they competed against styrene-acrylate copolymer dispersions, which are also finely divided dispersions based on emulsifiers. As both types are referred to as acrylate copolymer dispersions, they are differentiated by calling the one type straight acrylics and the other, styrene-acrylate dispersions.

4th Generation – "crosslinking" and "wet adhesion"

The polymer dispersions belonging to the first three generations dry physically. The 4th generation brought "crosslinking" and "wet adhesion". For use as wall paints, gloss emulsion paints had many defects, such as poor shear resistance, lack of alkyd-like flow behaviour, poor blocking resistance for application to doors and windows, and an absence of "wet adhesion" on old oil and alkyd paints. The 1980s saw the advent of environmental awareness, and legislation aimed at replacing solvent-borne coatings with their water-borne counterparts. The weaknesses that had been inherent in polymer dispersions as well as emulsion paints up to that point were overcome by four means:

"Crosslinking" (see Chapter 4.6.4), "core-shell technology" (see Chapter 4.4), and "wet-adhesion promoters" (see Chapter 4.6) were incorporated into new polymer dispersions, while the flow behaviour was influenced by a new additive: "diurethane thickener" [70].

5th Generation – "eco-friendlier"

The emergence of water-borne, silicone architectural coatings brought a combination of high water vapour transmission and excellent hydrophobicity that enabled façades to be coated with "hydrophobised porosity" [69] (Chapter 4.6.3).

Although emulsion paints are among those coatings with the lowest VOC content, health concerns have been raised about some constituents of polymer dispersions and water-borne coatings, leading to the goal of lowering the content of the problematic substances or finding replacements for them. While binder-rich, block-resistant gloss emulsion paints based on pure acrylic dispersions still require some film-forming agent, new low-odour styrene-acrylate or VAc-ethylene copolymer dispersions are now available for matt-drying façade and indoor paints [71, 72].

4.3 Composition of acrylates and their influence on performance

4.3.1 Parameters influencing binder properties during latex preparation

This chapter discusses the different production methods und their influences on emulsion polymers in greater depth.

4.3.2 Raw materials

As well as the monomers, ingredients such as initiators, emulsifiers, buffer systems and chain-transfer agents, and process parameters such as pressure, temperature and metering time, exert a considerable influence on the polymer characteristics and the properties of the binder. The following ingredients need to be borne in mind when a binder is being prepared:

- Main or basic monomers
- Auxiliary or functional monomers (stabilizers, crosslinkers, adhesion promoters, etc.)
- Emulsifiers
- Initiator system

- Chain-transfer agents/regulators
- Buffer substances
- Complexing agents
- Neutralizing agents
- Biocides/preservatives; and
- Defoamers

4.3.2.1 Monomer selection

Functional monomers

The basic properties of the copolymer, such as polarity, hardness and flexibility of the coating, are primarily determined via the choice of the standard monomers, such as n-BA, 2-EHA, EA, MMA or styrene [36, 73 - 75]. However, functional monomers, which as auxiliary monomers, usually make up only 0.5 to 10 wt.% of the overall polymer, also exert a considerable influence on the overall properties. The rheology, for example, can be controlled by copolymerizing (M)AA, (M)AM or using sulpho-functional monomers, such as acrylamidopropanesulphonic acid. The colloidal stability of the dispersion can be increased in this way, too. Crosslinking monomers, examples of which are divinylbenzene, bisacrylates such as ethylene glycol dimethacrylate, and epoxy-functional or N-methylol-functional monomers, enhance resistance to solvents and other chemicals. They also yield coating films of greater mechanical strength.

Monomers with amino, acetoacetoxy, phosphate, siloxane or urea functional groups, as well as (M)AA with their carboxylic acid groups, can improve the adhesion to the substrate as a result of specific interactions or chemical reactions.

Structure and properties

The main polymer properties, such as T_g, film mechanics and polarity, are influenced by the structure of the main chain and side-chains. Table 4.1 (page 133) shows how the main and side-chain structure influences properties such as water solubility and T_g. As the solubility in water increases, there is an increase in the polarity of the resultant polymers.

Polymethacrylates have more-rigid chains than the homologous polyacrylates due to the additional methyl groups and the resultant steric hindrance of rotation of the main chain. This greater rigidity imbues polymethacrylates with higher T_g, greater hardness, and reduced flexibility in comparison with the homologous polyacrylates. As the side-chain becomes longer, hardness and T_g decline and extensibility increases (up to 8 carbon atoms in the case of acrylates and about 12 carbon atoms in the case of methacrylates; with longer chains, hardness increases stem from increasing crystallinity) [74].

Table 4.2: Influence of growing side chain length on the properties of poly(meth)acrylates

	Influence on properties
Growing side chain length	Hardness decreases
	Flexibility and extensibility increase
	UV stability rises
	Gloss retention improves
	Tackiness increases
	Alkali resistance increases
	Water sensitivity decreases
	Alcohol resistance increases
	Hydrocarbon solubility increases

Increases in chain length are also accompanied by an increase in tackiness of the polymers, as shown by tack measurements of different homopolyacrylates (Figure 4.27, page 231).

The influence exerted by the side-chain is illustrated in Table 4.2. Specifically, poly(meth)acrylates bearing unbranched side-chains and formed from unbranched alcohols are softer and more extensible than those originating from branched isomers. This is also reflected in their different T_g values.

Glass transition temperature of copolymers

Copolymerization of the monomers affords a way of varying the properties of the polyacrylate dispersions over a broad range. For coating binders, it is usual to combine monomers of low T_g ("soft" monomers), such as BA and EHA, with monomers of high T_g ("hard" monomers) of the homopolymers, such as BMA and MMA, so as to yield a copolymer of T_g between 0 and 40°C. Only when a high level of low-temperature elasticity is required, such as for crack-bridging exterior coatings, are lower T_g values (down to -45°C) conferred in the copolymers by using appropriately high fractions of soft monomer. Interior wood coatings and special industrial coatings require a high level of coating hardness, and so the T_g values of the binders should be 40°C or more.

Copolymer variants

In the case of the styrene-acrylate copolymers, costs can be reduced and properties optimized by replacing some of the MMA commonly employed (to bestow the necessary hardness) for straight acrylic copolymers with inexpensive styrene monomer. This works because styrene (S) tends to copolymerize well with acrylates (relative to MMA) and because the two homopolymers have roughly equal T_g values. In the resultant polymers, the incorporated, non-polar styrene monomer, acting as a substitute

Table 4.3: Binder properties arising from use of styrene or methyl methacrylate as hard monomer

Property	S	MMA
Hardness	++	++
Light stability	+/- to -	++
Water resistance	++	+/-
Water vapour permeability	-	+
Chalking/gloss reduction	+/- to -	++
Dirt pick-up resistance	++	+
Saponification resistance	++	+/- to +
Pigment binding power	++	+/-
Film gloss	++	+
Price	+	-

for MMA, confers superior water resistance and alkali resistance and a higher PBP. As PS has a higher refractive index than PMMA, moreover, incorporation of styrene leads to higher coating gloss. However, due to the inherent UV absorption of the phenyl groups, a high styrene content in the polymer and low levels of pigmentation in the coatings can conduce to increased photo-induced binder degradation, chalking, loss of gloss and yellowing in the course of long-term weathering (for a comparison of properties, see Table 4.3 and Chapter 4.7).

Co-solvents

For reliable film formation at room temperature, even with "hard" standard copolymer dispersions which have T_g values higher than 20°C, it is common to use temporary plasticizers, i.e. solvents which evaporate after film formation. Unlike the permanent plasticizers which are likewise used to lower the MFFT, these solvents do not remain in the film. They are emitted to the environment at rates dependent on the ambient temperature, atmospheric humidity and boiling point, and on the resultant vapour pressure.

The solvents in emulsion paints are therefore frequently referred to as film-forming aids or coalescents [76a]. Besides white spirit, the main coalescents are water-miscible glycol ethers (butyl glycol, butyl diglycol, dipropylene glycol monomethyl ether, dipropylene glycol monobutyl ether) and their acetates, although the debate on VOCs and eco-label provisions (e.g., the EU's sunflower symbol, Germany's Blue Angel) has led to the increasing use of high boilers, such as "Texanol"[1], and esters of dicarboxylic acids, such as "Lusolvan" FBH[2], and tripropylene glycol monoisobutyrate.

The reason is that, according to the provisions of the European Union and the German paint industry association, the term "solvent" applies only when the boiling point is below 250°C (at 1 atm) [76b - 76d]. Nowadays, therefore, all film-forming auxiliaries which have boiling points above 250°C are by definition plasticizers.

[1] Eastman [2] BASF SE

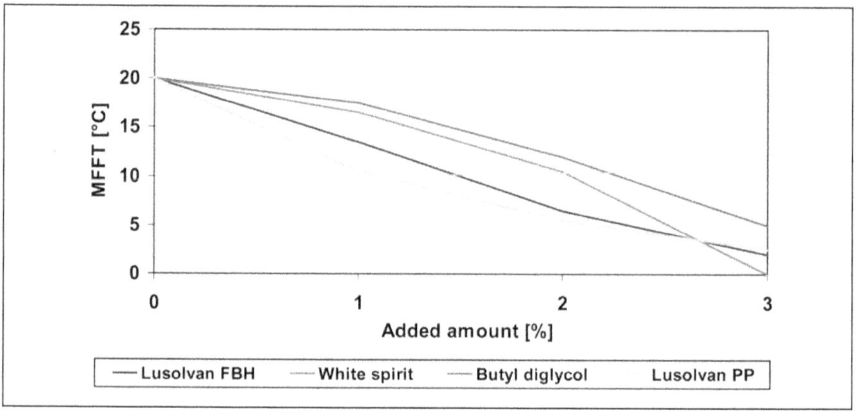

Figure 4.3a: Influence of the co-solvent on the MFFT of a typical acrylate/styrene binder

They can be used without restriction in emulsion paints labelled as "solvent-free", despite the fact that they do not remain permanently in the coating [76b, 76c].

The film-forming agent exerts a particular effect on the position of the MFFT. An important role is played by its compatibility with and ability to act as solvent for the latex particles [76e]. Hydrophobic solvents, such as white spirit or Texanol, are highly compatible with the polymer. Accordingly, to an extent depending on their affinity for the latex particle, they cause greater swelling and plasticizing than hydrophilic solvents, which predominate in the aqueous phase. Hydrophilic solvents, such as ethylene glycol or propylene glycol, have virtually no plasticizing effect. However, they slow the evaporation of water and so retard film formation. They are thus ideal auxiliaries for prolonging the coatability phase (open time or wet-edge time). For this reason, they are used to improve processing properties, while conferring frost resistance on the paints by lowering the freezing point.

The effect of various solvent additions on the course of the MFFT is illustrated by the binder "Acronal" 290 D[1], a styrene-acrylate dispersion with a T_g of 23°C and an MFFT of approx. 20°C (Figure 4.3a).

Service properties

High boilers, such as "Texanol"[1] and "Lusolvan" FBH[2], remain in the film, sometimes for weeks or months, with the result that the end properties of the coating, such as freedom from tackiness, good hardness and high blocking resistance, are not achieved until sometime after application (Figure 4.3b). In contrast to film-forming agents, true plasticizers remain permanently in the film, permanently altering its properties. As the film exhibits marked tackiness, the coatings have

[1] Eastman [2] BASF SE

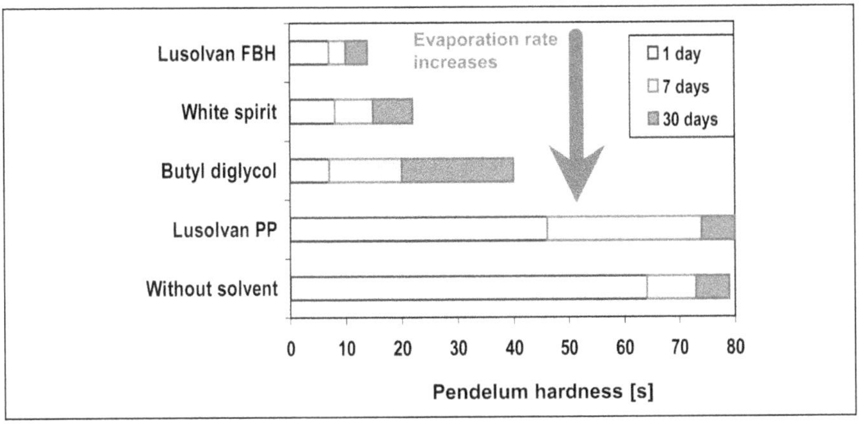

Figure 4.3b: Influence of co-solvents on the surface hardness of a typical acrylate/ styrene binder

a much greater tendency to dirt pick-up. Consequently, the use of permanent plasticizers is restricted to specialty formations, e.g. flexible, crack-bridging coatings. Furthermore, plasticizers in many cases have a tendency to migrate to the surface, which can further intensify the soiling tendency of such modified coatings throughout their lifetime.

In general, film formation by a polymer dispersion or a dispersion-bound paint is contingent on a sufficiently high processing temperature and a sufficiently long drying time. Only at temperatures above the MFFT does the system form a continuous, clear film which can withstand mechanical stress. If the water evaporates at temperatures below the MFFT, fissures or even powdery, turbid films are formed.

Other than by the choice of film-forming auxiliary, the drying rate and film quality are also affected to a considerable extent by the temperature, the relative atmospheric humidity, and the rate of air change [76f].

Environmental aspects

Increased environmental awareness among end users in Europe in particular has encouraged paint and coating manufacturers to increasingly offer low-solvent or solvent-free products.

For this purpose, raw materials manufacturers nowadays offer "internally plasticized" binders in their ranges, with a greater proportion of copolymerized soft monomers (for current product reviews see [76g]). For good film formation without added solvent in adverse weather conditions, the binder MFFT must not exceed

5°C; a MFFT less than 3°C is even better [76h, 76i]. According to a directive of the German paint industry association, "eco paints" made with such binders can be labelled "solvent- and plasticizer-free" if the residual VOC and plasticizer content is less than 1 g per litre of paint [76b, 76c].

The MFFT of a dispersion is usually determined by means of a *Kofler* bar (e.g. as per ISO 2115).

4.3.2.2 Auxiliaries

The emulsifiers or surfactants – usually combinations of anionic and nonanionic types are used in binders for architectural coatings – are responsible for providing adequate colloidal stability and also for the pigment and filler compatibility of the dispersions. This is ensured by charge stabilization or steric stabilization. The nature and quantity of the emulsifiers, however, also determine the particle size and thus the viscosity and film properties of the system.

Emulsifiers are amphiphilic compounds, normally composed of a long-chain, hydrophobic organic tail and a hydrophilic head. The organic tail is generally an alkyl (usually C12 to C18), alkylbenzene, alkyldiphenyl oxide or alkylphenol group (usually C8 to C9). For anionic emulsifiers, the hydrophilic, polar head is a sulphate, polyether sulphate, sulphonate, sulphosuccinic acid, carboxyl, phosphate or phosphonate group [77, 78] Anionic emulsifiers that are very frequently used in industry are sodium dodecyl sulphate ($C_{12}H_{25}O\text{-}SO_3Na$) and sodium dodecylbenzenesulphonate ($C_{12}H_{25}\text{-}C_6H_4\text{-}SO_3Na$). Studies have shown that they are also suitable for EP.

The hydrophilic part of non-ionic emulsifiers is played by uncharged long-chain polyethylene oxide (EO units: 8 to 50) or polyglycoside (in the case of alkyl polyglycosides) chains. Polypropylene oxide/polyethylene oxide block copolymers also serve as non-ionic emulsifiers in EP. Emulsifiers based on alkylphenol ethoxylates, which for decades have shown themselves to be good technical emulsifiers for EP, have attracted increasing criticism in recent years owing to their ecotoxicity (toxicity to fish, and disputed endocrine activity) [6].

For this reason, they are increasingly being replaced by alternative ethoxylated emulsifiers based, for example, on natural fatty alcohols or synthetic oxo alcohols, and by alkyl polyglycosides [80 – 82].

Protective colloids are natural or synthetic polymer emulsifiers (polyvinyl alcohol, starch, polyvinylpyrrolidone, hydroxyethylcellulose or polypeptides such as gelatin). They contribute to steric stabilization, especially with respect to the effect of the electrolytes, and help to improve storage stability. Moreover, as a result of their swelling capacity and hydration capacity, they also regulate the viscosity and flow behaviour of the dispersion and the formulated paints, but usually have an adverse effect on water resistance [83]

Initiators and chain-transfer agents

Water-soluble, thermally decomposing, free-radical initiators or free-radical formers, such as alkali metal (Na, K) and ammonium peroxodisulphates, are employed. However, hydrogen peroxide, organic peroxides and hydroperoxides, and azo compounds are also suitable. Free-radical formation takes place by homolytic scission of the peroxo groups or, in the case of the azo initiators, by elimination of nitrogen. The decomposition characteristics are normally chosen to allow polymerization at 75 to 95°C. An alternative is to use redox systems, in which an oxidizing agent is combined with a reducing agent for the purpose of initiating the polymerization. Redox polymerization of this kind requires only minor thermal activation, and so permits polymerization at relatively low temperatures (down to room temperature). Examples of redox systems are combinations of hydrogen peroxide with ascorbic acid, and hydrogen peroxide with reductive iron(II) or copper(I) salts.

The nature, quantity and metering strategy for the initiator and of the chain-transfer substances are critical for the molecular weight and the polymer architecture (e.g. branching and crosslinking); they also influence the residual monomer content at the end of the polymerization.

Commonly employed chain-transfer agents include mercapto compounds, such as thioethanol and long-chain mercaptans, e.g. n- or tert.-dodecyl mercaptan. Here, the free S-H group serves to transfer hydrogen to the growing polymer chains. The mercapto free-radicals are highly stable, however, and have limited ability to promote chain growth. All in all, therefore, the use of chain-transfer agents leads to a controlled reduction in molecular weight.

Buffers and neutralizing agents

Buffers, such as sodium carbonate, sodium acetate and sodium bicarbonate, and complexing agents such as ethylenediamine tetraacetate, can improve stability and ion compatibility during polymerization.

Where the EP is initiated with peroxodisulphates, or where a copolymerizable (M)AA acid is used for stabilization, the dispersion at the end of the process will be in acidic form.

Neutralization (commonly pH 7 to 9) with ammonia, amines or alkali metal hydroxides can boost the stability of the dispersion. It was common in the past to use ammonia, a volatile base, for this purpose. Problems with its odour as the dispersions are being processed and forming films has seen it being increasingly replaced by low-odour neutralizing agents. [76h, 76i].

Preservatives

Adding biocides, usually active substance combinations of methyl and chloroisothiazolinones, benzisothiazolinones, or of formaldehyde or formaldehyde donors, can prevent infestation of the dispersions by microorganisms, such as bacteria, fungi (moulds), yeasts and algae, during storage and transport.

Defoamers

Small quantities of defoamers are added to excessively-foaming dispersions to prevent excessive surface foam or microfoam during preparation, handling and transport. Defoaming and the various classes of defoamer are discussed in more detail in the literature [66].

4.4 Emulsion polymerization processes

4.4.1 Polymerization control

The process parameters, such as pressure, temperature and feed time, act in tandem with the chosen initiator, chain-transfer system and monomers to lay down the polymer architecture, i.e. the chain length, the degree of branching and the gel content (or crosslinked domains). The polymer architecture in turn has a great influence on important technical properties of the binder, such as film tack and pigment binding power (PBP).

The expertise of the leading dispersions manufacturers in controlling the properties of the binder stems from their ability to correctly combine the ingredients and judiciously balance the process parameters. As there are so many different possible influences, clearly two dispersions of identical monomer composition can still differ considerably in their properties. A good overview of market products at the beginning of the 21st century is given in [76g].

4.4.2 Multi-phase systems

Special polymerization techniques, such as staged or gradient polymerization, enable polymer particles to be made that have two or more polymer phases [84–92]. Such polymers are available in a very wide variety of morphologies. Known morphologies include customary core-shell particles and raspberry, strawberry or half-moon shaped particles, as well as particles that have inclusion structures or an inverted core-shell structure.

Multi-phase systems are playing an increasing role as binders in the architectural coatings sector, especially for gloss emulsion paints (Chapter 4.6.1.4), wood coatings (Chapter 4.6.1.5) and also solvent-free paints.

By combining a soft, film-forming polymer phase (polymer of low T_g, < 10°C) with a hard, non-film-forming polymer phase (polymer of high T_g, > 50°C) in one and the same particle, and by tailoring the particle morphology, it is even possible to achieve properties which appear per se to be contradictory. Such systems have a low MFFT and high elasticity, along with good freedom from tackiness, excellent blocking resistance, and good coating hardness (see [93]).

A further example of the use of staged emulsion polymerization to adjust properties is the incorporation of, say, a carboxylic acid-rich polymer phase on the surface of latex particles to regulate the rheology of the emulsion.

4.4.3 Seed polymerization

Ready-made fine (diameter < 50nm) emulsions acting as seed or seed latex afford a way of determining the particle number and thus the ultimate particle size with a relatively high level of precision at the very outset of EP [94a]. This technique, known as seed polymerization, ensures excellent reproducibility of the particle size distribution from one batch to the next and thus a highly consistent viscosity.

An alternative called the in situ seed process has become established in industry. In it, the polymerization proceeds phase-wise in a one-pot process. To start with, a portion of the monomer emulsion is polymerized under the addition of further emulsifier and a portion of the initiator. This step precedes the actual EP in the reactor.

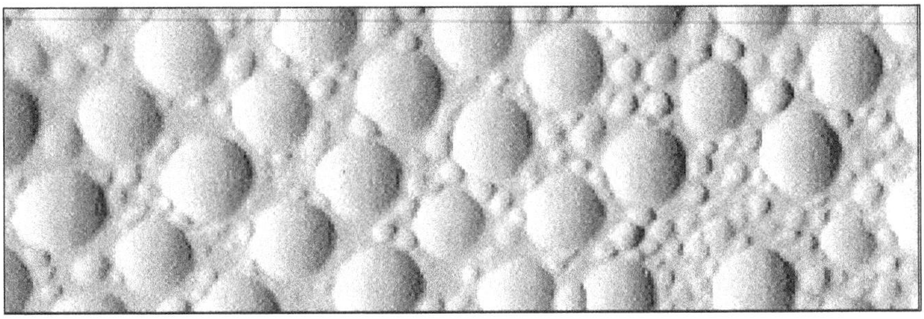

Figure 4.4: AFM view of a highly concentrated dispersion film with bimodal particle-size distribution

Directly after formation of the finely divided in situ seed dispersion (particle size usually less than 70nm), the main polymerization is carried out in the same reactor [66].

Other seed materials, such as inorganic materials like SiO_2 [94b] and TiO_2 [94c, 94d] can also be used and can form interesting nanoparticles with special properties [94e].

4.5 Combinations of acrylic dispersions with other binders

4.5.1 Combinations with other dispersions

Primary polymer dispersions are thermodynamically unstable systems. They can achieve kinetic stability via the steric hindrance provided by auxiliaries such as protective colloids, or water-soluble monomers such as (M)AA, and (M)AM.

Adding surface-active substances of the kind used in polymer dispersions of different base monomer, e.g. S/Bu or PVAc dispersions, to these thermally unstable systems alters the mixture over a certain time/temperature period and brings about extensive changes. Often observed in practical work, these can take the form of severe viscosity changes, such as a viscosity increase (until a gel forms) or the opposite, namely a marked drop in viscosity. These and other changes ultimately destabilize the dispersion systems completely. That is why it is not normal practice to mix equal quantities of a polymer dispersion based on polymer A with one based on polymer B if A and B are very different.

Acrylic dispersions serve as thickening systems in all kinds of applications. In many applications, primary polymer dispersion will often be rheologically modified with the aid of acrylic dispersions. Normally, additions of such acrylic dispersions are low, expressed in terms of the other polymer dispersion. However, in each case, the compatibility of the two polymer dispersion systems needs to be checked, as the changes observed in practical work emerge over the course of time. It is now more readily understood that the stability of formulated products over time should be checked not just at room temperature, but also at elevated temperatures. On the basis of the common rule that a temperature increase of 10K doubles the reaction rate, storage tests on formulated products at elevated temperatures such as 50 and 60°C permit better conclusions to be drawn about long-term behaviour at room temperature. The advantage is that it takes less time to achieve the results.

Combinations with PVAc dispersions

Ongoing improvements in vinyl acetate dispersions (PVAc dispersions) have seen them evolve into the most important class of binder for coatings applications [95]. For applications requiring water retention potential, PVAc dispersions are widely used.

By virtue of their chemical structure, PVAc dispersions continue to absorb water even after extensive drying; they absorb even more when cellulose derivatives act as the thickening system. An at least partial reduction in this hydrophilicity of PVAc-containing systems can be achieved through the use of an acrylic dispersion thickener.

Combinations with Ak/S dispersions

Straight acrylic dispersions and styrene-acrylate dispersions (Ak/S) are compared in detail in Chapter 4.7 and 4.6.1.3.

In this chapter, we will discuss a combination of Ak/S dispersions as binder in the formulated product with a thickening system based on acrylate thickeners. As the thickening system has to act in the aqueous phase, the greater hydrophilicity of straight acrylics, compared with Ak/S, must be seen as an advantage. Since acrylates are water-resistant compared to other polymer systems after drying (see Chapter 4.1 and 4.6), they have a beneficial effect as when serving as thickeners in Ak/S binder systems.

Combinations with S/Bu dispersions

Styrene/butadiene (S/Bu) dispersions mainly find application wherever their rubber-like performance is of interest [96]. In tyres, coatings and the like, it is their hydrophobic character which is the attraction. Clearly, tyres need to be hydrophobic in the rain, but this is a property which polymer systems for carpet floorings and corrosion inhibition coatings must also possess, albeit to a lesser extent. The hydrophobic profile of S/Bu dispersions must not be impaired by auxiliaries and so acrylate systems are chosen as the thickeners for S/Bu dispersions.

Combinations with PVC/PVDC dispersions

Dispersions based on polyvinyl chloride or polyvinylidene dichloride (PVC/PVDC dispersions) are generally used in applications where their hydrophobic polymer performance is desirable. This calls for a thickening system which does not adversely affect the water resistance. Acrylic thickeners are therefore the first choice.

4.5.2 Combinations with water-soluble binders (starch, cellulose, lignin and proteins/casein

Many (industrial) applications still make use of water-soluble binder systems such as starch, lignin, and cellulose, and proteins such as casein.

This can be explained on one hand by local availability and on the other by ecological considerations, especially of late. The aforementioned water-borne binders are natural binder systems, which are advantageous in recycling systems.

One of their disadvantages in water is their pronounced thickening action, which can easily lead to gel formation. This thickening action is due to the swelling behaviour of these binders in water. For better control over the rheology of water-borne systems containing these binders, a water-insoluble thickener polymer such as a water-borne acrylic thickener can prove advantageous.

4.6 Applications of acrylate primary dispersions

4.6.1 Emulsion paints

Emulsion paints have a dual function. On the one hand, through their colouring, they make a striking contribution to the aesthetics of the building or to the decoration of structural components, while on the other they afford the material protection against external influences, such as humidity, sunlight, or mechanical or chemical damage.

The majority of aqueous emulsion paints are complex mixtures of a very wide variety of chemical components, as shown by Table 4.4.

It is not uncommon for aqueous emulsion coatings to contain between 10 and 20 different ingredients.

The function of the binder is to give the coating the necessary cohesion, durability, weatherability, good mechanical properties such as flexibility and hardness, and to give the emulsion its advantageous processing properties. The binder embeds the colouring pigments and fillers in a stable matrix and joins them to the substrate. This distinguishes the finished coating from, say, classroom chalk, which does not contain a binder and so is easily washed off.

Besides water glass, which is a purely inorganic material and has a long history of use in silicate systems, modern binders for aqueous coatings are predominantly polymer dispersions.

Table 4.4: Components of water-borne emulsion paints

Main components	Additives/auxiliaries
Water	Dispersants and wetting agents
Binders	Thickeners/rheology modifiers
Pigments	Defoamers
Fillers	Preservatives/biocides
	Solvents/film forming auxiliaries

In Europe alone at present, more than 1,000,000 metric tons of aqueous polymer dispersions are processed into architectural coatings each year. The solvent-borne alkyd resin systems which formerly dominated the sector are increasingly being displaced by more environment-friendly, aqueous coating systems bound with polymer dispersions. In Germany, the production statistics of the paint industry association (VDL) [97] for 2009 record a total of 1,098,924 metric tons of aqueous emulsion paints for interior and masonry application alone.

Formulation of water-borne emulsion paints

This chapter presents the structure of water-borne paint formulations and also the effect and chemistry of the other paint ingredients [1, 4, 47, 98, 99]. It also describes common features and general requirements which apply to all types of paint and their binders. Particularities which apply only to one or another application are dealt within the respective specialised sections.

The typical ingredients of emulsion paints are set out in Chapter 4.1. Besides the polymeric binder and water as diluent, the main constituents of emulsion paints are pigments and inorganic fillers. In addition, there are a number of additives, such as solvents, dispersants, thickeners, preservatives and defoamers, which are needed for providing the paint with stability and favourable processing properties and which confer durability and protective properties on the coating.

In the preparation of paints, aqueous polyacrylate systems serve both as binders and as thickening polymers and dispersants.

Pigment binding power and critical pigment volume concentration

The most important parameter for characterizing a paint system is the pigment volume concentration (PVC) [100 - 108]. It is the mathematical ratio of the volume fraction of the pigments and fillers to the total volume of the dried coating:

$$\%\ \text{PVC} = \frac{\text{Volume of pigments and fillers} \cdot 100}{\text{Volume of binder} + \text{Volume of pigments and fillers}}$$

The higher the PVC, the less binder the paint contains. The PVC is quoted only for paints containing fine fillers. As plasters are formulated with coarse fillers, there is no point in stating PVC for them. The **critical** PVC (CPVC) is the PVC at which the binder in the paint film just fully wets the pigments and fillers and fills all of the interstices [100 - 108].

Table 4.5: Coating properties as a function of PVC

Property	PVC < CPVC	PVC > CPVC
Gloss	high	low
Porosity	low	high
Water absorption	low	high
Water vapour permeability	low	high
Elasticity	high (T_g-dependent)	low, i.e. film brittle
Hiding power	low	high (dry hiding effect)
Wet scrub resistance	high	low

Accordingly, the binder film is just about coherent and continuous. Above the CPVC, the coating film develops open pores; i.e., voids exist, and the binder merely provides bridges of attachment and points of adhesion between the pigment and filler particles.

The higher the CPVC attainable with a binder, the smaller the quantity of binder that is required to achieve the desired service properties. Consequently, it is the attainable CPVC which decides the economics of the binder. The capacity of a binder to cause the pigments and fillers of a paint to "stick together" to form a film having the desired service properties is also often called the PBP.

Many properties of the coating film change drastically when the CPVC is exceeded; for example, there are sharp increases in water absorption, water vapour and carbon dioxide permeability, hiding power and the brittleness of the coating, while the gloss and wet scrub resistance of the coating drop (Table 4.5). Adjusting the PVC is therefore an important tool for controlling the overall pattern of properties of emulsion paints.

The position of the CPVC is determined by the chemical nature and the particle size of the binder on one hand and by the pigments and fillers on the other. In general, with fine dispersions and especially with Ak/S binders, the CPVC is only reached at higher PVCs and when coarse dispersions [107, 108a] or other classes of binder are used (see also Chapter 4.6.1.1). The binder demand of the pigments and fillers, expressed by the oil number, i.e. the ability to absorb linseed oil, rises as the particle size falls and is heavily dependent on the chemical nature and the crystal structure. The CPVC of acrylic dispersions for standard fillers is usually between 45 and 60%.

The statement of the PVC is indirectly also a definition of the gloss range and application range of paints. The graph in Figure 4.5 shows the typical applications of acrylic paints and the associated degrees of gloss by PVC [109]. The principal application sector for Ak/S dispersions is that of highly filled paints, i.e. in the PVC

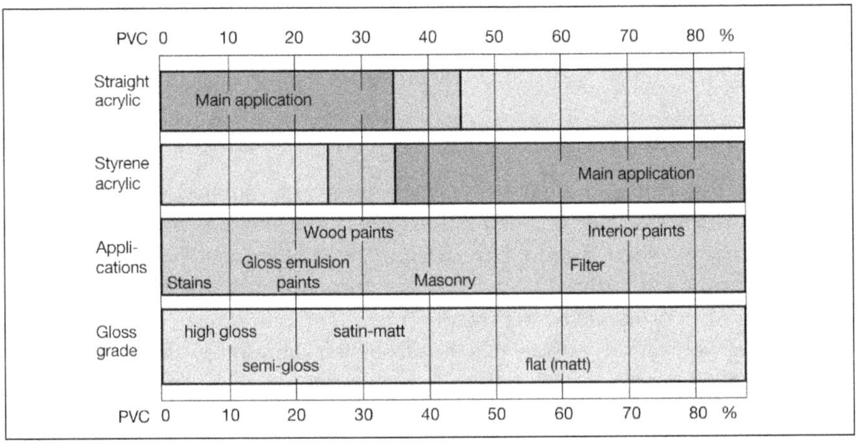

Figure 4.5: Applications of acrylate dispersions

range > 40%. Below a PVC of 30%, i.e. in the high gloss and satin gloss range, preference should be given to the use of straight acrylics on the basis of their better UV stability. In general, however, straight acrylics are used in the PVC range from 0 to 45% for wood coatings, gloss emulsion paints, coloured aggregates and high-performance masonry paints. Typical applications for paints containing Ak/S include masonry paints, fillers, synthetic resin plasters and interior paints. In the masonry paint segment, in particular, there is an overlap between the fields of use of the straight acrylic and Ak/S dispersions (see Chapter 4.6.1.3).

Interior and exterior paints

Apart from classifying paints by their PVC, another possible distinction made in practice is by type of application. The requirements imposed on interior paints are different to those for exterior paints. In recent years in particular, a lack of odour has become a very important property feature for interior paints, whereas with exterior paints the odour is not so important. In the latter application, weatherability is of greater interest and is covered in detail in the chapter on masonry paints. Primers are discussed in Chapter 4.6.1.1.

Standard formulations of matt interior and exterior emulsion paints are given in [66]. The essential differences between a matt interior wall paint formulation and a high-performance exterior paint formulation derive from the use of different types of binder, different grades of TiO_2 and different fillers, and from their different proportions in the formulations. On account of the requisite water resistance and elasticity of masonry paint, it requires more binder than does an interior paint.

The relatively binder-rich masonry paint, moreover, requires a higher TiO_2 fraction. This is because of the low filler content and the resultant lack of dry hiding power.

Saponification resistance

As already described, coating systems are frequently applied to cementitious substrates which have not yet fully carbonated, an example being the plaster on masonry or sites where repairs have been made. The cement component in the plaster is strongly alkaline (pH > 12). Therefore, polymer dispersions for such coatings must in principle be highly alkali resistant and stable to saponification. A lack of saponification resistance in the dispersion can severely curtail the useful life of a coating, as a result of chalking, cracking or loss of adhesion. In general, the ester groups present in polyacrylate and polyvinyl ester dispersions make them potentially saponifiable (hydrolysable).

The saponification test number can serve as a measure of hydrolytic resistance. In this measurement, 10g of a dispersion (for a solid content of 50 wt.%) is diluted with 30ml of water, adjusted to a pH of 7 and made to react with 50ml of 1N sodium hydroxide solution at 50°C for 24 hours. The consumption of alkali is determined by titration against 1N hydrochloric acid. A value of 50 (ml of hydrochloric acid) represents a perfectly saponification-stable polymer (for example, the Ak/S dispersion "Acronal" 290 D[1]), where no sodium hydroxide solution is consumed; lower values indicate a certain tendency to hydrolysis. For complete saponification, the value is 0.

Wagner tested the saponification numbers of different polymer dispersion types [110]. From his results, it is clearly the case that Ak/S and straight acrylic copolymers possess the best saponification stabilities. An essential condition here is that the copolymers have been prepared with long-chain acrylic esters which are difficult to saponify, such as n-BA and 2-EHA.

Water resistance of the polymer film

When films of polymer dispersions are stored in water, they absorb water and undergo blushing or whitening [111].

The water resistance of a dispersion film may be measured both by the rate of water absorption and by the quantity of water absorbed after a fixed period (usually 24 hours). Water penetration usually has a plasticizing effect and increases the extensibilty of the films, but lessens the mechanical strength and in many cases causes loss of adhesion to pigment and substrate. The reason for this is the reduction in the adhesion forces of the polymer as a function of the quantity of water absorbed. The aim is therefore to minimize swellability in water or water absorption (WA) of a coatings' binder.

[1] BASF SE

Water absorption (WA)

The extent to which a dispersion film absorbs water is determined by a number of factors:

- Chemical composition and polarity of the polymer;
- Type and quantity of water-soluble salts and emulsifiers (which, enclosed between the particles, produce an osmotic pressure);
- Type and quantity of water-swellable auxiliaries, e.g. protective colloids;
- Particle size;
- Film quality, drying conditions;
- Film thickness;
- Glass transition temperature;
- Temperature; and
- Salt content and pH of the water.

The WA of the polymer is determined above all by the polarity of the monomers used. Hydrophilic functional groups (e.g. carboxyl), which are solvated by water, increase the WA. The general rule is that the more hydrophilic the basic polymer itself, the higher is the WA under otherwise identical conditions. This can be seen from the WA values of a series of polyacrylate dispersions with virtually identical T_g.

The dispersion films absorb significantly more water as the chain length decreases and thus as the polarity of the soft acrylate monomer increases (in the order EHA < BA < EA), and also when there is a change from the hard monomer styrene to the more hydrophilic alternative MMA (see Figure 4.6) [112].

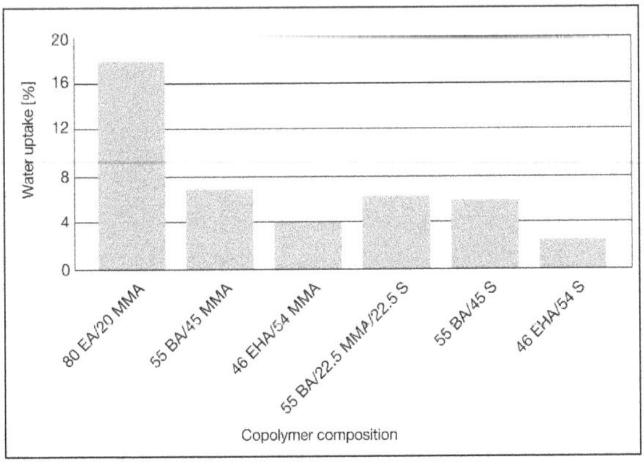

Figure 4.6: Water absorption of polymer films with the same auxiliary system and similar T_g (after 24h, values from [112])

Effects of auxiliaries

Emulsifiers and other water-soluble materials and salts (e.g. potassium sulphate as a decomposition product of the potassium peroxodisulphate polymerization initiator) form a network structure in the dispersion film and in part also accumulate at the film surface. This improves wettability by water, which at the same time can penetrate into the films as a consequence of capillary forces, and so cause whitening (blushing). The water-soluble materials pass into solution, and osmotic pressure is generated.

Depending on the elasticity and T_g of the polymer, the film yields to the pressure and so creates space for newly penetrating water. Channels are then formed through which further leaching of soluble constituents occur. Water-whitening is due to refractive index inhomogeneities caused by the penetration of the water into the interstitial phase. As a defect, it particularly affects transparent and semi-transparent coatings, such as clear varnishes, coloured stone plasters and wood stains.

As the quantity of stabilizer grows, there is normally an increase in the WA and in the tendency of the films to blush. The relationship between emulsifier content and water resistance of a dispersion has been clearly demonstrated by *Lamprecht* [111] and *Snuparek* [113]. *Snuparek* was able to lower the WA significantly by using dialysis methods to effect subsequent removal of the emulsifier from a model dispersion. The general rule when deciding on the quantity of stabilizer for binder preparation is to find a good compromise between the water resistance of the film on one hand, and sufficient colloidal stability of the dispersion on the other.

Effect of particle size

The WA level of the pure dispersion film after 24 hours is a first range-finding parameter for assessing the water resistance of the binder. From the effect of particle size on the long-term water resistance, however, it becomes clear that the state reached in terms of WA after 24 hours is usually still not one of equilibrium. Coarse dispersions, which usually exhibit poor film formation, display rapid water penetration. Fine polymers, which lead to a more continuous film, absorb water very slowly but, as there are higher barriers to the leaching of the water-soluble constituents, frequently attain higher final values after protracted immersion in water [113].

Test method

WA levels increase as the film thickness falls, the water temperature rises, and the salt content of the water falls. Consequently, standardized test conditions (e.g. DIN 53495) and the use of deionized water are vital if WA values are to be compared.

Wagner likewise addressed the WA of different types of polymer dispersions [110]. His results (Figure 4.7) reveal that straight acrylics and Ak/S copolymers generally swell to a lesser extent in water than do polyvinyl esters. Here again, however, depending on the styrene content, Ak/S copolymers are superior to the straight acrylics, as styrene is significantly more hydrophobic than the alternative hard monomer, MMA, which is used in straight acrylic systems.

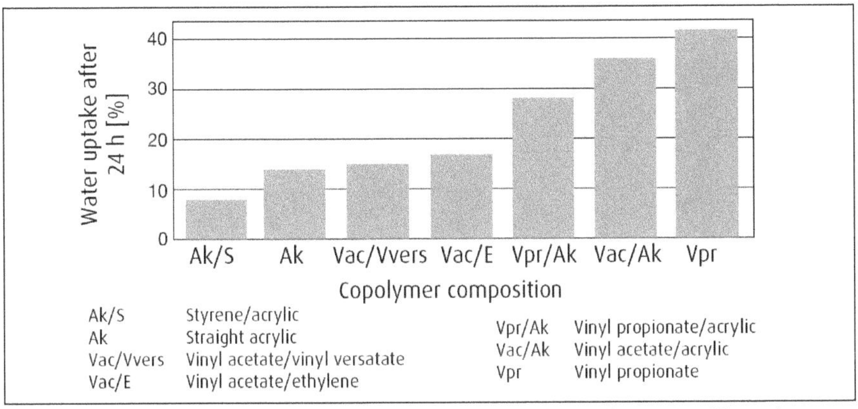

Figure 4.7 Comparison of water absorption values of polymer dispersion films of different composition [110]

When dispersion films are exposed to water a number of times with intermittent drying, the general observation is of increasing water repellency [108, 112, 114]. This can be attributed to the leaching of water-soluble fractions and an improvement in film quality as a consequence of progressive film formation. Figure 4.8 shows a graphic representation of the WA levels of the films prepared from two styrene-acrylate dispersions (Ak/S1 and Ak/S2), a straight acrylic dispersion (Ak), and a styrene-butadiene dispersion (SB), as a function of the number of water-exposure cycles. The test cycles comprised 24 hours of immersion in water followed by 48 hours of restorative drying at 50°C. The films had been dried at room temperature (to constant weight) for several days beforehand. The dry film thickness was approx. 500µm. For all of the dispersions, the WA decreased with the number of cycles. The greatest decrease was observed for the first two to three cycles. For dispersions Ak/S2, AK and SB, which had a high starting level of WA (initial WA values > 20%), the water repellency was much more pronounced than in the case of dispersion Ak/S1, in which the initial level was already low.

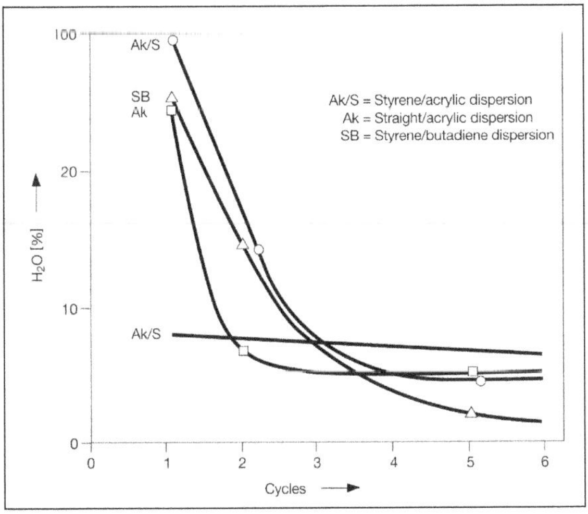

Figure 4.8: Water absorption of dispersion films after subsequent immersion/drying cycles [114a]

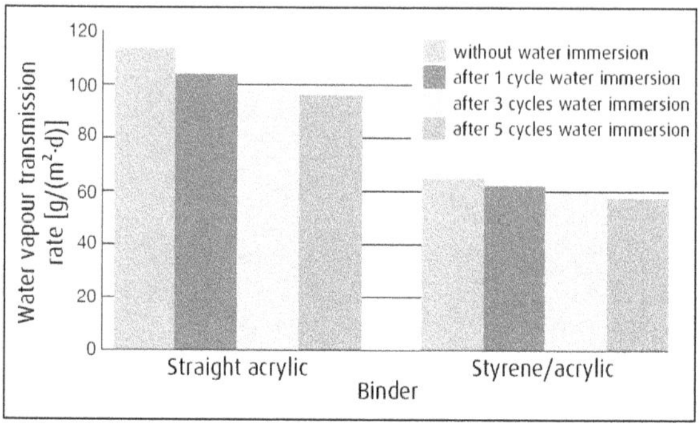

Figure 4.9: Water vapour transmission rate of polymer dispersion films

Water vapour permeability" of the polymer film

According to *Künzel*, WA and water vapour permeability (WVP) must be in balanced proportion [115a, 115b]. Penetration of water into the substrate must be prevented as a first priority by the good water resistance of the coating. Should moisture enter the substrate, however, then adequate permeability of the coating to water vapour should guarantee rapid restorative drying. Accordingly, the interplay of the WVP of the dispersion film (measured as per pr EN 1062-2, ISO 7783 or DIN 52615) and its water resistance are crucial. The WVP of polymer films was determined comparatively by *Kossmann* and *Schwartz* [109] as a function of the number of moisture cycles using a styrene-acrylate dispersion (Ak/S) and a straight acrylic dispersion (Ak) as examples (Figure 4.9).

Over all cycles, the film of the Ak/S dispersion is always less permeable to water vapour than that of the Ak dispersion. As the number of cycles rises, there is a decrease in the WVP of the Ak dispersion, whereas there is virtually no change in the case of the S/S dispersion.

Glass transition temperature

A major influence on the properties and, consequently, the application possibilities of acrylic dispersions is exerted by the T_g. Figure 4.10 shows the appropriate T_g range for various acrylic dispersions. Adhesives (Chapter 4.6.6) should be soft and tacky, while floor polishing agents (Chapter 4.6.9) must be hard and resistant. Architectural coatings come in between these two.

Acrylate thickeners

A broad range of acrylate thickeners is available for formulations [66]. Alkali-swellable acrylate thickeners (ASE and HASE) are polyacrylate dispersions which are rich in acid groups and which swell considerably when the pH exceeds the neutral point. The extent of swelling, and thus the thickening effect, depends on the following factors:

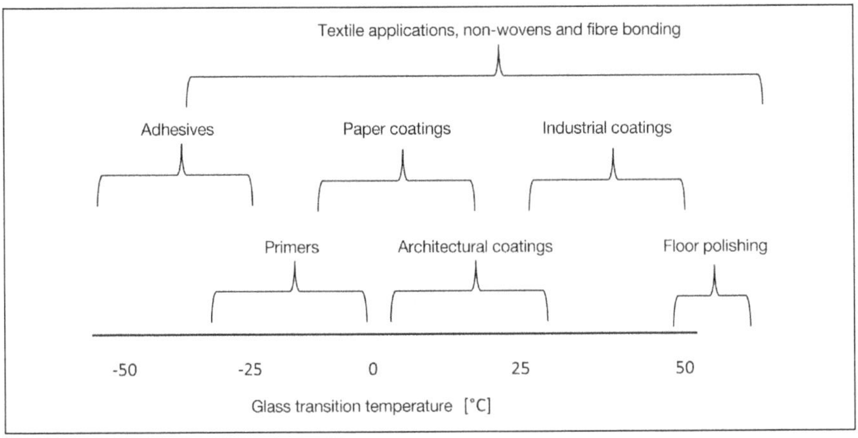

Figure 4.10: Glass transition temperature of polymers for various acrylate applications [116]

- Type of acid in the latex
- Amount of acid in the latex
- Distribution of acid in the latex
- Degree of neutralization
- Glass transition temperature (T_g) of the base polymer
- Crosslinking density; and
- Polarity

Besides the choice of acid monomer (usually (M)AA), quantity of acid (usually from 10 to 40 wt.%), and distribution of acid in the latex particle (between serum, surface and particle interior), the T_g of the polymer is important. Thus, soft copolymers of low T_g swell more extensively than hard, rigid systems of high T_g. Moreover, hydrophilic copolymers thicken more extensively than hydrophobic polymers for a given quantity of acid and T_g. For this reason, soft, straight acrylic dispersions containing EA are used predominantly, with (M)AA serving as acid monomers. The polymers, prepared in the acidic range (solid content about 40 wt.%) swell extensively when base is added from a pH of approx. 7 on. Usually, the thickener activity passes through a maximum as the pH increases and the quantity of acid grows. The viscosity maximum is reached at optimum swelling of the particles.

The effect of further constituents in a typical formulation, such as pigments, additives and fillers, on acrylic dispersions is discussed in the literature [66].

4.6.1.1 Primers

Prime coats are usually applied to materials such as wood, metal and especially mineral building materials. Here, we shall deal primarily with primers for mineral substrates. Wood primers are covered in Chapter 4.6.1.5.

One of the functions of the primer in the context of mineral building materials is to bind loose sections of the substrate together, such as fillers and pigments from old, weathered coatings, and sand from old plasters, and so to consolidate the substrate. A prerequisite for this is very good film formation by the binder in the primer. In the case of dry, porous surfaces, the primer is additionally required to prevent the solvent of the subsequent topcoat, e.g. water, from being lost by penetration and to stabilize the absorbency of the substrate.

Primers should further be able to suppress the emergence of dissolved salts from the substrate in order to provide long-term prevention of delamination of the topcoat as a result of sub-film migration of efflorescing salts. Both aims can only be achieved by effective water resistance and a water barrier effect. For the primer to achieve a sufficient depth effect, it should be capable of readily penetrating the substrate. In view of the high alkalinity of cementitious substrates (e.g. plasters), especially when fresh, it is also important for the primer to possess adequate resistance to hydrolysis. Furthermore, the high proportion of soluble ions, including polyvalent ions (e.g. Ca^{2+}, Mg^{2+}, Al^{3+} ions) in mineral building materials renders it essential for the primer and especially the binder to also be highly stable to electrolytes. These means prevent premature coagulation on the substrate surface.

Water-borne primers based on acrylic dispersions

In the past, primers were prepared primarily from solution polymers, since low-molecular weight polymers in a state of molecular solution usually penetrate deeply into porous substrates, as a result of which consolidation extends far into the substrate.

Nowadays, however, owing to increased environmental awareness among processors and end users, the use of water-borne, emulsion primers is on the increase [4, 97].

Because water-borne primers must be resistant to hydrolysis, only Ak/S or straight acrylic copolymer dispersions are offered on the market. Preference is given to soft dispersions which form films readily without solvents ($T_g < 10°C$), which have a certain tackiness and hence good binding power, and which provide good anchorage for the topcoat.

Penetration

Experience has shown that particulate systems find it more difficult to penetrate into porous building materials than molecularly dissolved systems. Nevertheless, modern-day ultrafine acrylic dispersions (particle diameters of 30 to 60nm and solid content of 30 to 42 wt.%) make it possible to obtain depths of penetration into mineral materials that are comparable with those of dissolved polymers. This is done by exploiting the fact that, for good penetration, the particle size of the polymer must be 1/10th of the pore diameter of the building material [117, 118a, 118b].

Requirements

In many cases, particle sizes of 100 or 200nm are enough to achieve good consolidation on large-pore mineral substrates. This is why a number of standard acrylic dispersions, such as "Acronal" 290 D[1], still find use as binders for pigmented primers in particular. The simplest primer is a highly diluted form of the water-borne topcoat paint.

Primers which must combine excellent consolidation with very good penetration capacity, however, are formulated essentially with the ultrafine specialty dispersions described above (e.g. "Acronal" A 508[1]) and are applied in pigment-free form.

Prerequisites for effective penetration are, as already mentioned, not only the fine particle size discussed above and the stability of the binder to electrolytes but also the use of the water-borne primer in a highly diluted form. Both measures prevent premature coagulation of the primer arising from the usually high concentration of soluble salts from the substrate.

Applying the water-borne primer at low solid (usually about 10 wt.%) also guarantees a low viscosity even after water fractions have been lost by penetration into the substrate, this being a mandatory requirement for effective penetration.

For adequate fineness combined with retention of stability to electrolytes, the stabilizer system of the dispersion must be optimized. Formulation with particle sizes much lower than 100nm normally requires relatively large quantities (2 to 10 wt.% expressed in terms of monomer) of strongly surface-active anionic emulsifiers during EP.

As the quantity of emulsifier increases, the particle size decreases. By optimizing the EP process, the initiator system and the emulsifier system, it is technically possible to achieve particle diameters of a minimum of 20 to 50nm for acrylic copolymers, at an acceptable solid content.

Formulation of primers

A transparent unpigmented primer with good penetration power contains essentially only the highly diluted, finely divided binder, some defoamer, film-forming aids if necessary, and a preservative for conferring adequate stability in storage.

The quantity of primer to apply varies from 25 to 100g m^{-2}, depending on the absorbency of the substrate.

[1] BASF SE

A typical primer formulation comprises:

- Water
- Finely divided acrylic dispersion (e.g. "Acronal"[1] A 508)
- Defoamer
- Biocide/preservative
- Film-forming agent (as a function of the binder's MFFT)
- Pigment preparation (0.5 to 1% expressed in terms of the overall formulation)
- Solid content approx. 10 wt.%

Pigmented primers should be richer in binder and of lower viscosity than the topcoat. They must in general form films and be closed-pored, i.e. the PVC should be lower than the CPVC. Pigmented primers mainly supplement hiding in the first application where the substrate is strongly coloured or offers a strong contrast. Pigmenting also has the effect of accentuating uniform application of the primer. Special primers for plasters are often formulated in the PVC range of 50 to 60%, i.e. just below their CPVC.

4.6.1.2 Exterior paints

Exterior paints are coatings systems which are applied outdoors. This class can be subdivided into masonry paints as well as flexible paints, wood coatings and textured finishes. Primers for these systems are described in Chapter 4.6.1.1.

The outdoor coating should show excellent weatherability. However, the coating is subjected to many stress factors, which ultimately lead to its degradation. These stress factors are:

- Physical influences
  - Temperature/change in temperature
  - Abrasion
  - Mechanical stress/change in stress

- Chemical influences
  - Photo-induced degradation – UV/visible radiation – photo-catalysis
  - Chemicals/dirt pick-up
  - Oxygen (atomic, hydroperoxides)
  - Dissolving and leaching of ions and/or pigment(s)
  - Water / wet-dry cycles (swelling/shrinkage)

- Biological influences
  - Mildew growth
  - Lichen and
  - Algal growth

In the following paragraphs, masonry paints will be discussed.

Masonry paints

Masonry paints can be subdivided into different classes:

1. Classical ("normal") masonry paints: Nowadays, the most important classical masonry paint is an emulsion-based type. Over the years, it has supplanted the former standard solvent- and/or oil-based coatings or lime paints.

A subclass of emulsion paints are

2. Tinting and full tone paints and
3. House paints.

Besides these, there are

4. Silicate paints, which are growing in importance (Chapter 4.6.2), and
5. Silicone paints (Chapter 4.6.3).

First, "normal" emulsion-based masonry paints will be discussed, and mainly white shades. For coloured paints, specific aspects are discussed in the chapter on tinting and full tone paints in [66].

Stringent requirements are imposed on masonry paints owing to their exposure to the weather. The impact on

- saponification resistance
- chalking resistance
- water vapour transmission rate (WVTR)
- water absorption (WA)
- adhesion
- gloss and its retention; and
- colour retention

need to be considered when choosing a binder. These requirements and various tests are discussed below.

Significance of natural weathering

It is a well-known fact that artificial weathering methods cannot fully reproduce the effects of natural weathering on a coating [119]. This is due to differences in the interaction between temperature, humidity and ultra-violet radiation. Nevertheless, it is still possible to gain meaningful results even from short-term natural weathering tests if certain principles are observed (see "Whitestone method" in Chapter 4.7) [120].

Solvent-borne masonry paints

Even today, the majority of masonry paints are formulated with the help of solvents or coalescing agents. A large number of different polymer emulsions exist for formulating masonry paints. In order to assess their performance potential, comparative laboratory studies on emulsion based paint films were carried out on two typical commercialised acrylic copolymer dispersions of the type straight acrylic (Ak) and styrene-acrylate (Ak/S) in the PVC range of 15 to 55%. The two copolymers have comparable water absorption, but the WVTR of the paint films of the Ak/S system is inferior to that of the Ak system (Figure 4.11) [121].

The mechanics of the paint film are, as expected, heavily influenced by the PVC, but also by water immersion and drying processes. Remarkable is the U-shape of the tensile strength curves (Figure 4.12). Untreated Ak/S paint films (i.e., those not immersed in water) were found to have the lowest strength, irrespective of PVC. The tensile stresses of the Ak and Ak/S control specimens (not immersed in water) pass through a minimum at 30% and 35% PVC, respectively. In order to explain the shape of the curves, it is necessary to consider two opposing effects. Firstly, emulsifiers in the binder, and thus in the paint, can have a plasticizing effect, an effect that lessens with increasing PVC. Secondly, the "hard" component in the formulation – pigment and extender – reinforces the polymer matrix (the more so the higher is the PVC), thereby increasing the film's stiffness [122] and tensile strength.

Immersion in water was found to increase film strength, most being due to the first wash/dry cycle [121]. This effect is more pronounced in the Ak/S system than in the

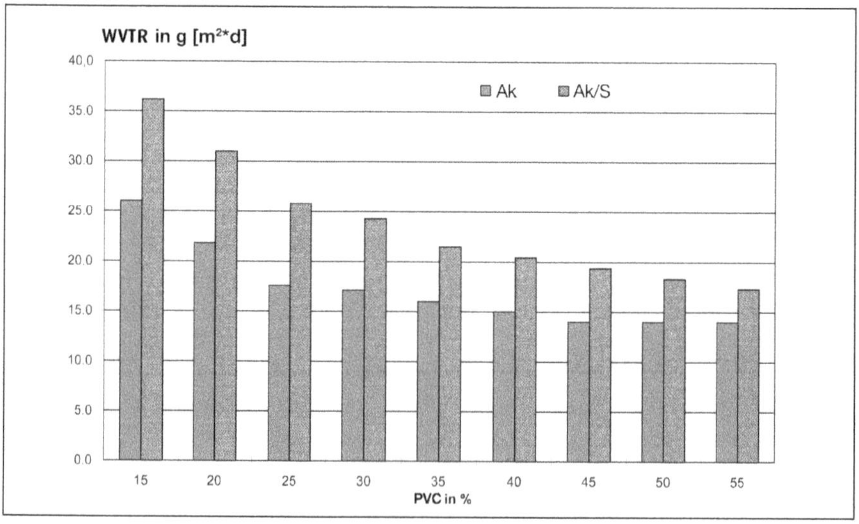

Figure 4.11: Water vapour transmission rate of masonry paints [121]

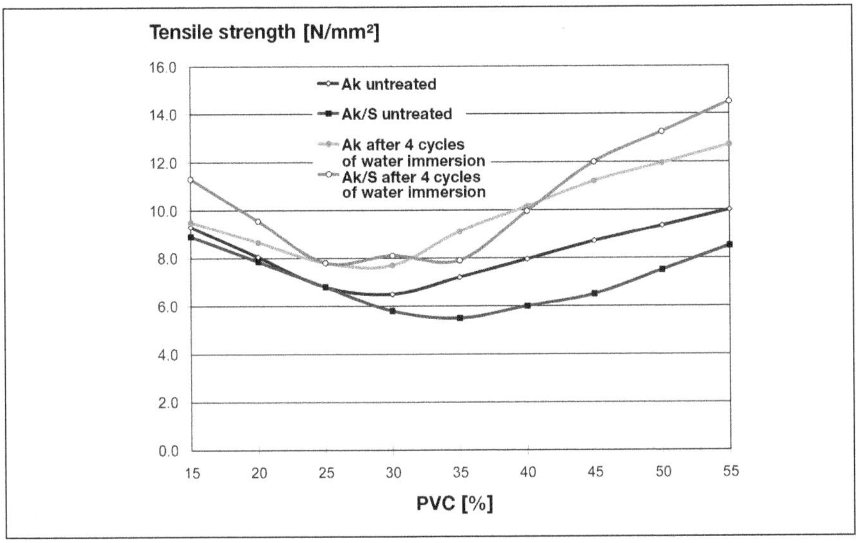

Figure 4.12: Results of tensile strength experiments [121]

Ak system. Also, the percentage increase in film strength appears to depend on the PVC. Increasing the PVC of the paint formulation causes a dramatic fall in the extensibility of the film. This property is further worsened by repeated immersion in water. With increasing PVC, the film's elongation is reduced independently of water immersion. Moreover, due to the higher quantity of binder, the elasticity of films with PVC 15% is higher than for those with a PVC of 55%. Repeated immersion in water increasingly reduces the capacity for elongation.

All these experiments demonstrate that, by washing out the water-soluble components of the paint formulation, the paint films' mechanical properties change – the film becomes more brittle and partly loses its extensibility. This phenomenon was described in a study by *Bradac* and *Novak* [123]. In general, it is assumed that the water-soluble constituents of dispersion films accumulate in voids which appear at the former boundaries of the polymer particles (see also Chapter 4.1). Although these voids are small in terms of their volume and of the surface area bounded by them, they can exert a considerable influence on the strength of the film. This observation can be regarded as being due to, for example, the emulsifiers' plasticizing effect [124] or their unfortunate ability to impede film formation [125] of emulsifiers. In addition, any forced drying of the wet film at 50°C – well above the polymer's T_g – helps to continue the film-forming process, leading to a more cohesive film. In principle, the results of the tests on Ak paints are comparable to those of the Ak/S paints. One difference worth noting is that the action of water does not increase the tensile strength of Ak films as much as it does with the Ak/S films, particularly at low pigment loadings.

Water toughens and improves the extensibility of Ak/S paint films more than that of Ak paint films. Outdoors, where paint films are subjected to the action of water in the form of dew and rain, the improvement observed in the tensile properties of Ak/S paint films is advantageous.

In artificial weathering tests, only little colour change was observed in the Ak/S paints, but it was still more noticeable than in straight acrylic formulations. However, the differences were not serious. Moreover, no significant difference was found with regard to the yellowing tendency of the two types of paint. The yellowing tendency Δ b* is, for the Ak/S as well as for the straight acrylic paints, in the same range and practically independent of PVC. These findings are in agreement with those of *Stevens* [126], who concluded that, at PVC of 20%, the presence of styrene had no effect on the yellowing tendency of styrene-acrylate paints. The large total colour difference values (Δ E*) occurring in the Ak/S formulations, particularly at 55% PVC, therefore result from changes in the values not of b*, but L* and a* [121]. A detailed discussion of outdoor weathering of masonry paints can be found in Chapter 4.7.3.

Table 4.6: Characteristics of the three Ak/S copolymers

Binder	% n-BA	% S	T_g [°C]	MFFT [°C]
Disp. Ak/S soft	60	40	8.4	< 1
Disp. Ak/S medium	50	50	27.0	18
Disp. Ak/S hard	40	60	43.3	40

Table 4.7: Formulations of the masonry paints [127]

Component	wt.-%	wt.-%	wt.-%
Water	61	55	58
Dispersing agent, polyacrylate	3	3	3
Polyphosphate	4	4	4
Sodium hydroxide	2		
Ammonia		2	2
Biocide	3	3	3
Thickener, hydroxyethyl cellulose as 2 wt.% solution	100	100	100
Solvent 1, white spirit (180 to 210°C)	0	13	13
Solvent 2, "Lusolvan" FBH[1]	0	7	7
Wetting agent	10	10	10
TiO_2 pigment from Table 4.8	194	190	190
Filler 1 (Calcium carbonate, 5µm)	246	240	240
Filler 2 ("Talcum" AT 1)	53	50	50
Defoamer	3	3	3
Disp. "Ak/S soft"	321		
Disp. "Ak/S medium"		320	
Disp. "Ak/S hard"			317
Total	1000	1000	1000

[1] *BASF SE*

Masonry paints formulated without solvent

The weatherability of masonry paints is heavily

influenced by the binder used as well by the other paint ingredients, as described in Chapter 4.7.3 with the example of pigments [109, 112].

In the following, the influence of the white pigment TiO_2 on the paint behaviour is discussed [127].

Paint formulations and pigments used

For this masonry paints were formulated with three binders. The formulations shown in Table 4.6 were used. The PVC was 45% for all paints (Table 4.7). Different rutile types of TiO_2 were also used in the paint formulation. Only white colours were studied. Commercially available pigments from four producers of titanium dioxide (Table 4.8) were teste for their influence on the outdoor stability of the paint.

Dependence of brightness on the binder and TiO_2 type

The measured brightness values of the paints are practically identical for many TiO_2 types. Paints based on "Ak/S soft" exhibit lower brightness values with all pigments from supplier 1 in comparison with the paints based on the two other binders. Assuming equally good and constant pigment dispersing in the different masonry paints and no flocculation after production of the paints, it can be expected that the paints' measured brightness should be, to a first approximation, the same for all three binders, even when the pigment varies. In principle, this can be observed, when the results of P11 and P14 are disregarded, because they seem to be artefacts. As the dispersing capacity of TiO_2

Table 4.8: TiO_2 types (SP = Sulphate process, CL = Chloride process)

Designation of TiO_2 pigment	% Al_2O_3	% SiO_2	Producer	Process
P1	3.6	9.6	Supplier 1	SP
P2	2.3	2.1	Supplier 1	SP
P3	3.1	1.1	Supplier 1	SP
P4*	3.1	0.15	Supplier 1	SP
P5	3.9	0.18	Supplier 1	CL
P6*	3.9	0.5	Supplier 1	CL
P7	2.7	0.7	Supplier 2	SP
P8	2.8	0.05	Supplier 2	SP
P9	2.7	0.9	Supplier 2	SP
P10*	4.3	0.1	Supplier 2	SP
P11*	3.3	0.12	Supplier 3	SP
P12*	3.4	0.6	Supplier 3	CL
P13	4.2	0.14	Supplier 3	CL
P14	5.6	2.07	Supplier 3	CL
P15	4.4	0.1	Supplier 4	CL
P16	4.4	1.5	Supplier 4	CL

** = Additionally contains ZrO_2*

pigments depends on their inorganic and organic surface treatment, it can be guessed that the brightness values of the paints can be correlated with data available for the surface treatment of the pigment used in the paints. With relatively good correlation, the brightness becomes more pronounced with increasing quantity of Al_2O_3 in the pigments. This result makes sense, because Al_2O_3 is used to improve the dispersing capacity of the pigment. Such interactions were also investigated by *Reck* und *Wilford*[128]. In the experiments described here, however, no good correlation can be found between the hardness of the binder and the wetting and dispersing capacity of the pigments. The above-mentioned deviations for some of the pigments can normally be easily adjusted via the quantity of dispersing agent used in the paint formulation.

Results of artificial paint weathering

The plot of results of colour measurements obtained after artificial weathering by a Xeno1200 tester for the paints based on the binder "Ak/S soft" is shown in Figure 4.13, as paints with the other Ak/S binders gave similar results.

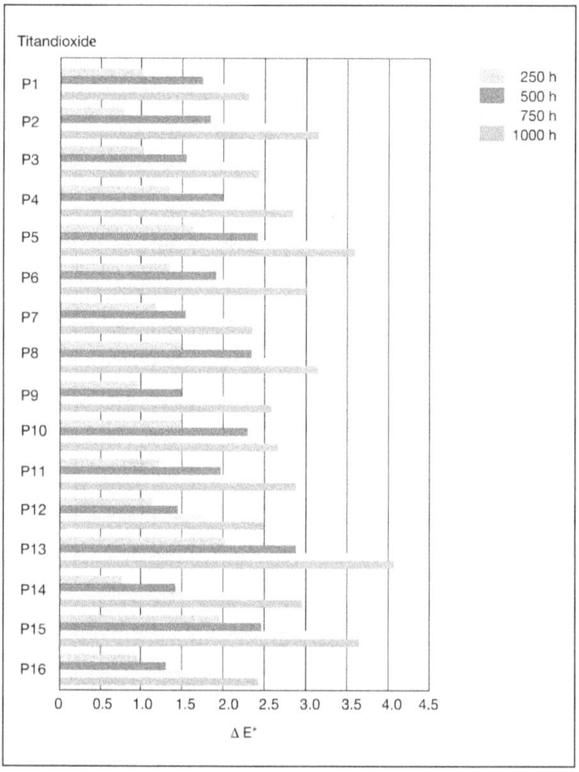

Figure 4.13: Δ E of masonry paints based on Ak/S soft in Xeno 1200 test*

Different TiO_2 types in the paints lead, even when the same binder is used, in some cases to strong colour changes, as expressed by Δ E*, which increase with irradiation time. Pigment P1, made by supplier 1 in a sulphate process, provides better protection against colour changes in all paints, irrespective of the binder, than pigments P5 and P6, made in the chloride process. However, in the case of pigments made by supplier 1 in the sulphate process, the paints also show differences in colour changes as a function of irradiation. On the basis of these results, pigment P1 from supplier 1 can be evaluated as the best. These results clearly show the influence of post-treatment of the TiO_2 pigment, which affects the UV stability of the pigment and its function of protecting the binder. Similar differences in colour changes were found for the pigments from the three other suppliers. With increasing post-treatment with silica (SiO_2), colour changes due to weathering decrease slightly. The mechanism of the photochemistry of TiO_2 types is discussed in the literature [129]. It is shown there that the OH concentration on the pigments' surface has an influence on photo-catalytic oxidation. This concentration is a function of the post-treatment of the pigment.

The colour changes Δ E* after 1000 h artificial weathering in the Xenol200 test do change, but there is no correlation with increasing hardness of the binder, i.e. increasing amounts of styrene.

Chalking

As time passes, the chalking tendency in the Xenol200 test increases. After 750h Xenol200 testing of all paints, class 5 chalking was observed, irrespective of the binder and the white pigment used.

Outdoor exposure of masonry paints

Colour changes wrought by two years' outdoor exposure were greater than the corresponding values for artificially weathered samples. This result can be explained by the fact that artificial tests are conducted without dirt. Obviously, dirt on the coatings surface changes the colour shade in outdoor exposure more markedly than in artificial weathering and therefore leads to higher Δ E* values. As already discussed in the artificial weathering tests, a good correlation cannot be found between the hardness of the polymer, i.e. between the styrene content of the polymer, and changes in Δ E* values in outdoor exposures [66].

Δ E* is the cumulative vector from the changes in brightness (Δ L value) and changes on the red-green and yellow-blue axis, Δ a* and Δ b* respectively. Figure 4.14 shows the Δ L values obtained after two years' outdoor exposure. Paints based on binder "Ak/S soft" have only slightly higher Δ L values than the two other binders. The measured differences in Δ L values are not detectable by the human eye. The finding that the Δ L values for paints based on "Ak/S soft" are higher than for the other binders is borne out by studies in which unpigmented free polymer films were submitted to outdoor exposure [130].

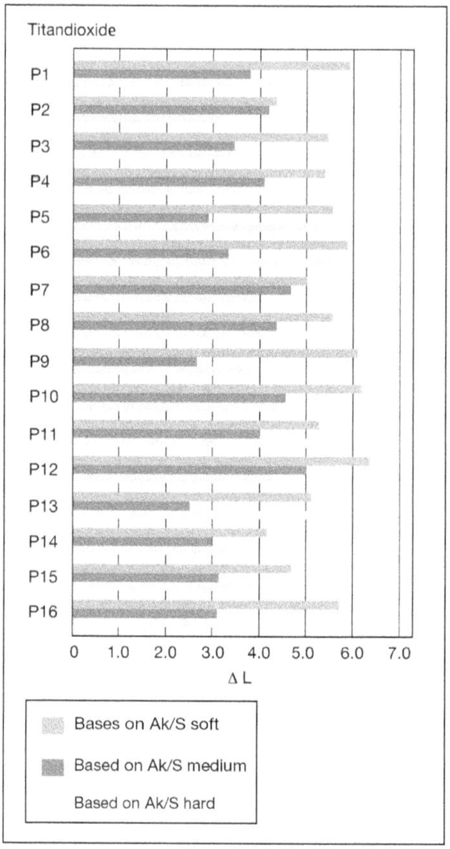

Figure 4.14: Δ L after 2 years' outdoor exposure* [127]

Summary

Acrylic-based masonry paints, formulated with binders of different MFFT, show an influence exerted by the white pigment TiO_2 on weatherability. Colour changes after 1000 hours' artificial weathering were lower than after 2 years' outdoor exposure. In contrast, artificial weathering leads to stronger chalking of the coating than natural weathering, a fact which once again confirms that dirt protects against degradation. A direct correlation between the hardness of the binders used in a paint and its dirt pick-up, chalking and degradation behaviour could not be found, neither by artificial nor by natural weathering. For chemically similar binders, which differ only in their main monomer composition, the influence of the pigment grade (rutile type) on the coating durability is, as a rule of thumb, stronger than that of the binder [127].

The stability of masonry paints in outdoor exposure is heavily influenced, as shown, by the raw materials used. Therefore, it should be always remembered that masonry paints, and in general, all exterior paints, should be carefully tested under natural weathering conditions before market launch.

Masonry paints formulated above the critical PVC

In some regions and for special applications, masonry paints are formulated above the critical PVC (CPVC).

Especially for renovation of old buildings as well as for coatings applied to lime, masonry paint should have an excellent WVTR. For this purpose, typical commercial binders – of similar particle size – from the group styrene-acrylates (Ak/S) and straight acrylics (Ak) were used to formulate masonry paints above the CPVC. Some characteristic data of the chosen products are shown in Table 4.9 and Table 4.10.

Table 4.9: Characteristics of the three Ak/S copolymers

Binder	% n-BA	% S	T_g [°C]	MFFT [°C]
Disp. Ak/S1	55	45	12	7
Disp. Ak/S2a	50	50	25	18
Disp. Ak/S2b	50	50	25	20

Table 4.10: Characteristics of the two Ak copolymers

Binder	% n-BA	% MMA	T_g [°C]	MFFT [°C]
Disp. Ak1	55	45	12	7
Disp. Ak2	50	50	20	13

Binders Ak/S2a and Ak/S2b have the same monomer composition. In production, different types of emulsifiers as well as different bases for neutralization after polymerization were used. Disp Ak/S2a was neutralized by ammonia, while Ak/S2b contains sodium hydroxide.

Irrespective of the binders' MFFT, all paints were formulated with solvent to avoid different drying times. The binders were used to formulate matt masonry paints. The PVC varied from 40 to 60%.

Results of laboratory tests

The WA of free paint films decreases with increasing MFFT of the binder. This occurs independently of the binder type. As expected, the WA of Ak-based paints is higher than that of corresponding Ak/S-based paints. Surprisingly, paints based on the dispersion Ak/S2b, which is neutralized by sodium hydroxide, have a lower WA and a lower capillary water up-take than Ak/S2a based paints. This holds for all of the PVC range tested. Further, the WVTR of paints based on binder Ak/S2b is better than that of paints based on binder Ak/S2a.

The values for WVTR and for the capillary water up-take of paints based on "soft" dispersions are better than those of paints based on "hard" dispersions. For paints formulated below the critical PVC (40 and 50%), surprisingly low WA and high WVTR values of paints based on the "soft" binder Ak/S1 are obtained in comparison with the "harder" binders based on Ak/S. This single result should, however, not be transferred to all "soft" Ak/S binders without verification. In the case of straight acrylic binders, a decrease in the MFFT of the dispersions leads to an increase in the paints' WA.

Free paint films absorb less water with rise in MFFT of their binders. For applied paints, this rule changes due to the fact that the determination methods differ. Free films are used for measuring the "normal" WA. Such films will never be used in practice. For the determination of the capillary water uptake and the WVTR, paints were applied to a mineral substrate or on *Kraft* paper.

Film formation on sandstone simulates application on to wall on a laboratory scale particularly well.

The water resistance of paints based on straight acrylic binders decreases as the PVC increases from 40 to 50%. As the values for capillary water up-take and WA show, such a decrease in water resistance was not observed in the Ak/S binders tested as the PVC was increased to 50%.

Films of paints with a PVC of 60% were too brittle to enable manipulation of a free film for mechanical testing and determination of WA; for both binder classes, the critical PVC was exceeded.

Paints based on Ak binders have a higher WVTR than the corresponding paints based on Ak/S binders, irrespective of the PVC. It is interesting that, for Ak/S-containing paints, there is only a slight increase in capillary WA when the PVC increases from 50 to 60%, while the capillary WA for the Ak-based paints grows tremendously. For such systems of low binder content, the capillary water up-take decreases with increase in the MFFT of the paints' binder.

Especially for masonry paints formulated above the critical PVC "hard" Ak/S binders should be preferred due to a better ratio of WVTR to capillary WA.

The differences between the tensile strength values for various binder classes are small for binders having the same MFFT. The elasticity of the free paint films, as expressed by elongation at break, is better for binders based on Ak/S than for those based on Ak [66].

Colour changes due to outdoor exposure

The colour changes Δ E* are independent of PVC. Therefore, only the results for paints with PVC of 50% are discussed. The Δ E* values decrease with increase in the MFFT of the binder. This corroborates previous statements about unpigmented polymers tested outdoors [130]. Colour changes in paints based on straight acrylic binders are more extensive than those in paints based on corresponding Ak/S binders. For straight acrylic paints, the decrease occurs slightly over time. The plots of the colour changes of the Ak/S paints are U-shaped for the first and second years. In the third year, the value of Δ E* of Ak/S2b-based paints is lower than that of Ak/S2a-based types. After a year's outdoor exposure, the change in L value for straight acrylic-based paints is higher than that for Ak/S paints. This is confirmed optically by the higher dirt pick-up potential of the straight acrylic-based paints in comparison with paints based on Ak/S binders (Table 4.11). With ongoing outdoor exposure, the L values of Ak paints increase due to chalking, while those of Ak/S paints exhibit a plateau after 2 years of weathering. Dirt pick-up, as an optical aspect, and the change in L value, favour Ak/S based binders. Comparison of the two binders Ak/S2a and Ak/S2b is interesting. Besides the fact that

Table 4.11: Dirt-pick up and chalking during outdoor exposure (evaluation based on DIN 53 280 with 0 = no chalking. 5 = heavy chalking)

	Dispersion					
Weathering duration		Ak1	Ak2	Ak/S1	Ak/S2a	Ak/S2b
1 year	**PVC**	**40**	**40**	**40**	**40**	**40**
Dirt pick-up	2	1.5	1.5	1.5	1.5	
Chalking	0	0	0	0	0	
2 years						
Dirt pick-up	3	2	2	2	2	
Chalking	1	1	2	1	2	
3 years						
Dirt pick-up	3	2.5	2	2	2	
Chalking	1.5	2	3	3	3	
1 year	**PVC**	**50**	**50**	**50**	**50**	**50**
Dirt pick-up	2.5	2	2.5	2	1.5	
Chalking	0	0	0	0	0	
2 years						
Dirt pick-up	2.5	2	2	2	2	
Chalking	3	1	3	3	3	
3 years						
Dirt pick-up	3.5	3	3.5	3.5	3.5	
Chalking	3.5	2	4	4	4	
1 year	**PVC**	**60**	**60**	**60**	**60**	**60**
Dirt pick-up	2.5	2.5	2	2	2	
Chalking	0.5	0.5	0.5	0.5	0.5	
2 years						
Dirt pick-up	3	2.5	2.5	2.5	2.5	
Chalking	3	3	4	4	4	
3 years						
Dirt pick-up	4	3.5	3.5	3.5	3.5	
Chalking	4	3	5	4.5	5	

the two have the same monomer composition, Δ E* as well as the changes in L values for Ak/S2b are better, compared to the values for Ak/S2a. The results show a lower dirt pick-up tendency for Ak/S2b in comparison with Ak/S2a.

Chalking due to outdoor exposure

The paints' tendency to chalk increases with increase in PVC. Chalking worsens with duration of exposure. Paints based on straight acrylic binders tend to chalk less than those based on Ak/S. In all cases, paints based on the binder Ak2 chalk more than those based on the binder Ak1. No such differences are observed for the corresponding cases for the Ak/S binders (Ak/S1 and Ak/S2). The outdoor behaviour of binders Ak/S2a and Ak/S2b is very similar (Table 4.11).

Summary

Masonry paints based on acrylic emulsions of different MFFT and neutralization were formulated in the PVC range of 40 to 60% and subjected to artificial and outdoor weathering.

The following conclusions can be drawn:

- Free paint films absorb less water with increase in MFFT of the binder, irrespective of the binder type. For applied paints, the reverse applies.
- A binder, neutralized by sodium hydroxide, need not exhibit less water resistance in comparison with a binder neutralized by ammonia.
- Colour changes in outdoor experiments indicate an influence exerted by the type of binder. The straight acrylic binders yield higher ΔE^* values than the corresponding Ak/S copolymers, i.e. the dirt pick-up potential is higher. Neutralization of the binder is found in outdoor weathering to exert an effect.

In these outdoor experiments, no clear correlation could be found between the hardness of the tested binders and their dirt pick-up, chalking and degradation behaviour [131].

On account of the porosity of supercritical paint films, the CO_2 permeability of such masonry coatings is higher than that of paints formulated below the CPVC. Therefore, this type of coating is suitable for old walls and paints containing lime. Thanks to the good WVTR, this system can be described as breathable. Any water in the substrate can evaporate, without causing the coating to blister or peel. An advantage over silicate paints (Chapter 4.6.2) is the fact that supercritical masonry paints are only slightly alkaline, whereas silicate paints (due to the water glass) are highly alkaline. This alkalinity of the water glass of silicate paints can lead to efflorescence, which cannot occur with masonry paints that are formulated well above the CPVC.

Further, masonry paints formulated above the CPVC have good obliterating power due to the "dry hiding" effect – despite a low quantity of binder and pigment (for more details, see Chapter 4.6.1.3).

In South America, in Southern regions of Europe and in Scandinavia, especially, masonry paints formulated above the CPVC are used, sometimes for cost reasons, and sometimes due to technical requirements. In all these regions, these paints are applied to lime-containing substrates. It must be emphasized, however, that the durability of such coatings is not the best. However, not everywhere is there a desire for the exterior coating to last decades. The decision as to which is the best PVC range for the paint always requires a cost/benefit analysis.

For old walls, technical requirements are clearly more important. In humid and relatively cool regions, the water level of paints formulated below the CPVC can become a problem because the water vapour transmission rate is lower. This can be resolved with a masonry paint formulated above the CPVC.

Formulations based on styrene-acrylate dispersions

Proceeding from the discussion of the parameters which influence outdoor weathering of masonry paints, Table 4.12 shows a typical formulation which is well below the CPVC (PVC 50%) and is based on an Ak/S binder (at a solid content of 66%).

Table 4.12: Typical masonry paint formulated well below CPVC (PVC 50%) based on an acrylate/styrene binder

Component	wt.-%
Binder, e. g. "Acronal" S 716[1]	100
Water	100
"Natrosol" 250 HR[2] (thickener)	1
Pigment disperser NL[1] (dispersing agent)	4
Sodium polyphosphate, as 25% solution (buffer)	4
Preservative, e.g. "Acticide" FI[3]	2
Defoamer, e.g. "Agitan" 280[4]	4
"Isopar" G[5] (solvent)	10
"Dowanol" DPnB[6] (solvent)	10
Titanium dioxide rutile (pigment)	175
"Omyacarb" 5 GU[7] (extender)	255
"Naintsch" SE-Micro[8] (extender)	60
Binder, e. g. "Acronal" S 716[1]	245
Defoamer, e. g. Agitan® 280[4]	3
"Collacral" PU 85[1], as 5% solution (thickener)	12
Water	15
Total	1000

[1] BASF SE, [2] Hercules, [3] Thor, [4] Münzing, [5] Exxon, [6] Dow, [7] Omya, [8] Naintsch

Formulations based on straight acrylic dispersions

To aid comparison of pure acrylates and Ak/S as classes of binder, a typical formulation of a masonry paint based on straight acrylic is shown in Table 4.13. It is formulated with a solid content of 58% at a PVC of 41%, well below the CPVC. The reasons for the lower PVC of the straight acrylic-based paint and the formulating concepts have already been discussed elsewhere in this chapter.

4.6.1.3 Interior paints

In the broadest sense, interior paints are all those paints

Table 4.13: Typical masonry paint formulated well below CPVC based on a straight acrylate binder

Component	wt.-%
Water	137
Pigment disperser A[1] (dispersing agent)	2
Ammonia, conc. (base)	2
Preservatives, e.g. "Acticide" FI[3]	3
Sodium polyphosphate, as 25% solution (buffer)	4
"Natrosol" 250 HHR[2], as 2% solution (thickener)	50
"Collacral" PU 85[1] (thickener)	4
Defoamer, e.g. "Agitan" 280[4]	1
White spirit (180 to 210°C) (solvent)	12
Butyl diglycol (solvent)	12
Propylene glycol (solvent)	16
Titanium dioxide rutile, e.g. "Kronos" 2043[7] (pigment)	155
"Omyacarb"[8], 5µm (extender)	175
Talcum, 5µm (extender)	55
Defoamer, e.g. "Agitan" 280[4]	2
Binder, e.g. "Acronal" 18 D[1]	370
Total	1000

[1] BASF SE, [2] Hercules, [3] Thor, [4] Münzing, [5] Exxon, [6] Dow, [7] Kronos, [8] Omya

which are used indoors, i.e. shielded from the weather and UV radiation. In the narrower sense, the term interior paints refers to matt paints of the kind used to coat interior plaster, woodchip wallpapers and the like. Gloss or wood paints, although they may be intended for use in the interior sector, are not included in the description of interior paints below.

Interior paints represent by far the largest segment within the water-borne paints market. The range of products offered in the market is virtually limitless and technically covers the range from distempers made with water-soluble sizes (which are not considered below), very highly extended, ultra low-price emulsion paints through to standard grades and high-quality varieties, such as emission-free and solvent-free interior paints (known in Germany as ELFs). High technical quality, where interior paints are concerned, primarily means ease and simplicity of use and economy through high coverage.

Technical requirements on interior paints

As there is no outdoor exposure, the binders used in interior paints, in contrast to those for exterior paints, do not need to be resistant to either UV or weathering. This means that suitable binders for interior paints include all of the copolymer types offered in the market, i.e. Ak/S dispersions, straight acrylics, VAc copolymers, and high-pressure polymers of VAc and ethylene. In contrast to exterior paints, interior paints need not afford any protection against humidity either. Consequently, the PVC of interior paints can be formulated well above the critical CPVC. The resultant open-pored nature of the paint film is more of an advantage than a disadvantage, by virtue of what is known as dry hiding.

What are the requirements imposed on an interior paint?

For its production to be as economic as possible, the paint must produce the desired service properties with the smallest possible quantity of binder. This means, in turn, that the principal requirements of the binder are a high PBP and an excellent price/performance ratio.

The interior paint must possess the following application and service properties:

- high wet and dry hiding power;
- great ease of use, without defects;
- sufficiently long open time;
- high-volume application in one operation ('one-coat paint');
- no mud cracking;
- ready compatibility with tinting colours; and
- wash and/or scrub resistance.

Pigment binding power and critical pigment volume concentration

As described in Chapter 4.6.1, the binder must to a certain extent cause the pigments and fillers of a paint to "stick together" to form a film having the desired service properties. The film must have adequate adhesion to the coated surface. There is a correlation between PBP and the CPVC. Ways of determining the CPVC are discussed in Chapter 4.6.1. The higher the CPVC attainable with a binder, the lower is the quantity of binder required to achieve the desired service properties. In other words, it is the attainable CPVC that determines the economy of the binder.

Factors influencing the CPVC

The position of the CPVC can be influenced via the formulation – that is, the selection and combination of the most important ingredients – of an emulsion paint. These formulation constituents, namely polymer dispersion, pigment and filler, are discussed in this context below and illustrated with examples.

Influence of the polymer dispersion

The principal factor influencing the CPVC is the particle size of the binder. As the particle size falls, the CPVC rises. Model investigations show that for a given chemical composition, the CPVC shifts upward by 5% on passing from a coarse to a fine dispersion [132].

Table 4.14: Influence of the polymer dispersion on CPVC [133]

Polymer dispersion	Filler				
	Particle size [µm]	Socal P2[1] CPVC by Gilsonite test	Socal P2[1] CPVC by Film stress method	Microdol 1[2] CPVC by Gilsonite test	Microdol 1[2] CPVC by Film stress method
Acrylic/styrene	0.1	52%	50%	64%	62%
Ethylene-vinyl acetate-vinyl chloride	0.1	50%	50%	57%	58%
Vinyl acetate-vinyl ester	0.1 to 1.5	46%	48%	59%	57%
Vinyl propionate-acrylate	0.2 to 3	38%	40%	49%	48%
Vinyl acetate-maleate	0.3 to 5	42%	44%	57%	54%

[1] Solvay [2] Omya

Comparison of chemically different dispersions of approx. the same particle size show that the CPVC rises by about 63% in the case of VAc copolymers, by more than 66% in the case of high-pressure polymers, and by about 68% in the case of Ak/S copolymers. Therefore, the position of the CPVC can be influenced over a range of 15 PVC units by means of the polymer dispersions (Table 4.14).

In summary, the correct choice of binder and of chemical composition and particle size can lead to a considerable rise in the CPVC and can therefore play a crucial role in the economics of the formulation. It must also be noted that finely divided Ak/S dispersions can yield a very high CPVC which eludes any other class of polymer.

Of course, it should not be overlooked that (as will now be shown) the choice of pigment and filler has considerable influence on the CPVC.

Influence of the pigment

The influence exerted by different grades of TiO_2 with different surface treatments at a pigment/filler ratio of 40 : 60 can take the form of a shift in PVC of approx. 7 units. This means that even pigments with markedly different surface treatments influence the position of the CPVC within a relatively narrow range [133].

Influence of the filler

Normally, mixtures of fillers are used in emulsion paints. In quantitative terms, fillers constitute the bulk of a formulation. In order that conclusive statements

Table 4.15: Fillers used

No.	Type	Upper particle size [μm]	Average particle size [μm]	Oil number
1	Natural $CaCO_3$	50	10.0	13
2	Calcite	40	7.0	14
3	Calcite	25	5.0	15
4	Calcite	10	3.0	18
5	Calcite	12	2.7	18
6	Calcite	7	1.5	18
7	Calcite	3	0.7	21
8	Precipitated $CaCO_3$	10	0.3	26
9	Natural dolomite	45	7.0	11
10	Talcum	50	10.0	33
11	Talcum	50	10.0	32
12	Talcum	50	7.0	37
13	Talcum	20	5.0	32
14	Microtalc	10	3.0	46
15	Mica, ground	100	27.0	37
16	Mica	30	8.0	48
17	Kaolin, fine	30	1.7	43
18	Kaolin, very fine	80% < 1		48
19	Chalk	20	3.0	17
20	Chalk, ground	4	1.0	21
21	Precipitated aluminium silicate		0.035	120

may be made about their influence on the CPVC, it is necessary to discuss the individual effect of each filler on the CPVC. The fillers considered are those mentioned in Table 4.15. To illustrate the effect of the filler as clearly as possible, the discussion will be based on a pigment/filler ratio of 10 : 90.

Figure 4.15 shows the results [134]. The different plots of the three families of curves are the result of the influence exerted by the three particle sizes.

The CPVC decreases as the particle size of the filler decreases. In these studies, a direct link to the oil number of the filler was found. In other words, an increase in particle fineness and in oil number is accompanied by a decrease in the CPVC. The coarser fillers 4 and 9 yield CPVCs of between 62 and 65%. Using these three particle-size families, differences of up to 15 PVC units are attainable in the CPVC [133].

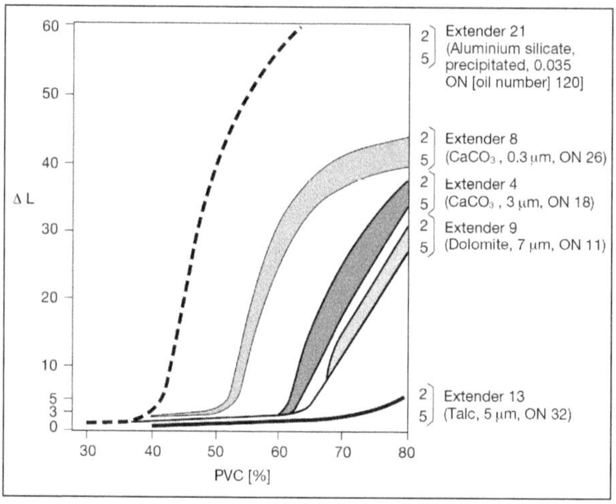

Figure 4.15: Effect of fillers 4, 8 and 9 - and also 21 and 13 - on the position of the CPVC by the Gilsonite test (polymer dispersion 1 with Kronos 2043 (2) and Kronos 2300 (5) at a pigment/filler ratio of 10 : 90)

Wet and dry hiding power

The hiding power is the ability of an emulsion paint to obliterate large differences in colour and lightness in the substrate. The hiding power should be as high as possible so that the paint can be effective at a low film thickness. The hiding power is dependent on the difference in the refractive indices of the materials used (binder, pigment, fillers): the higher this difference, the higher is the hiding power. TiO_2 is the deciding factor in hiding power.

Moreover, a comparison of the refractive indices of $CaCO_3$ and TiO_2 clearly shows why fillers provide no hiding power. Consequently, cheap paints containing very little TiO_2 and large quantities of filler are a poor choice, because the requisite hiding power can only be achieved by labour-intensive application of several coats.

In terms of hiding power, PVAc dispersions are marginally better than Ak/S copolymers or straight acrylics due to their slightly lower refractive index. The difference between the hiding power in the wet and dry states refers to the visual impression obtained when the paint is used. In the liquid state, the dispersion per se has greater hiding power than after it has dried to form a transparent film.

Besides its refractive index, the pigment's particle size is an important factor in the hiding power. For TiO_2, the optimum particle size for hiding power is around 0.2µm.

Hiding power is normally measured on the basis of films of paint applied to contrast paper (with black/white patterning). The contrast ratio is determined via the degree of light reflection from the white and black areas. The hiding power is considered adequate when the degree of reflection from the black areas reaches 98% of the value obtained from the white areas.

Application properties

For application, the paint must meet the following principal requirements:

- no-splashing
- no dripping
- sufficiently long "open" time, to prevent overlap marks; and
- no "shifting" when applied by roller.

Ease of application is heavily dependent on the auxiliary system in the dispersion and on the thickener system employed in the paint. Many paints have a relatively high intrinsic viscosity, and are usually thixotropic as well. This prevents the paint from dripping from the roller or brush and also suppresses the tendency of the paint to spatter when applied by roller.

Thanks to its thixotropy, the paint becomes temporarily less viscous under the shearing that is generated on application, and so spreading and flow of the paint are facilitated. After spreading, and thus after shearing, the viscosity rises again, which prevents the paint from running. Thixotropic behaviour also prevents "shifting" of the paint. Shifting refers to the slippage of the paint roller on a freshly applied coat of paint: the roller no longer rotates, and the paint becomes almost impossible to spread. From experience, the tendency of a paint to spatter, drip and shift can be determined only by practical tests.

High-volume application in one operation (one-coat paint)

For the purpose of minimising labour costs, the paint needs to be applied in one operation and have adequate hiding power, even on critical substrates, at a rate of 200 to 250g m^{-2} (one-coat paint). This requires the paint to have sufficiently high pigment content. Whereas formerly 20% TiO_2 was considered the minimum, paints containing 17 to 18% TiO_2 and less are common nowadays. However, the pigment content is not the only factor which is critical to the quality. The auxiliary system in the paint formulation must also permit application in a sufficiently thick coat without shifting or running.

The hiding power of the paint can be tested as described above; the other properties must be assessed in practice.

Open time

In the case of physically drying interior paints, the "open" time is the time within which the still wet coat of paint may be painted over with a second coat without the boundary between the two coats being visible after drying. With interior paints, the open time can normally be readily adjusted by adding glycols or high-boiling solvents. In the majority of cases, the addition of film-forming auxiliaries removes the need for adjustment to an appropriate open time. Since interior paints are not exposed to weathering, there is no need for a balancing act between sufficient open time and early rain resistance.

For emission-free and solvent-free paints, it is not possible to add solvents and the like, and so in this case adjusting the open time may be somewhat problematic. In some cases, the open time may be adjusted via the solid content and also via the nature and quantity of the cellulose thickener (retention effect on water).

The open time is tested by measuring the time after which a wet paint coat can be painted over without overlap marks. The optimum open time is 10 to 12 minutes on primed substrates offering little absorbency.

Mud cracking

Shrinkage cracks ('mud cracks') are short and irregular cracks in the paint film (Figure 4.16 [135, 136]). They are the result of changes in volume during drying.

Shrinkage cracks usually occur when the paint dries too quickly. This may be due to one of the following factors:

- the paint contains an unfavourable combination of fillers, with excessively small particle sizes;
- an excessively absorbent substrate or an overly diluted paint can likewise induce shrinkage cracks; highly absorbent substrates must therefore be primed;
- uneven application of the paint leads to drying in a coat which is too thick, and so leads to shrinkage cracking.

This quality feature is examined by applying the paint with a coating bar that has a wedge-shaped gap (from 50 to 2,000μm). Shrinkage cracking of the resultant wedge-shaped coat is assessed after drying. For interior paints, cracks should not appear at a coat thickness of less than 400μm.

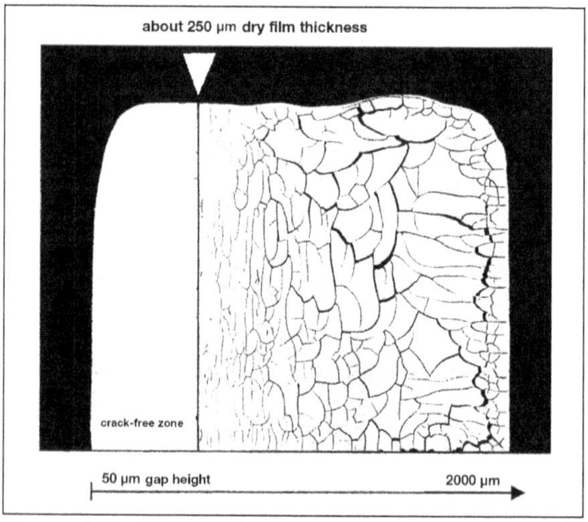

Figure 4.16: Mud cracking; the paint has cracks from 250μm on and thus fails to reach the lower limit of 400μm [135, 136]

Compatibility with tinting colours

Interior paints are often tinted with full tone colours or with colour pastes. The paint must therefore be readily compatible with the tinting product, as otherwise differences in shade may appear when a can is being used up. From experience, compatibility depends largely on the auxiliary system used in the tinting product.

Interior paints are often tinted with full tone colours or with colour pastes. Therefore, the paint must be readily compatible with the tinting product, as otherwise differences in shade may appear when a can is being processed. From experience, compatibility depends largely on the auxiliary system used in the tinting product.

Compatibility testing must be carried out on a case-by-case basis. For this, the "rub-out" test is appropriate. It involves comparing the shade of the dry rub-out of a paint with a "normally" dried paint on an absorbent substrate.

Wash and scrub resistance

Wash and scrub resistance refers to the ability of an emulsion paint film to withstand mechanical abrasion. It is a measure of its mechanical stability, which in turn is necessary for removing small spattering marks and the like without at the same time rubbing off the paint, and is critically influenced by the quantity of binder in the paint.

Testing of the wash and scrub resistance is standardized in DIN 53 778, Part 2. A coat of paint produced under precisely described conditions is scrubbed in the wet state with a brush under specified conditions. The paint is considered wash-resistant if after 1,000 scrub cycles the black substrate cannot be recognized (in other words, it has not been scrubbed right through at any point). The paint possesses scrub resistance if it withstands 5,000 scrub cycles.

In terms of wash and scrub resistance, Ak/S dispersions are superior to the other types of copolymer, primarily owing to the hydrophobicity of the polymer. Within any one polymer group (produced by the same polymerization technology), the wash and scrub resistance depend on the hardness of the polymer: the harder the polymer, the higher are the wash and scrub resistance. The majority of binders used in interior paints have MFFTs of approx. 18 to 20°C. Lower MFFTs adversely affect the wash and scrub resistance, while higher MFFTs cause the coats of paint to become too brittle.

Scrub testing to DIN standards is laborious – the coats need to be dried for 4 weeks before testing – and the reproducibility is not particularly good. Assessment of the scrub damage is impossible to render objective and is therefore heavily dependent on the individual assessor. It is certainly possible to shorten the drying time for comparative measurements, by drying for example in a conventional oven at 50°C. From experience, the scrub value after 1 week's storage has been found to shift insignificantly only in the case of the emission-free and solvent-free paints.

Owing to the relative lack of reproducibility, the method was developed further, and was published in September 1997 as international draft standard ISO/DIS 11 998. In this method, a coat of paint is weighed accurately, subjected to 200 wet scrubbing with a standardized scouring cloth, and dried, after which the abrasion is determined gravimetrically. Although the reproducibility of this method appears to be better, discussion is still ongoing as to whether the differentiation between paints of different scrub resistance is sufficient.

A comparison of the measurement techniques of DIN 53 778, Chapter 2, and the ISO proposal 11 998 Part 2 is provided by *Büppelmann* [137] who establishes a way of correlating the two measurement methods.

Emission-free and solvent-free interior paints (low VOC)

Numerous binders for interior paints have an MFFT of approx. 20°C. This value represents a good compromise between wash and scrub resistance, open time, and the quantity of film-forming auxiliary required. The addition of such auxiliaries is generally necessary for flawless film formation of the paint even at processing temperatures as low as about 5°C. The film-forming auxiliaries render the binder flexible and thus allow the binder particles to coalesce to a film during drying. After drying, the film-forming auxiliary has no further function. Film-forming auxiliaries are high-boiling solvents which slowly evaporate after film formation. There are a large number of such auxiliaries, which differ in their effect on the binder. Despite these differences, it should be kept in mind that the film-forming auxiliaries have a certain residence time in the film after drying, and so the surface properties change over time.

Owing to the requirement to reduce emissions generally, and especially to reduce the odour nuisance arising during processing, the last 10 years have seen the development of binders which form a film even at 5°C without the addition of film-forming auxiliaries. This presupposes a binder, which is plasticized by a non-volatile plasticizer or, better still, a binder that is already flexible by virtue simply of its polymer composition. By its very nature, a binder with film formation at 5°C will have a lower level of wash and scrub resistance than a binder having an MFFT of 20°C. As a result, initially paints containing flexible binders had poorer wash and scrub resistance values than those containing conventional binders. In the meantime, this disadvantage has been largely eliminated through improved binders and modified formulations.

Consistent formulation of an emission-free and solvent-free paint must avoid all volatile substances (except for water). This means not only omitting film-forming auxiliaries but also using solvent-free auxiliaries and tinting products. Solvents with a boiling point above 250°C are currently still classified as non-evaporating, or emission-free.

An intermediate stage between the traditional paints and emission-free and solvent-free paints is represented by the odourless paints. In these paints, the ingredients are only odourless, i.e. they are not emission-free paints.

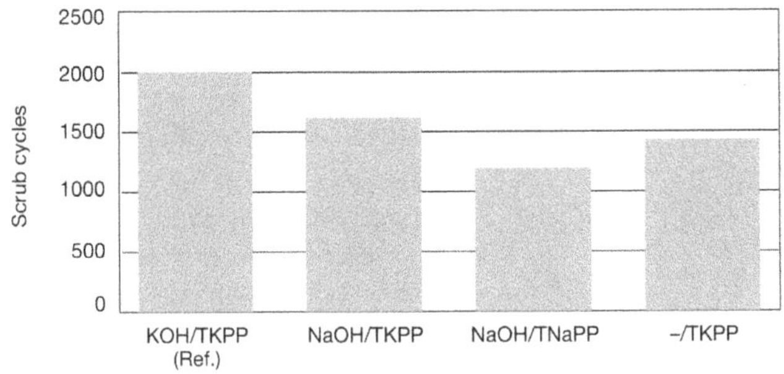

Figure 4.17: Scrub resistance as a function of additive (interior paint, PVC 79%)

Formulation scheme for interior paints

It is accordingly possible to work on the basis of the formulation scheme set out in Table 4.16.

Detailed information on formulating interior paints is described by *Brettner*[138], who also makes a direct comparison between different binder classes. He also provides confirmation that Ak/S copolymers confer outstanding scrub resistance.

The sensitivity of the property profile of interior paint to influences can be illustrated by a solvent-free interior paint based on a commercially available Ak/S binder. By varying the base and the phosphate in the paint formulation, the scrub resistance can be altered to a relatively large extent (Figure 4.17).

4.6.1.4 Gloss emulsion paints

Gloss emulsion paints are high-quality decorative coating systems which are marketed in different levels of gloss. The term gloss describes the intensity of the reflection of the light (i.e. mirror effect) at the coating surface. The measurement is made at a certain angle (usually 20, 60 or 85°, in special cases 45° also) of the incident light (Figure 4.18).

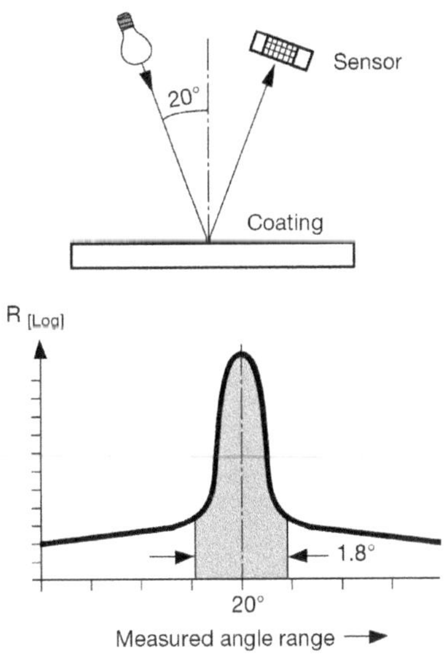

Figure 4.18: Gloss measurement
above: measurement principle
below: angle distribution of reflection intensity R, gloss measurement range indicated in grey

Table 4.16: Formulation scheme for interior paints

Components	wt.-%	Activity
Water	326	Dispersing phase; amount corresponding to the desired solids content and the geometry of the dispersing vessel
Hydroxyethyl cellulose (e.g. "Natrosol" 250 HR[2])	6	Protective colloid thickener; extends the open time through water retention; addition: max. 1%, as the water-soluble thickener increases the water sensitivity of the film. Increases matt appearance of the film surface
Tetrapotassium pyrophosphate, 50%	2	Wetting agent and dispersant; also complexes the water hardness and metal ions in the water; may also be replaced by polyphosphate ("Calgon" N[4])
Pigment disperser S[1]	4	Polyacrylate-based dispersant; attaches to the surface of the pigments and fillers and stabilises them by electrostatic repulsion
Preservative e.g. "Acticide" FI[3]	3	Usually a formaldehyde donor. Preserves the paint during storage
Defoamer	2	During dispersion, the air adsorbed on the particle surface is released, and air is also introduced by stirring. The purpose of the defoamer is to produce rapid and effective removal of the air
White spirit (180 to 210°C)	12	Much-used combination of 2 film-forming auxiliaries (white spirit and "Lusolvan" FBH) reduces MFFT, not necessary if binders have MFFT < 5°C
"Lusolvan" FBH[1]	13	
Tio$_2$ rutile (pigment)	115	
"Omyacarb" 2 GU[5] (filler)	205	The pigment and the fillers should be added in the order of increasing particle size. This facilitates the removal of the adsorbed air (small particles adsorb more), in view of the fact that the viscosity increases in the course of addition and so air removal becomes increasingly slow
"Naintsch" SE Micro[6] (filler)	60	
"Omyacarb" 5 GU[5] (filler)	160	
Defoamer	2	Platelet-shaped fillers such as SE Micro or talcum prevent mud cracking. Should further promote the removal of air from the mixture before the dispersion is added
Styrene/acrylic binder e.g., "Acronal" 290 D[1]	90	This amount produces good scrub resistance; wash resistance would be achieved with 70 parts.
Total	**1000**	

[1] BASF SE, [2] Hercules, [3] Thor, [4] Benckiser, [5] Omya, [6] Naintsch

The light reflected from the coating is compared with the light reflected from a smooth black glass surface (refractive index 1.523 or 1.567). The ratio of the light reflected from the coating and the glass surface, multiplied by 100, gives the gloss [142].

Table 4.17: Requirements imposed on gloss emulsion paints

Processing properties	Surface and film properties
Good levelling	High gloss and low haze values
Long open-time	Hardness and scratch resistance (for interior)
High brushing consistency/resistance	Long-term elasticity (for exterior, on wood for resistance example)
Low foam generation	Blocking resistance
	Chemical resistance
	Dirt pick-up resistance
	Good adhesion, including wet adhesion
	Water resistance (for exterior)
	Weathering stability (for exterior)
	Yellowing resistance
	Good hiding power

The old DIN 53 778 classified coating systems by their gloss (with a tolerance) and the measurement angle as high-gloss (64 ± 5 at 20°), gloss (≥ 60 at 60°), satin gloss (31 ± 5 at 60°), satin matt (45 ± 3 at 85°) and matt (7 ± 1 at 85°). According to the new DIN EN 13 300, this classification will be gloss (60 at 60°), medium gloss (< 60 at 60° and 10 at 85°), matt (< 10 at 85°) and very matt (< 5 at 85°).

Aqueous gloss emulsion paints based on acrylic dispersions, known as acrylic paints and formulated as satin gloss to high gloss paints, are a true environmental alternative to the traditionally solvent borne air drying alkyd resin paints. The latter are still used by DIY enthusiasts and in particular by professional decorators in very large quantities (in Germany alone: approx. 100,000 metric tons per year).

The requirements which are imposed on gloss emulsion paint therefore also essentially concern the same processing, surface and film properties familiar from alkyd paints (Table 4.17). In this context, however, the requisite property profile is again dependent on whether the products are to be used exclusively indoors or whether they will later be exposed to the elements.

The field of gloss emulsion paints embraces applications which differ greatly with the substrate and the desired effect. The following text will present only a few general points on aqueous gloss emulsion paints, and will focus in particular on high gloss paints.

Gloss and haze

The gloss of a coating is a visual perception which arises when surfaces are viewed. It involves an optical light reflection phenomenon, which together with a high distinctness of the resultant mirror image, makes a key contribution to the aesthetic appearance of the coated surface to the viewer. Classic examples of high-gloss coatings are motor vehicle finishes. Microstructures, of the kind that can occur in the coating owing to poor pigment dispersion, for example, give rise to scattered light of low intensity in the region of 1 to 3°, close to the natural main reflection. This gives the surface a milky appearance which detracts from the brilliance and the distinctness of the mirror image. This unwanted effect is referred to as bloom or haze [143].

The overall gloss of a coating is a function of the following factors [142, 144]:

- Viewing angle
- Refractive index of the coating material; and
- Roughness of the surface.

A high level of gloss, with little haze due to scattered radiation, is realized primarily by good surface smoothness or, in other words, by an absence of surface micro-roughness in the coating film.

Gloss emulsion paints are therefore normally formulated with high binder content and a relatively low pigment and filler content (see general formulation in Table 4.18).

Table 4.18: Typical gloss emulsion paint formulation

Component	Parts by weight per 100 parts of paint
1. Preparation of the mill base	
Water	3 to 6
Dispersant	0.5 to 4 (usually a polyacrylate-type)
Neutralizing agents	0 to 2 (ammonia, NaOH or 2-aminopropanol)
Titanium dioxide (rutile type)	12 to 25
Fine Fillers	0 to 15 (at best fine calcites, 1 to 2μm)
Defoamer	0.2 to 0.4
2. Reduction	
Dispersion (50%)	45 to 60 (straight acrylic or styrene acrylic)
Defoamer	0.1 to 0.2
Associative thickeners	2 to 4; normally a combination of high shear (for brush drag) and low shear thickeners (for good levelling)
Solvents	0 to 10; amount of film-forming auxiliaries dependent on the MFFT of the binder, glycols (for prolonging the open-time)
Preservative	0.1 to 0.2 (for in-can preservation)
Other additives	0 to 2.5; levelling agents, matting agents, water repellants, waxes, etc.

For high-gloss systems, only the white pigment TiO_2 – but no fillers – is used.

This is a consequence of the fact that, despite an increase in the refractive index of the paint, an increase in PVC goes hand in hand with a reduction in gloss. This gloss reduction occurs due to the increase in surface roughness, which occurs slowly at first and then more rapidly until the CPVC is reached. For this reason, the PVC values for gloss emulsion paints are usually in the range of just 15 up to a maximum of 25%. This offers the best compromise between attainable gloss level and required hiding power.

Theoretical approaches to light scattering indicate that surface roughness values of just 0.05 to 0.1 μm lead to a considerable loss of gloss [144]. Accordingly, inorganic fillers, which are normally coarse (particle diameters greater than 1 to 100μm), have an even greater gloss-reducing effect on the coating than the finely divided TiO_2 (particle diameters 250 to 300nm). For this reason, fillers are used only in satin and semi-gloss paints (up to a PVC of 28%), but not in high-gloss paints. Additionally, only very fine, wet-ground or precipitated grades of filler, with high whiteness, are suitable for satin gloss paints. In some cases, finely dispersed, organic white pigments (opaque dispersions) are used in satin to semi-gloss paints as a partial substitute for TiO_2.

At given PVC, the attainable gloss level of emulsion paints is critically influenced by the binder and the TiO_2, and also to a great extent by the choice of solvents, thickeners and dispersing resins. An essential prerequisite is that the additives be highly compatible with the binder. Advantageous effects are also exerted by high-boiling, hydrophobic and strongly plasticizing film-forming auxiliaries. Finally, besides the choice of solvent, the drying conditions (e.g. temperature and ambient humidity) also have a distinct influence on the attainable gloss level of emulsion paint [145].

Binders for gloss emulsion paints

The binders for gloss emulsion paints are usually relatively fine-sized acrylic dispersions (with solid content from 45 to 50 wt.%), which in some cases also include a small styrene fraction (up to about 30 wt.%). The T_g is set at levels of 15 to 50°C depending on the primary intended application and on weatherability requirements. The hardest polymers with the highest T_g are used in particular for scratch-resistant and blocking-resistant high-gloss paints for interior use. Accordingly, the systems usually need a high content of film-forming auxiliaries (up to 10 wt.% in the paint formula).

To satisfy the German "Blue Angel" requirements for eco-friendly paints [146], the solvent content must not exceed 10%. Additionally, the residual monomer level must be < 500ppm, which does not normally cause any problem in the case of straight acrylic systems, owing to the usually high level of conversion of the (meth)acrylate monomers in the EP.

Nowadays, binder manufacturers offer (for gloss emulsion paints as well) multiphase polymers with a greatly reduced solvent demand, which lead to coating systems that combine good blocking resistance and hardness with excellent film elasticity (for further details see [139, 147] and Chapter 4.6.1.5). As an alternative to these, there are many ongoing investigations into the preparation of blends from soft, film-forming and hard, glasslike dispersions. These are regarded as a further possible solution on the path to low-solvent or completely solvent-free gloss emulsion paints [148 – 150].

Owing to the low PVC of the high-gloss acrylic paints and the associated high polymer content close to the coating surface, particularly stringent requirements are imposed on the blocking resistance of the binder film (tested e.g. as follows: 200 to 400µm wet layer thickness of the coating, 5 to 24h drying time at room temperature, 24h blocking at up to 400g m^{-2} pressure at room temperature up to 60°C). Blocking resistance is adequate when either the T_g of the binder is high enough or when a transparent nanoscale filler is introduced by admixing or incorporating non-film-forming polymer fractions into the binder [147]. Blocking and chemical resistance of the acrylic gloss paint may also be improved by introducing a crosslinking component. For this reason, acrylic dispersions for gloss emulsion paints or varnishes are in many cases formulated to be self-crosslinking.

Binders and gloss

Water-borne acrylic paints still cannot match the high gloss levels of solvent-borne alkyd systems. The lower gloss is attributable essentially to the poor surface smoothness of the acrylic coatings. This in turn is primarily a consequence of the particulate nature of the binders. A role is also played by the associated specific film-forming mechanism, and by the more pronounced pseudo plasticity of the aqueous acrylic paints, which is manifested in the flow curves. This pseudo plasticity leads to poor levelling and hence the lack of a smoothing effect at the surface.

For high gloss values, the acrylic binder should preferably have a fine particle size (< 100nm) [143]. This generally guarantees good coalescence and surface smoothness during film formation. If the dispersion is too fine, however, the more pronounced pseudo plasticity limits levelling of the paint, which again is detrimental to the gloss of the film. For high gloss paint binders, the optimum particle diameter is in the range 80 to 200nm. The film quality – and hence gloss – can be somewhat improved by lowering the molecular weight of the binder. Increasing the refractive index of the polymer likewise leads to higher gloss levels. The n_D^{25} values of the polymers span the range 1.46 for PVAc through to 1.62 for polyvinylidene chloride (PVDC). Theoretically, a high PVDC content would necessarily give the highest degree of gloss. A number of years ago, therefore, consideration was in fact given to raising the gloss level of aqueous acrylic coatings by incorporating VDC fractions into polyacrylates (n_D^{25} values of only 1.46 to 1.48) [143]. The refractive index of acrylic polymers can also be raised by copolymerizing them with halogen-free styrene (PS, with a refractive index of 1.59). In some cases, therefore, styrene fractions of up to 30% (but not higher, due to weatherability reasons) are incorporated into high-gloss paint binders. For a given particle size and stabilization, Ak/S dispersions which have a high styrene content can be used to achieve film gloss levels which are higher by up to 10 units than those achieved using their straight acrylic counterparts.

As the refractive index of the binder increases, however, there is a concurrent slight drop in the hiding power of the paint, since the refractive indices of binder and pigment are approaching each other.

Investigations by *Zimmerschied* [151] have shown that the choice of pH and neutralizing agent (NH_3, NaOH, KOH, AMP, etc.) used in preparing the binder may also have an effect on gloss and on other gloss emulsion paint properties (e.g. rheology and blocking resistance).

The gloss can be further improved with special multi-phase polymers or comparable blends. These yield a low-molecular weight polymer fraction which is alkali-soluble [143].

TiO_2 and gloss; effect of the dispersing operation

To bring about a reduction in surface roughness, dispersing of the pigment should be as thorough as possible with the goal of establishing a high level of gloss. Such dispersing removes coarse pigment agglomerates and aggregates which reduce the gloss [143]. Besides effective wetting of the pigment and mechanical disruption of the pigment agglomerates through long and intensive shearing, it is necessary for the dispersing resin to provide optimum protection to the pigment primary particles of the gloss paint against aggregation and flocculation [152]. This protection must be provided during both storage and formation of the acrylic paint film.

The selection of the type and quantity of dispersant for emulsion paints therefore requires particular care. Dispersing resins which have proven suitable for acrylic paints are special, low-molecular-weight polyacrylic acids, often hydrophobically modified.

For high-gloss systems, the pigment manufacturers nowadays offer special surface-modified grades and also fine or special monodisperse grades of TiO_2. These are easier to wet and disperse, a fact which induces lower surface roughness in the finished coating. Standard rutile grades now have particle diameters of 250 to 300nm.

As the particle size of the TiO_2 decreases, gloss increases further, while the hiding power of the paint decreases [153]. For modern high-gloss paints, finely dispersed rutile is used at PVC levels of 15 to 20%.

Properties of acrylic gloss emulsion paints

As regards their surface properties, gloss emulsion paints are rapidly approaching the quality level of conventional alkyd resin paints, but still show certain weaknesses, especially in respect of the processing properties (Table 4.19). This is evident in particular in the wet edge time, which in many cases is too short for applying paint to large areas. On account of the more pronounced pseudo plasticity, poorer levelling properties can also be expected from such water-borne coating materials.

In the case of dissolved resins, as with the alkyds, the freshly dried coating film can be dissolved by fresh paint, enabling the film to be over-painted without brush or

Table 4.19: Comparison of properties of water-borne acrylic and dissolved alkyd resin paints

Properties	Water-borne acrylic paint	Conventional alkyd paint
Processing	moderate to good	good
Levelling	moderate to good	very good
Tendency to sag	low	low
Tendency to foam	low to high	usually low
Hiding power	moderate	good
Yield or spreading rate	moderate	good
Drying	rapid	slow
Dust-dry after ...	1 hour	at least 3 to 4 hours
Gloss	moderate to good	very good
Gloss retention	very good	moderate to poor
Yellowing tendency	very low	medium to high
Long-term elasticity	very good	moderate to poor
Hardness and scratch resistance	moderate	very good (after through-drying)
Adhesion (dry)	good	good
Adhesion (wet)	moderate to good	good
Blocking resistance	good to very good	very good (after through-drying)
Weatherability	very good	moderate to good
Water resistance	moderate to good	good to very good
Reparability/ Sandability	poor	good

roller marks. In other words, conventional paint generally has a long open time. In contrast, emulsion paints form films rapidly and irreversibly before complete water loss, which means that they dry more quickly and are over-paintable more quickly than the solvent-borne alkyd paints but have only a relatively short open time. By optimizing the solvent combination with high-boiling coalescents and glycols, such as propylene glycol, it is possible to extend the open time of gloss emulsion paint to several minutes, as is described in Chapter 4.1. However, this is not always long enough for an unskilled to paint large surface areas (e.g. doors) without producing an uneven or streaked appearance.

On account of the absence of autoxidative crosslinking and the associated long-lasting thermoplasticity, acrylic paints are better than their alkyd system counterparts in respect of long-term elasticity, yellowing resistance and durability. The thermoplasticity however, becomes a problem owing to the poor sandability of acrylic-based systems. This is manifestly disadvantageous in particular when it comes to repairs and renovations. The earlier problem, namely poor adhesion of gloss emulsion paints on aged alkyd coatings, has been solved satisfactorily by incorporating special monomers, known as wet adhesion promoters, into the acrylic binders (for further details, see Chapter 4.6.1.5).

The particulate nature and the associated special film-forming mechanism of the acrylic dispersions lead in the high gloss range to somewhat lower gloss levels (values of 50 to 70 at 20° for water-borne acrylic paints compared with values of up to 90 or more at 20° for solvent-borne alkyds) and to higher haze values (customary values for acrylic paints > 100). However, in recent years, marked improvements have been made, as regards both the binders and the high-gloss emulsion paint formulations.

Interaction with associative thickeners

The differences in the rheology of alkyd paints and emulsion paints are addressed in Chapter 2 of [66], along with the mode of action of associative thickeners at the latex surface. For high-gloss paints in particular, good approximation of the flow curves of acrylic paints to the Newtonian flow behaviour of alkyd paints is of great importance. This is accompanied by good levelling, high gloss, and good brush drag. The outcome is high film build and substantial freedom from spattering when the paint is applied by brush or roller.

In a manner similar to wood stains, therefore, gloss emulsion paints are formulated exclusively with expensive associative thickeners (PU thickeners or HASE thickeners), i.e. without the cellulose ether or acrylate thickeners (ASE) which would further reinforce the pseudo plasticity. In order to keep down the costs of the thickener, the binder should be designed for very good interaction with the associative thickener. Key factors in controlling thickening with the associative thickeners are the quantity of binder in the formulation, the particle size, the emulsifier content, and the surface characteristics of the binder, plus the nature and quantity of the solvents used [154 – 161].

As the particle size and quantity of emulsifier in the latex decrease, there is usually an increase in the viscosity of the paints for a given quantity of associative thickener. This is a consequence of the competition between the hydrophobic blocks of emulsifier and thickener molecules at the latex surface. A decrease in particle size and a decrease in emulsifier coverage, however, are also accompanied by an increase in the free latex surface area which is available for the associative thickener. This leads to new linkage points for the three-dimensional network of the associative thickener, thereby enhancing its thickening effect.

Hydrophobic solvents, such as white spirit and esters of long-chain carboxylic acids, can considerably strengthen the response of associative thickeners. When formulations for solvent-free paints are being devised, therefore, it is usually necessary to adjust the quantity of thickener. Water-miscible, hydrophilic, film-forming auxiliaries, such as butyl diglycol and even propylene glycol, can likewise exert an adverse effect on the interaction of the associative thickeners with the binder. This can lead to a considerable reduction in viscosity.

4.6.1.5 Wood coatings

Coatings for wood, a natural "living" material, and the binders designed for them are subject to a number of specific requirements which differ markedly from those for coatings on mineral substrates, such as interior wall paints or masonry paints [162, 163].

Special features of wood as a building material

In order to gain a better understanding of the properties generally required in a wood coating, it is appropriate first to give a brief outline of the special features of wood as a substance and as a building material [163, 164]:

- Variability in its basic structure and chemical composition
- Water swelling and hygroscopicity
- Anisotropy of the wood structure
- Absence of dimensional stability (swelling and shrinking with humidity changes, tendency to crack)
- Instability to UV light (colour change, greying, lignin degradation); and
- Sensitivity to microbial attack when the humidity content > 20%

Essentially, wood consists of cellulose (40 to 60%), lignin (15 to 40%), hemicelluloses (15 to 20%), and residual organic (sugar, starch, peptides; 2 to 8%) and inorganic constituents (mineral substances). The cellulose molecules give the wood its high tensile strength (reinforcement), the lignin constitutes the connection (cement or matrix) between the cellulose chains, and the hemicellulose molecules strengthen the chemical bonds between the cellulose and lignin molecules. The properties of wood are characterized not only by the three main constituents but also by accompanying ingredients, such as alkaloids, dyes, fats, oils and starches. They are partly responsible for the colour, aroma and durability of the wood. Difficulties with the surface treatment of wood can often be traced back to these ingredients.

As a hygroscopic material, wood absorbs humidity from its surroundings and can also release it again. The quantity of humidity bound by the cellulose is in equilibrium with the humidity of the surrounding atmosphere.

Changes in wood humidity content cause it to swell and shrink. These volume changes ("movement" of the wood) are different in the three sectional directions of the wood (radial, tangential and longitudinal). For this reason, wood is said to be anisotropic. Depending on the type of wood and the change in humidity content, wood swells and shrinks radially (in the direction of the wood rays) by up to 10 times as much and tangentially (along the annual rings) by up to 17 times as much as in the longitudinal direction (in the fibre direction or direction of longitudinal growth). Due to the anisotropic swelling and drying characteristics, cracks in the wood and the coating are especially likely at the annual rings and at the boundary between heartwood and sapwood.

Additionally, excessive wood humidity (> 20%) promotes fungal infestation and rotting. Fungi which destroy or discolour wood, such as moulds or the blue stain particularly prevalent in coniferous woods, detract from the aesthetics and the durability of wood and wood coatings. This becomes a particular problem if the wood is used as a construction material outdoors, where it is exposed in some cases to considerable humidity (driving rain, snow, dew, condensation, etc.)

Wood can also absorb UV light. Exposure to UV radiation from some light, and the absorption of this radiation by aromatic chromophoric groups, cause the photochemical breakdown (depolymerization) of certain wood constituents, principally lignin. This leads initially to discolouration in the case of light woods, followed later by greying, bleaching and destroying phenomena, and is accompanied by cracking at the surface and loss of adhesion of the coating on the loose cellulose framework. The latter is due to the leaching of water-soluble lignin breakdown products under weathering.

In view of the special features outlined above, it is important for maintaining the functional reliability of wooden structures to use coating systems which are tailored to the variety and quality of the wood and to the stresses to which it will be exposed under the various ambient conditions.

Classification of wood coatings

Water-borne wood coating materials are characterized according to whether they are used outdoors or indoors, according to their function, and according to the degree of transparency [163, 165]. For outdoor use, there are wood preservatives which contain antifungal additives and can penetrate well into the wood, as well as waterproofing compounds which contain no active substances but instead offer merely physical surface protection. Indoors, the main types of coating are furniture and parquet coatings. Functionally, wood coatings can be differentiated generally as primers, fillers, and topcoats. Topcoats can be subdivided in turn into transparent and semitransparent systems (clear varnishes and stains) and pigmented systems (waterproofing paints and emulsion paints). A distinction can be made between high-grade coatings for dimensionally stable components, such as windows and doors, with particular requirements in respect of freedom from tack, blocking resistance and humidity protection, and simple coating materials for components where dimensional stability is not so important, such as panelling, balustrades, fences or pergolas.

A fundamental requirement is to use primers, fillers and topcoats whose properties are matched to one another.

Primers/impregnating stains

The primer or impregnating stain is intended to make the connection between wood and topcoat and also to guarantee a deep-acting protection against fungal attack and blue staining by the introduction of a biocide (e.g. copper, chromium salts, borates, organic fungicides). The idea is to provide an effective and firm foundation so that the subsequent coatings adhere well and do not flake off. As the wood fibres stand up strongly upon contact with water, the primer must be sandable very soon after drying. In order to avoid cracking, however, it should not have excessive hardness.

Characteristics of wood primers are the low solid content (usual $< 15\%$), and the low viscosity, both features guaranteeing improved penetration and hence effective consolidation of the wood surface. For coniferous woods at risk of blue staining, such as pine, spruce or fir, special unpigmented primers containing fungicide are used, known as blue stain preventatives or preservatives. Suitable binders include conventional alkyd resins because of their capacity for penetration into wood, as well as alkyd emulsions or very fine, relatively soft polyacrylate dispersions.

Fillers are used when the aim is to obtain closed-pore surfaces and to compensate for unevenness or damage in the wood. This intermediate coating is best compared with a plaster compound and should be sandable after drying. For this reason, somewhat harder acrylic binders are commonly employed for that purpose.

Stain blockers

For the coating of woods rich in tannin, such as red pine, iroco and merbau, but also for oak or resinous conifers such as pine or fir, a stain blocker is needed because the constituents of these woods have a tendency to bleed and so to discolour the covering coating by forming stains, spots or snail effects. The discolouration is mainly caused by the tannins or, generally, by phenolic or quinoid organic components, which may emerge in particular under alkaline conditions coupled with exposure to humidity [166]. The stain blocker must ensure permanent, effective protection against bleeding.

In the past, hydrophobic solution polymers were chiefly used for stain blockers, which worked on the water barrier principle. Nowadays, increasing use is being made of aqueous coatings based on special straight acrylic or Ak/S dispersions. Appropriate dispersions should possess good water resistance and have an active mechanism which suppresses the emergence of the coloured, usually tannin-containing wood constituents. Such a mechanism operates by immobilizing the phenolic wood constituents with polyvalent metal ions (e.g. zinc, zirconium, chromium or tin compounds) or by crosslinking the low-molecular weight colorants in the substrate. To date, however, even with the best water-borne systems, it has only been possible to achieve a good barrier effect with the formulation if the pigment volume concentration remains relatively low (< 40%) and ZnO is employed as active filler.

There are at present no useful water-borne solutions capable of hindering the transport of resin fractions in coniferous woods through the coating.

Exterior topcoats

The following chapter deals with topcoats exposed to water. Exterior topcoats are performed with stains, waterproofing paints, gloss emulsion paints (Chapter 4.6.1.4) and clear varnishes.

All of these types of coating are subject to the following basic requirements:

- good weatherability;
- high elasticity, including long-term elasticity (no embrittlement);
- effective wood penetration;
- good adhesion, including wet adhesion (and adhesion to old coatings, e.g. aged alkyd);
- good water resistance (e.g. low water swellability);

- high water vapour permeability;
- blocking resistance (for dimensionally stable substrates);
- hail resistance;
- environmental friendliness (compliance with eco-label criteria, low biocide content, low VOC);
- durable UV protection to prevent lignin breakdown;
- protection against destructive fungi (adjustment of wood humidity content < 20%);
- compatibility with sealing profiles and sealants;
- alkali resistance (resistance to cleaners and cementitious compounds, e.g. mortar);
- uniform foam-free coat structure;
- easy coloration;
- easy repair; and
- advantageous processing properties (good levelling, long open-time, non-dripping).

The specific requirements imposed on a wood coating for components that must be dimensional stable, such as windows, can be found in the old guidelines of the Rosenheim Institute of Window Technology [167] and in the more recent works by the Brunswick-based Wilhelm Klauditz Institute [168] for modern wood coatings. The literature [169] contains brief information about the most important requirements and test methods. Key components in the requirements for wood coatings which come into contact with weathering are described in the new European draft standard pr EN 927-1 to 5. It lays down the details for testing the weatherability of wood coatings in comparison with a standard alkyd resin formulation, and also classifies waterproofing systems by their durability, water resistance and water vapour permeability.

Wood stains

Wood stains are transparent or semitransparent wood coatings. They contain transparent pigments (e.g. transparent, ultrafine Fe_2O_3) in a quantity sufficiently low to permit the structure of the wood to remain visible.

The stain is intended on one hand to protect the wood surface against weathering and on the other to highlight and intensify the attractiveness of the wood grain and wood colouring [170].

The transparency of the wood stain may cause problems, since wood, as already described, is attacked by UV light. Lignin has an absorption maximum in the UV-B region of approx. 280nm, celluloses and hemicelluloses absorb at well below 200nm and, like other wood ingredients, have additional absorption in the UV-A region up to 400nm. For this reason, the wood stain is given the task of preventing the UV component of the overall radiation from reaching the substrate. The rule is that, the thinner and more translucent the coating, the more intensive must be the UV protection of the wood substrate. Accordingly, the pigment content of the formulation, at least, must not be too low if good shielding is to be obtained. However, there are other influences to bear in mind, too. In the case of the water-borne acrylic coatings, for instance, an additional problem is their high UV permeability and yellowing resistance compared with those of alkyd systems. To achieve the necessary UV protection with acrylate-containing transparent coatings as well, use is made, for example, of UV-absorbing, very finely dispersed transparent pigments (e.g. ultrafine TiO_2 or ZnO). The same aim is also pursued by adding organic UV absorber substances from the groups of the hydroxyphenyl benzotriazole, hydroxybenzophenones, hydroxyphenol-s-triazines or oxalanilides to the formulation of stains and clear varnishes.

The UV absorbers are frequently used in combination with free-radical scavengers, which have the capacity to intervene in the photo-oxidation process of lignin breakdown. These scavengers are essentially sterically hindered amines, known as HALS (hindered amine light stabilizers), the majority being tetraalkyl-substituted piperidines [171].

As far as the pigmentation is concerned, good UV protection of the substrate – without UV absorbers in the wood stain formulation – necessitates at least 0.5 wt.% (better still 1 to 2 wt.%) of transparent, ultrafine Fe_2O_3 (usually combinations of yellow, red and black Fe_2O_3 with average particle diameters $\ll$ 0.1µm). Micronised pigments of this kind, while offering little hiding power, do have a "glazing" effect, i.e. they bring out the wood structure, enhance the brightness of the surface and at the same time prevent the passage of the damaging UV radiation of wavelength < 400nm to the wood substrate.

In principle, the concentration of UV absorber in a coating which provides effective protection should be chosen so that it can capture more than 95% of the UV radiation. For a good compromise between cost and activity, the quantity of UV absorber used in the case of clear varnishes is usually about 1 to 3 wt.%, expressed in terms of the formulation as a whole.

In terms of their solid content, the semitransparent wood coatings may be subdivided into low-build and high-build wood stains (Table 4.20).

Table 4.20: Formulation of water-borne wood stains

Constituents	Low-build stains Fractions in [%]	High-build stains Fractions in [%]
Binder (e.g. "Acronal" LR 8960 [1], polymer solids)	10 to 20	25 to 45
Solvents/coalescents	0 to 10	0 to 10
Pigments	0 to 5	0 to 5
Additives*	0.5 to 3	0.5 to 5
Water	60 to 80	35 to 60
Solids content	**< 30**	**30 to 60**

(Thickeners, levelling agents, fungicides, matting agents, defoamers)*
[1] BASF SE

Low-build stains

Low-build stains are characterized by low viscosity and low solid content (< 30%). Materials of this kind produce dry film thicknesses of up to 25μm when applied in three coats. Low-build wood stains are suitable primarily for components where dimensional stability is not an issue. Per se, the humidity protection they offer is limited owing to the comparatively low film thickness. On account of their low viscosity, however, they usually penetrate the wood more deeply than the more concentrated high-build stains and therefore exhibit good adhesion. They are normally applied without a primer.

Thick-film stains

These are stains of high binder content and relatively high viscosity, normally formulated for thixotropy – that is, to inhibit dripping. The targeted dry film thicknesses are up to 120μm. This target can usually only be achieved by spray application. The Rosenheim Institute for Window Technology recommends a dry film thickness of at least 60 to 80μm for wood-stain systems in window structures.

Water-borne high-build stains based on acrylate are highly durable and the majority of them offer good humidity protection for the wood. Because of their high solid content and their pronounced viscosity, however, their depth of penetration is usually low. This leads to poor mechanical anchoring in the wood, leading frequently to problems with the adhesion of the coating to the wood surface upon exposure to humidity.

High-build stains are useful essentially only for dimensionally stable components which have already been primed, such as windows and doors, and the first coat is normally applied industrially.

Opaque coatings; waterproofing paints

The pigment fraction ensures that wood protection paints and gloss emulsion paints provide the wood substrate with adequate protection against UV light. There is therefore no need for expensive, additional UV absorbers. Moreover, because of their UV protective effect, opaque wood paints are more durable than the transparent systems. Since, however, increasing levels of pigmentation are detrimental to the elasticity and extensibility of a coating, pigmented wood coatings are usually only formulated in the PVC range from 20 to a maximum of 35%. The wood paints are predominantly formulated with just pigments (TiO_2 and colour pigments), with the use of finely ground fillers, such as talc or $CaCO_3$, being kept to a minor level. For opaque coatings in window systems, dry film thicknesses of 100 to 120μm are recommended.

Good wood paints have a sufficiently high level of water vapour permeability (diffusion-equivalent air layer thickness, s_d, is < 0.5m). It is especially important to observe this parameter for the coating of outdoor timber constructions, where rotting of the façing timbers by accumulated humidity beneath the coating must be prevented. If, on the other hand, the coating system is too open to diffusion, there is a risk of water entering the wood via the gas phase.

Binders for wood coating

The established solvent-borne, medium oil and preferably long oil alkyd resins can still be found on the market today for use in all of the above-mentioned types of exterior wood coatings. Interest is increasingly being directed, however, to the more environmentally friendly water-borne systems with alkyd resin emulsions and especially acrylic dispersions [170 - 172] as binders. In certain cases, hybrid systems – i.e. combinations of alkyd emulsions and acrylic dispersions – also serve as binders for wood coatings.

Table 4.19, page 198, depicts the most important differences in properties between the conventional alkyd-based wood coatings and the modern water-borne systems for this purpose that are formulated on the basis of acrylic copolymer dispersions. From such evaluations it can be said that only tailor-made acrylic dispersions can meet the totality of complex requirements imposed on a wood coating which is exposed to the elements.

Straight acrylic dispersions

Dispersion binders for water-borne wood coatings in the exterior sector nowadays are primarily water-borne straight acrylic dispersions with at most low styrene content. Protective-colloid-stabilized polyvinyl acetate-maleate dispersions, which were formerly prized for wood paints owing to their good water evaporability and flexibility, are now of little importance in the European coatings market owing to their moderate water resistance, poor blocking resistance and severe propensity for soiling.

The fineness of modern straight acrylic dispersions (particle diameter < 120nm) ensures a certain degree of penetration of the coating material into the porous wood substrate and thus adequate anchoring of the coating. At the same time, the fineness promotes the wet transparency during application (i.e. haze-free, transparent staining) of the formulated stains, i.e. the user can already gauge the appearance of the finished coating while it is still being applied.

Nevertheless, water-borne acrylates still have marked weaknesses in terms of penetration and wet adhesion relative to conventional solvent-borne alkyds or modern high-solid alkyd systems [173 - 175]. The wet adhesion differences between acrylic and alkyd systems were quantified in a study by *De Meijer* and *Militz*

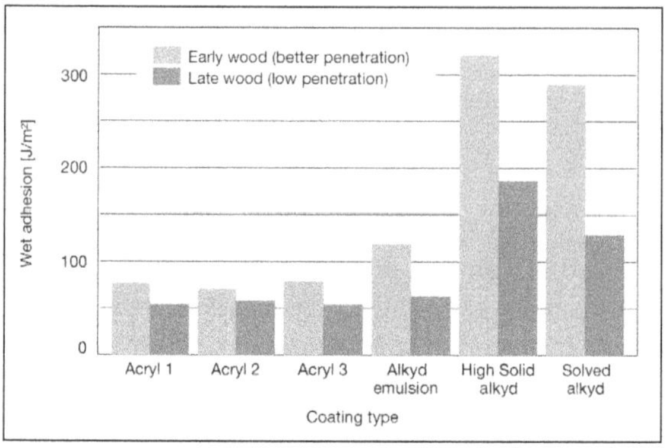

Figure 4.19: Wet adhesion of different coatings on pine sap wood [175]

(Figure 4.19) [175]. They used a tensile testing machine to peel off a defined adhesive tape stuck to damaged and water-exposed coatings (a modified tape test).

Stability features

In contrast to the alkyds, drying of water-borne coatings formulated on straight acrylic dispersions is merely physical in nature. Changes in glass transition temperature and elasticity over their normal lifetime of roughly 5 to 10 years [176, 177] are insignificant. As even coated wood shows movement under cyclical changes in temperature and humidity content, however, the dispersion chosen should permit sufficiently high stretching of the films, especially at low temperatures, so as to prevent cracking during cold seasons. However, even with acrylic coatings, the slow release of high-boiling plasticizing coalescents may cause the elasticity to deteriorate over the lifetime of the coating. Therefore, the T_g of the acrylic polymer (according to *Schmid*) should be < 10°C, and preferably about 0°C [177]. Such soft thermoplastic polyacrylates, however, do not yield sufficiently tack-free and blocking-resistant films when used to coat dimensionally stable substrates, in contrast to the crosslinked alkyds. In other words, window frames or doors coated with such paints may stick to the frame upon direct contact when exposed to pressure and heat. This is a particular problem with dark paints. In direct sunlight in the summer, such coatings can reach surface temperatures of up to 60°C or more.

Multi-phase acrylics

Since the pigments and fillers in wood stains or wood paints with low levels of filler contribute little or nothing to the hardness and blocking resistance of the coating surface, the search for alternatives was inevitable. It started in early to mid-1980s, when the first systems with acrylic-based multi-phase latex particles were developed for wood coatings.

These acrylic dispersions, often referred to by specialists as core-shell polymers, contain both soft, film-forming and hard, non-film-forming polymer fractions. [178, 179a, 179b] They permit the formulation of wood coatings which possess high elasticity combined with very good blocking resistance and dirt pickup resistance. The hard phase acts as a nanoscale, transparent, organic filler and provides reinforcement of the soft film and at the same time surface shielding of the soft-phase material. Both effects increase the blocking resistance. The elasticity of the system as a whole, however, is not adversely affected if there is a certain excess of soft-phase material of optimum particle morphology.

Crosslinked systems

In many cases, self-crosslinking acrylic dispersions are also used to improve the blocking resistance and hardness; this also benefits water and chemical resistance [180]. As windows are often provided with sealing profiles or sealants, the coating must be compatible with the plasticizers of such products and with the adhesives that are used in this context. Here again, it is useful to have a crosslinking component in the dispersion.

Moreover, a good alkali barrier effect is important for wood stains, because new buildings are plastered after the windows have been installed. Owing to the high alkalinity of the plaster, this may cause brown spotting on the wood substrate.

Wet adhesion

Experience has shown that problems arise when aged alkyd resin coatings are repaired with water-borne acrylic systems. Under conditions of humidity and frost, many simple acrylic coatings have difficulty adhering to the aged alkyd resin coats. Blisters develop, and in the wetted state, the new coating detaches from the alkyd. When the first aqueous acrylic systems were used in Scandinavia 15 to 20 years ago, this led to damage, some of it substantial. In order to ensure adequate wet adhesion of dispersion coatings in such cases, expensive specialty monomers are nowadays copolymerized for these purposes [181 - 183]. The monomers in question are usually based on (meth)acrylates, maleates, alkyl or vinyl ester compounds which bear amino, acetoacetate, cyanoacetate, urea, thiourea or cyclic urea groups. Attachment of an acrylate coating to the old paint is assured by these monomers by virtue of the formation of hydrogen bonds and by acid-base interaction with the alkyd substrate.

An alternative to incorporation of monomers, such as dimethylaminoethyl methacrylate, into the polymer, is a polymer-analogue strategy of incorporating amine functionalities into an acrylic dispersion. The basis for this strategy is the reaction between carboxylic acid groups and the reactive cyclic amines ethylenimine or propylenimine. In formulating dispersions with wet adhesion promoters, however, it should generally be borne in mind that surface-active additives, such as wetting agents, emulsifiers or thickening polymers, can considerably disrupt the adhesion mechanism in adverse cases.

Table 4.21: Water absorption of different wood coating systems (initial levels of water absorption from [177])

Binder system	Water absorption
Oil/alkyds, solventborne	5 to 10% by volume
Alkyd emulsions, water-borne	about 20% by volume
Alkyd acrylate hybrids, water-borne	10 to 20% by volume
Acrylates, water-borne	20 to 30% by volume

Moisture protection

By virtue of their pronounced hydrophobicity, solvent-borne alkyd resin coatings (Table 4.21) are initially able to provide very good protection for the wood against humidity [177, 184]. In many cases, however, they lose this positive property during their lifetime owing to the decrease in elasticity caused by weathering and the associated tendency to crack. The enduringly flexible, water-borne acrylic coatings, which in the beginning (before weathering) are fairly hydrophilic due to the formulating additives (thickeners, dispersants, etc.) and emulsifiers, exhibit low water absorption after the hydrophilic constituents of this kind have been leached out by rain-water. Ultimately, then, they permit better long-term protection against moisture than the conventional alkyd systems.

Even with acrylic systems, the initial WA should not be too high (< 25% by volume), since excessive incorporation of water into the coating film as a result of weathering can cause wrinkling and cracking. The cracks are primarily due to the formation of the surface skin as the water-swollen film dries (blancmange or pudding effect) [177].

Failure to implement structural wood protection, e.g. in the form of roof overhangs or protection against water splashes in the ground region, combined with the poor water barrier effect of some water-borne acrylic coatings on house facings, has led in some cases to severe fungal infestation of the wooden surfaces (wood rot syndrome). This phenomenon was observed a number of years ago in Norway and Sweden in particular. Now that these design effects have been recognized and the water resistance of the acrylic binders and their formulations improved, these problems are regarded as having been substantially resolved.

The high WVP of the acrylates (normally higher by a factor of 1.5 to 2 than that of alkyd systems), moreover, prevents a build-up of humidity under the coating. This guarantees freedom from infestation and long-term adhesion of the coating. The water vapour diffusion resistance of a coating has so far usually been determined in accordance with DIN 52 615. This so-called μ value is a dimensionless material constant of the coating. Related to it is the diffusion-equivalent air layer thickness s_d, which is the product of the coat thickness and the μ value. For a straight acrylic system, the s_d value is usually approx. 0.3m for a coat thickness of 100μm.

For unpigmented wood coatings, on grounds of stability, the acrylic binder should have a pH of 7 or 8. Levels of more than 8.5 to 9 may cause unwanted dark discoloration on critical woods, such as oak, and should therefore be formulated only in pigmented coatings.

Processing properties

The remaining weaknesses of water-borne acrylic coating materials primarily concern their processing properties. Owing to the rapid and irreversible physical drying, the open time is frequently too short for application to large areas. Levelling is often poor because of pronounced pseudo plastic flow. There have therefore been various attempts to prolong the open time, e.g. by increasing the water retention capacity of the polymer or by including high-boiling film-forming auxiliaries and coalescence retardants in the formulation. The levelling properties can be improved with special associative thickeners. These essentially increase the viscosity under high shear, but also lower the yield point (Chapter 4.6.1.4). Moreover, levelling can be improved further with levelling aids based on silicone surfactants.

Hybrid coatings

Blends of alkyd resin emulsions and acrylic dispersions are available on the market which make mention of the positive properties of alkyd resin coatings and dispersion coatings. The alkyd resin fraction is intended to improve the depth of penetration and thus the anchoring of the coating in the wood, while the acrylate component is intended to guarantee long-term elasticity. A fairly recent study, however, has found that the alkyd resin component starts floating during drying and therefore has limited ability to improve the adhesion to the wood substrate [185]. Nevertheless, the wet adhesion of water-borne hybrid systems is generally somewhat better than that of straight acrylic systems, although it does not reach the level of dissolved alkyd systems.

Hybrid binders dry both physically and oxidatively. However, nowadays they are used almost only in opaque coatings because of their poor film transparency. Nevertheless, the autoxidatively crosslinking alkyd resin fraction ensures easy sandability and repairability at the end of the coating's life. Straight thermoplastic acrylic systems, while longer lasting, must be stripped completely or sandblasted beforehand if the repair coat is to be impeccable: these are tedious operations.

One problem which affects all alkyd systems exists in connection with the use of the cobalt(II) drying accelerators (siccatives) which are still necessary for the autoxidation reaction, as cobalt compounds are regarded as toxicologically harmful. In addition, the water-soluble cobalt(II) salts may be precipitated as the inactive cobalt hydroxide [186] during storage of the water-borne paints or stains. At the present time, nevertheless, a large number of fundamental studies are underway into water-borne hybrid systems where the intention is to combine alkyd and polyacrylate in one particle by the processes of conventional emulsion polymerization or of modern mini-EP [187, 188].

Interior wood finishes

For interior use of wood coatings, the main concerns are aesthetic and decorative effects, such as highlighting the natural attractiveness of wood, as well as the natural concern to protect the wood against physical or chemical damage [189]. The requirements imposed on wood coatings in the wood flooring and furniture industry (usually clear varnishes) therefore differ markedly from those on waterproofing systems. The strong physical and chemical stresses necessitate stringent requirements on hardness, scratch resistance and stability.

Classic, solvent-borne acid-curing coating materials and nitrocellulose coatings, as well as 2-pack polyurethane systems, have been found to meet these requirements. Nevertheless, they are being increasingly supplanted by more eco-friendly waterborne coatings (based on acrylic dispersions) or by UV coating systems which offer 100% solid (Chapter 5). The EU-VOC directive agreed in 1999, which envisages a reduction in VOC emissions from industrial coatings within the European Union, is expected to promote the trend towards the use of water-borne systems.

The majority of acrylic dispersions for wood coatings are very fine (average particle diameter from 50 to 100nm) and they are usually formulated with a high T_g (usually from 30 to 60°C) in order to achieve the desired hardness.

The use of such polymers makes it possible to achieve film formation in the formulated coating by using solvent as well (from 3 to 10%). The solvent requirement can be further reduced, which is desirable on ecological grounds, by means of a heterogeneous structure of the acrylate particles, similar to that in the case of the wood stains, with no loss of surface hardness or blocking resistance [190]. Important performance properties, especially hardness, elasticity and chemical resistance, can be regulated via the copolymer composition.

Formulating aspects

Since the chemical resistance requirements imposed on wood coatings in the old furniture DIN 68 861 Chapter 1 B and new DIN EN 12 720 are very stringent, virtually all wood-coating binders are formulated to be self-crosslinking [180, 191, 192]. An alternative is afforded by external crosslinking [193, 194]. For carboxy- or hydroxyl-functional dispersions, this is effected with polycarbodiimides, polyaziridines, epoxy silanes or water-dispersible polyisocyanates. The incorporation of carbonyl-functional monomers into the polymers opens up the possibility of crosslinking by condensation with diamines or polyamines, or else with dihydrazides or multifunctional semicarbazides. None of these crosslinking systems, however, can win across the board in toxicology, activity, pot-life, freedom from yellowing, and price. A problem in respect of crosslinking continues to be the fact that the wood substrate, owing to the risk of thermal discolouration and gas emissions, only allows drying and curing temperatures of up to 80°C.

UV water-borne systems [180, 195], i.e. radiation-curable polymer dispersions, might be the solution of the future (see Chapter 5). They are produced, for example, by adding polyfunctional acrylic monomers such as tripropylene glycol triacrylate to acrylic dispersions. An alternative approach is polymer modification of the acrylic dispersions with free double bonds.

PUR combinations

Water-thinnable, highly reactive 2 C systems, starting from hydroxyl-functional acrylic dispersions and water-dispersible polyisocyanates, are being trialled for furniture and especially wood flooring. Such systems have already become established in the automotive sector for the coating of metals.

The addition of expensive, water-borne polyurethane dispersions (secondary dispersions) to acrylate systems leads to an improvement in the abrasion and scratch resistance while at the same time enhancing the film toughness. Usually, from 60 to 80 parts by weight of polyacrylate are combined with 20 to 40 parts by weight of polyurethane as binder. The polyurethane dispersions normally provide plasticization of the acrylate component owing to their water swellability and because of the residual solvent content (e.g. of N-methylpyrrolidone) which they have by virtue of their preparation. Therefore, very hard acrylic dispersions of high T_g can also be used for such combinations.

In many cases, however, it is found that there have been instances of deterioration in water resistance and chemical stability when polyurethane dispersions are added. Furthermore, hazing in the film can detract from the transparency and aesthetics of the coating. Therefore, for some years, a number of acrylic/polyurethane hybrid systems have been offered on the market in which the polyurethane component is introduced as a protective colloid first and an acrylic dispersion is then grafted on. Accordingly, both polymers are present in one particle [196]. Whether these systems offer overall advantages over plain mixtures is still a matter of dispute among experts [197]. In some cases, however, improvements have been found in the accentuation of the wood substrate and in the resistance to, e.g. ethanol [198].

Further formulating constituents

Besides the choice of binder and crosslinking system, the overall formulation of the water-borne coating material, such as the choice of matting agent, and the solvents, has an important part to play in the level of properties that can be achieved (Table 4.22) [199].

Table 4.22: Parquet coating formulation; matt clear varnish

Component	Parts by weight
Mixture 1:	
Solvents; propylene glycol-n-butyl ether	40
Propylene glycol	10
"Solvenon" DPM[1] (dipropylene glycol monomethyl ether, isomer mixture)	20
Additives/defoamers, wetting agent, levelling agent, thickener, matting agent	25
Water	30
Butyl glycol	20
Mixture 2:	
Acrylic dispersion ("Luhydran" A 848 S[1]; 45% by weight)	400
Polyurethane dispersion (40% by weight)	400
Waxed dispersion ("Polygen" WE 11)	45
Defoamer/levelling agent	10
Procedure:	
Introduce mixture 2 and add mixture 1 with stirring	

[1] *BASF SE*

4.6.2 Polymer dispersions in silicate systems

In recent years, polymer dispersions have become essential ingredients in the production of silicate emulsion paints and plasters. They are added to improve the stability and shelf life of the silicate systems. The choice of appropriate polymer dispersion and correct formulation of the paint as a function of the polymer dispersion used critically affect the properties of the end product.

Unlike the 2-component silicate systems, which must be mixed on the building site prior to use, 1-component silicate systems or silicate dispersion systems are ready for use. Dispersions are added to them as organic stabilizers, with DIN 18 363 restricting the total quantity of organic constituents to a maximum of 5%. Accordingly, there are essentially two binders in 1-component silicate systems, namely

- an inorganic binder, the water glass, and
- an organic binder, the polymer dispersion.

The two binders differ in their behaviour, both in the liquid paint and in the applied film. On application, water glass reacts chemically to form an inorganic silicate matrix, into which the pigments and fillers of the paint are incorporated.

Polymer dispersions, on the other hand, when applied and dried, participate in a purely physical process to form a film, which covers the pigments and fillers.

Within this paint system, however, the polymer dispersion not only functions as a stabilizer but also makes it possible to improve a number of parameters, the most significant of which are:

- moisture protection;
- chalking resistance; and
- adhesion.

4.6.2.1 Moisture protection

Moisture protection is an important criterion for masonry coatings. In silicate paints, it is even more important as the system includes a chemically reactive component, namely water glass. The "water balance" of a silicate coating can be readily described by two parameters:

- water absorption (WA); and
- water vapour transmission rate (WVTR).

On account of their morphology, the WVTR of silicate systems is excellent and easily meets the requirements presently imposed on diffusible coatings.

Of even greater importance in our latitudes, however, is the WA. In order to minimize capillary WA, silicate systems require particularly careful formulation and application. The use of organic binders has a measurable influence on the WA and the WVTR. This influence varies from one binder to another, however, as a function of the principal monomers, auxiliary monomers, emulsifiers, additives, and preparation process. Both the WA and the resistance to hydrolysis are therefore suitable criteria for selecting the appropriate type of dispersion for silicate dispersion systems.

4.6.2.2 Resistance to hydrolysis

On account of its high proportion of water glass, a silicate system is strongly alkaline, with the consequence that the polymer dispersion to be used must be very resistant to alkalis and thus stable to hydrolysis. As will be demonstrated later, a dispersion that is not resistant to hydrolysis can render the entire system unstable and therefore useless. The test method for determining the saponification number is described in Chapter 4.6.1.1. For practical purposes, there are only two suitable dispersion types for silicate systems: Ak/S copolymers and pure acrylic copolymers [110].

4.6.2.3 Water absorption (WA)

As discussed in Chapter 4.6.1, polymer dispersion films stored in water absorb the water and turn white. The more hydrophilic the polymer, the greater is its WA under otherwise identical conditions [121]. Also the WA analysis shows that the only suitable binders for use in silicate systems are Ak/S and pure acrylic copolymers (see Figure 4.7 on page 163).

Conclusion

In silicate dispersion systems, preference should be given to the use of Ak/S copolymers, whose resistance to hydrolysis, and low level of WA, makes them superior to all other types of polymer [104].

4.6.2.4 Interactions between dispersion and water glass

The interaction of a polymer dispersion with potassium water glass can be demonstrated both rapidly and forcefully in the water glass compatibility test, which is conducted as follows: the dispersion is mixed with an equal quantity of water glass and the mixture is stirred by hand using a spatula. Neither the dispersion nor the silicate solution should coagulate or flocculate [200].

This test illustrates the whole problem of silicate dispersion systems, as coagulum and/or flocks are formed with virtually all polymer dispersions that are used. To examine the reasons, it is necessary to look more closely at the chemistry of the two components.

The properties of the dispersion are critically co-determined by the auxiliary monomers [201 - 203]. These are added to the polymer in small proportions to confer defined performance properties. The most important are carboxy-functional comonomers such as (M)AA. The improvements that they effect include an increase in the stability of the dispersion (i.e. stability to shear and to electrolytes, and storage stability), enhanced viscosity control, and improved film adhesion.

The addition of alkali (in the case of silicate dispersion systems, the alkali is a potassium water glass with a pH greater than 11) gives rise to strongly dissociated carboxyl groups, which cause mutual electrostatic repulsion and extensive solvation. As a result, the polymer becomes more hydrophilic in proportion with the hydrophilicity of the nonionic main fraction of the polymer itself. The result is increased swelling of the polymer, a rise in WA, and an increase in the viscosity.

Figure 4.20 shows the viscosity characteristics of two Ak/S copolymers upon addition of alkali. The dispersion designated Ak/S1 is a 50%, high-viscosity, standard commercial product. Type Ak/S2 is a 50%, medium-viscosity dispersion developed specially for silicate dispersion systems. It can clearly be seen that the viscosity of the standard type Ak/S1 undergoes severe fluctuations when the pH changes. A decrease in viscosity with associated shrinkage of the polymer particles at a high pH is presumably linked to greater shielding of the ionised groups. In addition, however, there is evidently a dilution effect exerted by the 20% KOH. Dispersion Ak/S2, which is optimized for silicate systems, does not show any fluctuations in viscosity when the pH changes.

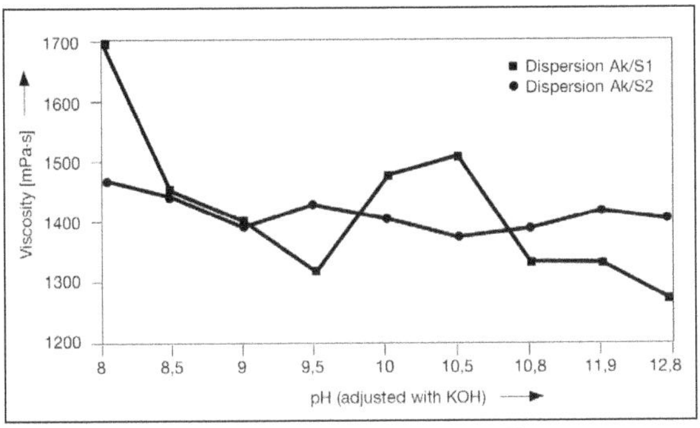

Figure 4.20: Viscosity as function of pH [200]

In addition to the effect on viscosity, which in extreme cases may even cause destruction of the disperse structure (known as alkali shock), the carboxyl groups present in the polymer also affect the WA. In the alkaline range, these groups are more highly dissociated and hence more osmotically active than in the acidic range. In the alkaline pH range, therefore, a markedly higher WA will be found. Dispersion Ak/S2 shows virtually no dependence on the respective pH. In the case of the standard type Ak/S1, on the other hand, the WA increases almost four-fold in the pH range of the silicate systems.

Any discussion of the effect of water glass should bear in mind that alkali silicates are not true chemical compounds, but rather are complexes consisting of defined alkali silicates, silicas, and alkali oxides [204]. All solutions of alkali silicates include monomeric silicate ions, polysilicate ions, and colloidal silica micelles. The prevailing dynamic equilibrium determines the type, bonding, and size of the ions and micelles. This equilibrium depends on the nature of the silicate, on the ratio of SiO_2 to K_2O, on the concentration, and on ambient conditions, such as temperature, pH, the presence of other substances, etc. Thus acids and acid salts, even carbonic acids and hydrogen carbonates, may liberate silica, which is produced as silica sol, gel or precipitate in accordance with the ambient conditions. The use of weak or dilute acids favours the formation of silica sols, while strong acids promote the formation of gels.

The very different pKa values of (M)AA on one hand and of silica and disilica on the other and the consequent differences in their reactions with the alkali present constitute the key to the stability (or instability) of the entire water glass/dispersion system.

It is therefore found that not only the polymer dispersion but also the water glass can be irreversibly damaged. Neither component can then fulfil the purpose for which it was intended in silicate dispersion systems. Accordingly, the selection of the polymer dispersion, and the requirements imposed on that dispersion are of central importance.

4.6.2.5 Demands imposed on an optimum dispersion

The remarks made so far can be boiled down to five requirements which the optimum polymer dispersion for use in silicate dispersion systems must meet:

1. It should be an acrylic/styrene copolymer.
2. It must be odour neutral, i.e. free from solvents, plasticizers and residual monomers, and from ammonia as well. Freedom from ammonia means that the dispersion should be neutralized with KOH.
3. It must be compatible with potassium water glasses, as determined by the water glass compatibility test.
4. It must be stable on storage, i.e. it must not undergo any change during storage or cause thickening.
5. It should have a low MFFT, so as to minimize the quantity of coalescent used in the formulation and to increase "film flexibility" [110].

4.6.2.6 Silicate dispersion system

Wagner [200] tested different parameters across a wide range of formulations. It was found that both the number of scrub cycles and the viscosity as a function of the storage period were more constant the higher the polymer fraction was in the formulated system. Moreover, both the WVTR values and the values of an equivalent air thickness layer remain approximately constant throughout the investigated range: 3 to 10% polymer fraction. In summary, no serious effect exerted by the polymer dispersion fraction on the WVTR is detectable. The WA coefficient w is not noticeably affected until the quantity of added polymer dispersion exceeds 6%. Finally, the resistance of the silicate paints to chalking and degradation on weathering increases with increase in the polymer fraction in the formulation. The optimum fraction of polymer dispersion (expressed in terms of solid in the entire formulation) is 4.5%.

Order of addition

The order in which the components are added, expressed in terms of the polymer dispersion, will now be considered. Advantageously, before the dispersion is added, the thickener (dissolved in water), the dispersant, the pigment, and the defoamer should be combined to form an initial charge which is subsequently dispersed for sufficient length of time. In the case of the polymer dispersion Ak/S2, it is advisable to add a small quantity of coalescent, despite the inherently low film-forming temperature, in order that better swelling of the polymer particles may occur. After the fillers, the water glass is added at the end of the formulation. In any case, it should always be added after the dispersion. Otherwise, if the dispersion were to be added slowly to the highly alkaline initial charge, the dispersion might experience the aforementioned alkali or electrolyte "shock". Indeed, even if the dispersion has successfully passed a water glass compatibility test, uncontrolled reactions, possibly going as far as the formation as coagulum, can always occur as a result of incorrect order of addition [110].

Silicate dispersion plasters

In principle, the aforementioned comments about silicate emulsion paints apply equally to silicate dispersion plasters. When formulating the plasters, however, there are a few additional points to consider.

Generally, a plaster may be regarded as high-build paint with a coarse structure. The greatest difference from silicate paint proper, besides the coarser fillers, lies in the ratio of binder to filler as a function of the structure that must be achieved. In silicate emulsion paints, this ratio is typically from about one to three, and in silicate dispersion plasters from about one to twelve, the binder being the total quantity of polymer dispersion plus potassium water glass (each considered as 100% solid). As a consequence of the high filler and solid content, the problem which arises is to introduce enough liquid to adequately disperse the pigment and filler. For this reason, the polymer dispersion, plus thickener and dispersant, could be introduced initially and the pigment could then be dispersed.

One consequence of this is that, even more so than in the silicate paints, the dispersion must be extremely stable to shear. Practical experience shows that between 3 and 4% of dispersion (expressed in terms of total solid) is best, but the precise amount varies with the desired plaster structure. The water glass should be introduced into the formulation alongside the dispersion of the pigment but before the fillers are added. Note that highly concentrated potassium water glass (28 to 30% strength) must be used so as not to exceed the overall liquid demand of the formulation [200].

Suggested formulation for a silicate emulsion paint

The relationships described above between polymer dispersion and water glass will now be translated into a suggested formulation for silicate emulsion plaster (Table 4.23) [200].

Table 4.23: Stating point formulation for a silicate dispersion paint

Component	wt.-%
Water	291
Dispersing and wetting agent	2
Thickener	5
Silicate stabiliser	5
Pigments	65
Fillers	417
Defoamer	3
Polymer dispersion	95
Film forming agent	5
Silicate binder	90
Viscosity adjustment	3
Hydrophobising agent	10
Total	**1000**

Table 4.24: Formulation of a grooved silicate dispersion finish

Component	wt.-%
Water	77
Polysaccharide	1
Dispersants and wetting agents	2
Defoamers	2
CMC thickener, 2%	34
Polymer dispersion	64
Hydrophobicity agents	3
Pigment	25
Silicate binder	98
Cellulose fibres	5
Calcite, 40µm (filler)	236
Calcite, 130µm (filler)	364
Plastorit, 300µm (filler)	44
Film-forming auxiliaries	5
Quartz shingle (filler)	40
Total	**1000**

Suggested formulation for a silicate dispersion plaster

As described above, especially in this chapter, silicate dispersion plasters have special features. The translation of that information into a general formulation is given in Table 4.24 [200].

4.6.3 Polymer dispersions as binders in silicone resin systems

One field of use which has gained considerable importance in recent years is that of silicone resin paints and plasters or, to put it more correctly from a technical standpoint, silicone resin emulsion paints and dispersion plasters. Their principal field of application is in the protection of façades.

These systems comprise a polymer dispersion as an essential constituent, namely as the primary binder. The choice of appropriate polymer dispersion and correct formulation of the paint or plaster critically affect the properties of the end product.

Polymer dispersions in silicone resin systems

An exterior architectural coating should have a low WA but also be permeable to water vapour so that the substrate dries out rapidly if wet through [205].

There are more qualitative rules, derived from climatic data and mathematical models, on the relationship between WA and water vapour diffusion [206]. Furthermore, semi-quantitative relationships have been established for central European conditions, based on outdoor exposure trials [115, 207]. Coatings with a WA coefficient w of less than 0.1kg m^{-2} $h^{-0.5}$ (classified as "impermeable to water") and, at the same time, a diffusion resistance s_d of less than 0.1m (classified as "microporous and permeable to water vapour") are suitable [208].

So what is the situation when silicone resin and polymer dispersion are combined? As a result of its deliberate "supercritical" formulation, water vapour can easily diffuse through the coating, but liquid water finds it difficult to penetrate, as the pores have been rendered hydrophobic by the silicone resin. In other words, silicone resin dispersion systems attempt to satisfy the demand for minimum diffusion resistance coupled with a low WA coefficient, by producing "hydrophobised porosity" [69, 209 – 212].

So why not formulate with a silicone resin as sole binder if the surface is porous in any case? The reasons are as follows. The PBP of silicone resins is too low to ensure adequate immobilising of pigments and fillers. This has a strongly adverse effect inter alia on weatherability. The reason for combining the silicone resin with a polymer dispersion, therefore, is a deficiency in the silicone. The added dispersion, however, not only functions as a binder in these paint systems but also leads, moreover, to an improvement in a number of quality features, the most important of which are

- storage stability;
- chalking resistance; and
- adhesion.

The PBP, WA and, a further important performance property, weatherability (including the tendency to pick-up dirt) are therefore criteria used for selecting the type of polymer dispersion in a dispersion/silicone resin system.

The effect of the polymer dispersion on the properties of the system as a whole differs from one dispersion to another, varying with the principal monomers, auxiliary monomers, emulsifiers, auxiliaries, and preparation process [213].

Pigment binding power

A detailed discussion of PBP can be found in Chapter 4.6.1.1 and in [213].

The CPVC can be determined by measuring a number of film properties. The most widely established methods of determining the CPVC of emulsion paints are the following:

- measurement of the hiding power on black/white contrast cards;
- measurement of the water vapour permeability;
- measurement of the film porosity by coloration and determination of the difference in lightness;
- qualitative [132] or quantitative [214] measurement of the "internal stress" on PVC films; and
- measurement of the reduction power.

A feature common to all of the methods is the preparation of a PVC series. This is created by mixing a defined emulsion paint mill base with the appropriate quantity of dispersion for the respective PVC. Determining the respective film property then makes it possible to determine the position of the CPVC, namely, as the exact point at which the property in question undergoes a marked change.

As is evident from a look at the literature, the CPVC and its position are hotly debated. In some cases, theoretical considerations contrast sharply with practical experience gained with different polymer dispersions [132, 215 - 218]. *Wagner*, for example, used eight standard commercial polymer dispersions which differed in principal monomer composition, stabilizing system (i.e. emulsifier or protective colloid dispersion), and average particle size [213]. The position of the CPVC was determined by three different criteria (hiding power as contrast ratio, WVTR and s_d values, and film porosity by the Gilsonite test (measurement of difference in lightness)). The three methods of determining the CPVC yield values which show very good agreement; only in the case of the relatively coarse dispersions VP-1 and VPA-1 are there discrepancies. The finer dispersions yield higher CPVC values. The Ak/S and straight acrylic copolymers investigated have a clear lead over the other competitors.

Silicone resin emulsion and CPVC

The best way to answer the question about the nature of the binder in the silicone resin emulsions is to consider their PBP. To this end, *Wagner* [213] selected three different silicone resin emulsions and attempted to determine the CPVC position using the methods mentioned above.

On the one hand, these results provide good confirmation of the correctness of the measurement techniques employed. On the other, however, they permit only one conclusion: on the basis of the binder properties which are clearly present and detectable, when calculating the PVC, the silicone resin emulsions tested must be considered together with the polymer dispersion as binders.

Weatherability

Studies reveal a marked influence exerted by the T_g of the polymer on dirt pick-up. The higher the T_g, the lower is the dirt pick-up; however, at higher T_g, the difference in polymer structure, provided the T_g was the same, increasingly became less significant. At a T_g of 10°C, there is no difference between the dispersions used [219].

Clearly, the difference in dirt pick-up is attributable, at least progressively, to differences in the severity of chalking. More severe chalking, accordingly, is said to go hand in hand with lower dirt pick-up, as removal of the topmost layer by chalking or breakdown will also see the removal of any dirt deposited thereon.

Corresponding chalking studies have shown that the difference in dirt pick-up cannot be the result of a "self-cleaning" effect. Rather, it is also necessary to bear in mind that, in addition to the thermoplasticity of the polymer, the PVC is also able to influence the dirt pick-up and chalking tendency.

For systems such as silicone resin paints, which are formulated at significantly higher PVC than exterior architectural paints, it has been found that the dirt pick-up tendency falls as the PVC increases, i.e. as the proportion of thermoplastic material decreases [214, 219].

Demands imposed on an optimum dispersion

In summary, for use in silicone resin systems, the optimum polymer must meet five requirements:

- It should be an acrylic/styrene or a straight acrylic copolymer.
- In the binder, the particles should have an average diameter of less than 0.2µm (so as to maximise the pigment binding capacity).
- The dispersion should be odour neutral, i.e. free from solvents, plasticizers and ammonia, and should also have low residual monomer content.
- It must be compatible with the silicone resin emulsion to be used, i.e. it should not coagulate and should generate the least possible, if any, turbidity in the film.
- It should have low water absorption.

Formulation of silicone resin paints

The suggested formulation for solvent-borne silicone resin paints presented in Table 4.25 is the fruit of many years' experience [213].

Table 4.25: Formulation of a solvent-containing silicone resin paint

Component	wt.-%
Water	213
Polyphosphate (dispersant)	0.5
Polyacrylate (dispersant)	3.5
Thickener	1.5
Preservative	1.5
Film-forming auxiliary	10
Defoamer	1
Titanium dioxide, rutile (pigment)	110
Aluminium silicate 0.04µm (filler)	15
Mica 1µm (filler)	30
Calcite 3µm (filler)	220
Calcite 7µm (filler)	135
Polymer dispersion	100
Defoamer	1
Silicone resin emulsion	100
PU associative thickener, premixed with	4
Butyl diglycol	16
Water	38
Total	**1000**

4.6.4 Elastic coating systems

An important function of exterior architectural coatings, besides decoration, is effective protection against moisture. Often over decades, façades develop cracks which are usually irrelevant to the status of the construction but which promote the penetration of moisture. Façades of this kind must be renovated with systems having the ability to cover these cracks. The main requirements of suitable systems are

- high elasticity, even at low temperature; and
- low dirt pick-up tendency.

The following discussion will therefore include the structure of binders which meet these requirements, the role of surface tack as a cause of dirt pick-up, and the possibilities for reducing it.

Nowadays, the protective function is at least as important as the aesthetic function. Protecting against climatic effects is primarily protection against moisture, which must be prevented from penetrating into the building substance in the form of rain, dew, high atmospheric humidity, and melted water. In the case of reinforced concrete, the coating must also protect against the diffusion of carbon dioxide, which through carbonation would otherwise lower the pH in the concrete and so diminish the steel reinforcement's protection against rust [220].

4.6.4.1 Effective protection against moisture, according to Künzel

There are two main ways in which moisture can penetrate the substance of a building:

1. diffusion of liquid water through the intact coating; and
2. inward seepage through defects in the coating, e.g. fine and ultrafine cracks.

The first process, capillary water uptake, is relatively slow, its rate increasing with the open-pored nature of a coating. A quantitative measure of the capillary water uptake is the WA coefficient (W); for many building materials, it corresponds to the slope of the line obtained by plotting the weight increase on immersion in water as a function of the square root of the time. The WA coefficient for exterior architectural coatings based on polymer dispersions is approximately 0.1kg $m^{-2} h^{-0.5}$, and is thus much smaller than for customary construction materials (concrete: about 1 to 2kg $m^{-2} h^{-0.5}$; lime sandstone: 5 to 7kg $m^{-2} h^{-0.5}$).

The second process can be very rapid; due to the relatively high WA coefficients of the building materials, fairly large quantities of water enter the building substance in this way, at least in the region of the defect.

The subsequent drying of the building substance always occurs by evaporation, predominantly through the coating. This process is also slow, but increases in rapidity with the open-pored nature of the coating. One measure of this is the diffusion resistance number, μ. It is the resistance of a building material to diffusion divided by the resistance of an equal thickness of air at rest to diffusion, and is considered to be a constant of the material.

In considering the diffusion process, it is also necessary to take account of the thickness of the layer(s), because the product of diffusion resistance number and layer thickness, the equivalent air layer thickness s_d, is more suitable for the purposes of practical assessment. In other words, s_d indicates how thick a layer of air must be in order to have the same diffusion resistance as the tested coating, at a given thickness. As they are applied only in very small coat thicknesses, exterior coatings based on polymer dispersions have markedly lower equivalent air layer thicknesses than standard building materials [220].

For effective protection against moisture, the WA coefficient and equivalent air layer thickness of the coating must be as small as possible. In practice, both parameters of a coating tend to be mutually contradictory; a low WA generally also implies low evaporation, i.e. high equivalent air layer thickness. Accordingly, *Künzel* [115, 207] quantified the conditions for effective moisture protection on the basis of measurements made in the particularly adverse climate of the Alpine foothills. Protection is afforded when the WA coefficient and the equivalent air layer thickness meet the following criteria:

$$W \leq 0.5\text{kg m}^{-2}\text{h}^{-0.5}$$

$$s_d \leq 2\text{m}$$

$$s_d \cdot W \leq 0.2\text{kg m h}^{-0.5}$$

These criteria impose considerable technical requirements on a coating system and its binders. The industry nowadays provides many types of binder which can be used to formulate coating systems that meet these criteria. The product class which has attained the greatest technical sophistication is that of aqueous polymer dispersions [220].

4.6.4.2 Principal requirements imposed on coating systems for renovating façades

A problem when renovating such façades is that the width of the cracks varies with temperature, as a result of changes in the volume of the substrate [221].

Accordingly, appropriate coatings must be sufficiently elastic to follow the temperature and moisture-related changes in crack width. Otherwise, the development of new cracks and the propagation of existing cracks, even in the renovated façades, cannot be ruled out and, indeed, is highly probable. This imposes even greater demands on the elasticity. Naturally, the coating should also fulfil its aesthetic functions over a long period, i.e. the façades should still look clean after many years.

Therefore, the principal requirements imposed on crack-bridging coatings are as follows:

- suitable mechanical and elastic properties in the climate-specific temperature range of approximately -10°C to 40°C; and
- high resistance to dirt pick-up over the long term.

Conventional binders for exterior architectural paints are unsuitable for crack-bridging because at low temperatures they are not sufficiently extensible or elastic. The proposed solution of using binders permanently plasticized with nonvolatile plasticizers is problematic since the plasticizer greatly increases the dirt pick-up tendency. Plasticizers may also have toxicological problems. Binders plasticized solely by virtue of the monomer composition, i.e. "internally" plasticized binders, would likewise attract criticism in respect of dirt pick-up.

In the text below, a number of chemical factors are examined which are responsible for the mechanical properties and dirt pick-up resistance, especially of binders based on aqueous polymer dispersions. Knowledge of these factors makes it easier to understand the structure of modern binders for crack-bridging coatings [220].

4.6.4.3 Mechanical properties of dispersion films

Glass transition temperature (T_g)

The mechanical properties of a dispersion film vary with the temperature. At low temperatures, the film is glassy and brittle. Above a certain temperature, it becomes soft and elastic (Figure 4.21), to an extent which is dependent on the composition of the polymer [222]. The T_g can be determined, for example, from the temperature dependence of the modulus of elasticity. At temperatures below the T_g, the modulus of elasticity is high, corresponding to very low extensibility. On heating, the modulus falls in the region of the T_g by several orders of magnitude. As will be shown, the precise course of the modulus of elasticity above the T_g depends on the degree of crosslinking. In any case, the extensibility there is substantially higher than below the T_g. It is possible to set a desired T_g with precision via a judicious combination of monomers. Obviously, binders for a system which is to retain its crack-bridging capacity even at -10°C must have a T_g below -10°C. The appropriate binder for this purpose would then have to contain primarily soft monomers, whose homopolymers have T_g values below -10°C [220].

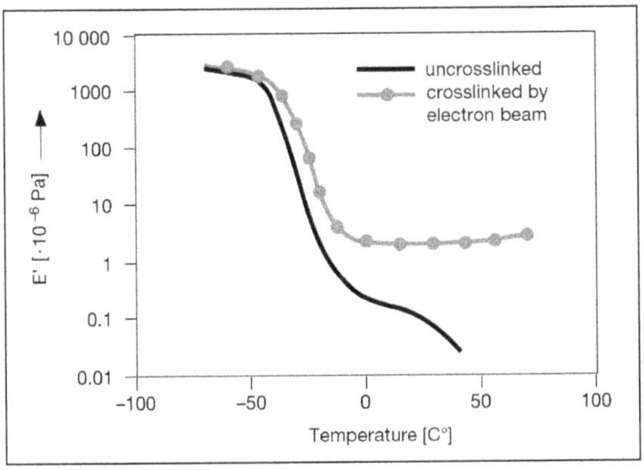

Figure 4.21: Storage modulus E' as a function of temperature: n-butyl acryl ate homopolymer

Tensile stress-elongation test

A low T_g is not an absolute criterion for the mechanical properties of the binder for a crack-bridging system. Equally important is that the extension (elongation) must be highly reversible, i.e. the material must be elastic. A system which becomes thermoplastic even upon slight elongation, i.e. an irreversibly deforming system, without exhibiting any resilience, would be unsuitable. The extensibility of polymer dispersion films always has a thermoplastic component and an elastic component. The elastic component can be increased at the expense

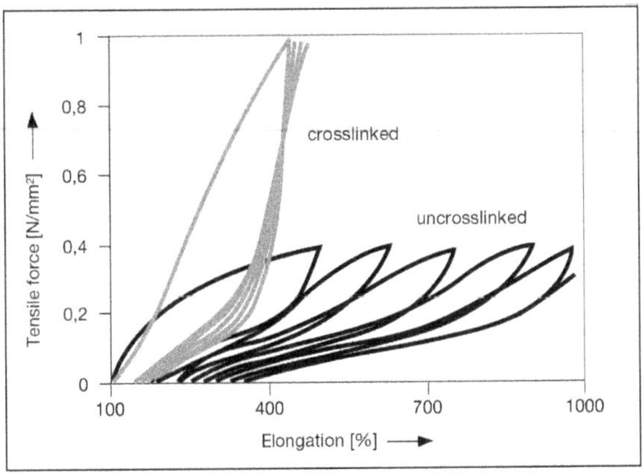

Figure 4.22: Cyclic tensile stress-elongation test

of the thermoplastic component by crosslinking the binder. This leads to an increase in the modulus of elasticity above the T_g. Viewed mechanically, the film above the T_g appears more elastomeric, i.e. its behaviour is substantially independent of temperature. The T_g itself is increased only insignificantly by the crosslinking.

The change in mechanical behaviour as a result of crosslinking becomes clear from the periodic tensile stress-elongation test, once with an uncrosslinked dispersion film and then with the same system but with added crosslinker (Figure 4.22). When the tensile force is recorded as a function of elongation, the uncrosslinked sample shows irreversible deformation of more than 200% after 5 elongation cycles and thus has lost any resilience. The crosslinked sample, although it still has a small thermoplastic component which is active in the first cycle, shows virtually no further thermoplastic deformation in the subsequent elongation cycles.

Type of crosslinking

Roughly, three types of crosslinking can be distinguished in a dispersion film (Figure 4.23):

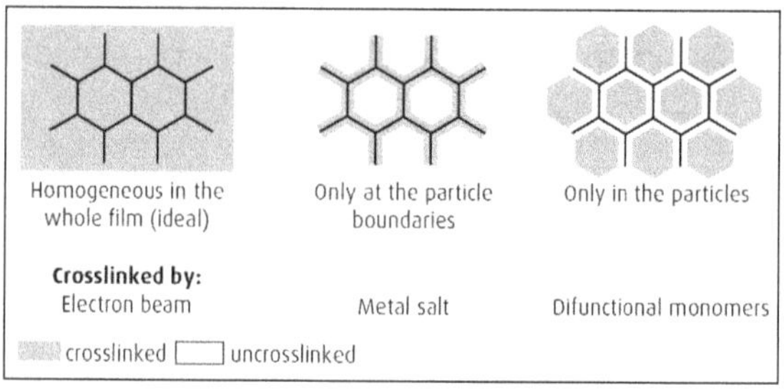

Figure 4.23: Types of crosslinking

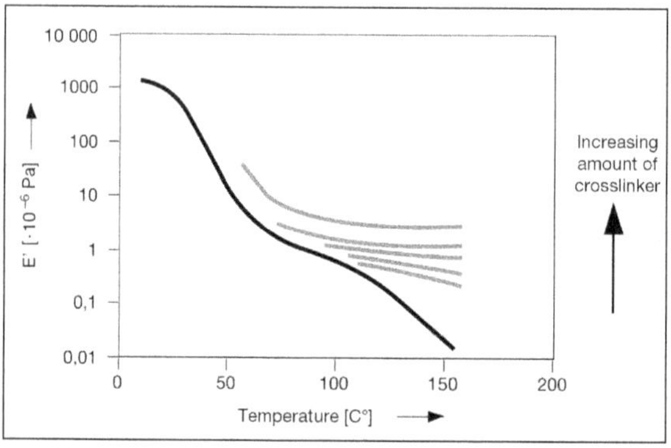

Figure 4.24: Storage modulus E' as a function of temperature: polymer of butyl methacrylate crosslinked with diacrylate

1. Substantially homogeneous crosslinking, attainable, for example, by subjecting the film to an electron beam. In accordance with the modulus of elasticity above the T_g, the film exhibits conventional elastomeric behaviour (cf. Figure 4.21 on page 227).

2. Homogeneous crosslinking within the particles, but restricted only to some of them and not progressing beyond the particle boundaries. This crosslinking is obtained, for example, with difunctional monomers during polymerization. Thereafter, crosslinking ceases. The behaviour of the film in the modulus of elasticity

range above the T_g becomes more elastomeric as the quantity of difunctional monomer increases (Figure 4.24).

3. Crosslinking which becomes effective only in the course of film formation and extending across particle boundaries; the particle cores remain more or less uncrosslinked. Crosslinking within the film is very heterogeneous, and is obtained, for example, by adding salts of polyvalent metal ions or thermally induced curing of dispersions containing N-methylacrylamide. Above the T_g, the film exhibits limited elastomeric behaviour (Figure 4.25).

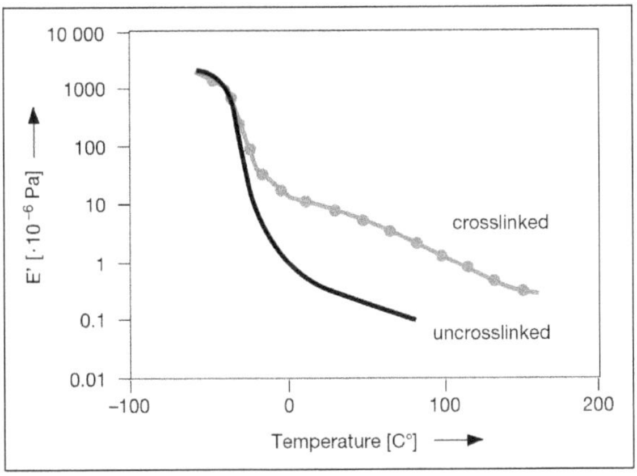

Figure 4.25: Storage modulus E' as a function of temperature: copolymer of BA/AA crosslinked with metal salt

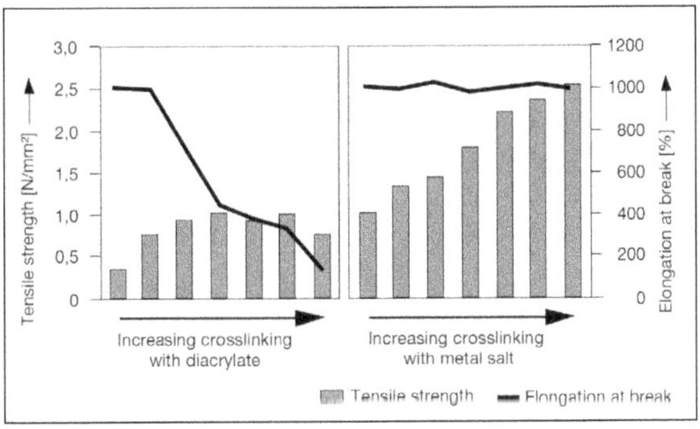

Figure 4.26: Tensile strength and elongation at break on crosslinking with diacrylate and metal salt

Regarding the dirt pick-up propensities of each of the variants, the following remarks may be made:

Type 1 crosslinking is not realizable in practice.

Type 2 has a decisive disadvantage which is not visible from the modulus of elasticity measurements alone but which does become evident when the tensile strength and elongations at break of corresponding films are considered. As a result of the homogeneous crosslinking within the particles but not across the particle boundaries, the cohesion of the film is very low and tends to decrease as crosslinking progresses. Thus there is also a continuous reduction in the elongation at break as the quantity of crosslinker grows, with the tensile strength passing through a maximum. The extensibility of a coating system crosslinked in this way will not be sufficient for bridging cracks (Figure 4.26 left).

Films produced by type 3 crosslinking with a metal salt exhibit high film extension irrespective of the quantity of the salt; the tensile strength increases with the concentration of metal salt (Figure 4.26 right).

As crack-bridging is associated with high extension, the third type of crosslinking, which becomes effective during or after film formation, is the most suitable for crack-bridging systems.

Summary of mechanical requirements: The mechanical requirements imposed on a crack-bridging system can be met by a dispersion having:

- a polymer composition with a T_g below the temperature range within which the crack-bridging is to be effective; and
- crosslinking which becomes effective only during film formation, i.e. during drying, and so improves the elasticity [220].

4.6.4.4 Dirt pick-up resistance

Crosslinking and T_g are critical influences on the mechanical properties of crack-bridging systems. They are followed in importance by the dirt pick-up resistance. It determines how long a façade remains optically attractive, i.e. the length of time until the next renovation coat is needed.

A surface will become soiled more rapidly with dust, for example, the more the dust adheres to it, i.e. the tackier the surface is. The cause of tack may be the low-molecular weight constituents of the formulation, which migrate to the surface (during drying). Moreover, soft binders have a more or less tacky surface and soil to a greater extent [223]. To a first approximation, the softer the polymer concerned, the greater the tack.

Good mechanical properties – even at low temperatures – and low tack are therefore a priori mutually contradictory properties of a dispersion film. In practice, mechanical properties have absolute priority, and so soft binders must be used. For a given binder, the tack should be reduced as far as possible by taking appropriate measures [220].

Tack

By tack is meant a kind of stickiness which causes one material which comes into brief contact with another, e.g. dust, to form an assembly whose separation requires a measurable force. Admittedly, tack is not a precisely defined material parameter such as, say, the modulus of elasticity.

In the chapter below, the factors which determine the tack of a binder will be looked at somewhat more closely.

A comparison of rubber with, honey, for instance, shows that softness is not necessarily synonymous with tack. Both substances are soft but rubber is virtually tack-free whereas honey has high tack. The reason lies in the flowability of honey. This flowability brings about high deformability of the surface, so that at the time of contact there is a relatively large contact area. Although soft, rubber is not flowable, so only a small effective surface area is involved in contact, compared with the case for honey. In other words, it is the deformability of a substance at the moment of contact which determines the tack.

Tack measurement

A quantitative measurement of the tack is a prerequisite for understanding the parameters which determine it at the molecular level. Since tack, as mentioned, is not precisely defined, its measurement necessitates precisely defined boundary conditions [224]. In other words, the more deformable a substance appears for a short contact time, the higher is its tack.

Figure 4.27 shows the results of measurements of the temperature-dependent tack of a series of acrylate homopolymers. The tack passes through a maximum at approximately 60°C above the T_g, and the absolute level of each maximum increases in correlation with the side-chain length of the respective monomer. P-EHA, therefore, has greater tack than, for example, P-EA, owing to its higher elastic deformability in the short-term range. The average molecular weight between two points of entanglement, i.e. the mesh size of an entangled network, can be approximated. It increases with increase in length of the side-chain and, within the polyacrylate series, reaches its maximum in the case of P EHA.

Summary: These results show that tack can be effectively reduced by entanglement or crosslinking.

Surface crosslinking

However, there are limits on the extent to which tack can be reduced by crosslinking. Effective reduction requires such a high crosslinking level that the extensibility of the film would become too low for a crack-bridging system, i.e. the film would be

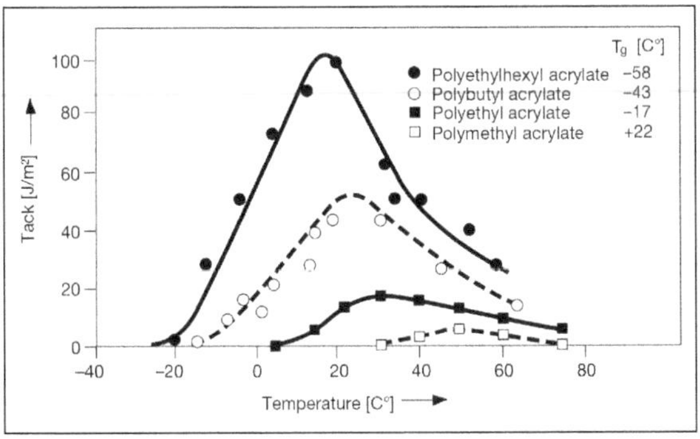

Figure 4.27: Tack and glass transition temperature of polyacylates

brittle. However, as tack acts only at the surface, a high level of crosslinking throughout the film is not necessary. It would be enough to restrict the high crosslinking density to the topmost layer. This can be effected by means of light-induced crosslinking with the aid of photoinitiators.

Under the influence of light, the photoinitiator breaks down into free-radicals, which in a subsequent reaction can cause the polymer chains to crosslink with one another (see Chapter 5). Even if the initiator is distributed uniformly throughout the film, on exposure it becomes effective only in the topmost layer, as the process of crosslinking is not a chain reaction; for each molecule of photoinitiator, only one crosslinking site is formed. Moreover, fillers and pigments prevent the penetration of light into lower-lying layers. Therefore, the resultant crosslinking of the uppermost layer reduces only the tack, and has no effect on the extensibility of a crack-bridging system.

Summary of findings relating to dirt pick-up resistance: In general, good mechanical behaviour and dirt pick-up resistance are mutually contradictory properties.

Dirt pick-up resistance, which is indirectly proportional to the tack of the surface, may be improved by crosslinking. Excessive crosslinking, however, has an adverse impact on the mechanical properties. Light-induced reaction makes it possible to limit crosslinking, which is critical for improved dirt pick-up resistance, to the topmost layer of the film. Consequently, film extension overall is not greatly affected [220].

Dirt pick-up resistance has not yet been standardized. The sole reason is the absence of a way of detecting the dirt pick-up of a masonry coating in a standardized manner within acceptable periods of time. Parameters governing actual dirt pick-up, such as weather, site, nature and degree of air pollution, for example, are highly complex and are also irregular. Many attempts have been and still are being made to elaborate corresponding tests, but to date there is still no test which, within acceptable periods of time, reliably simulates the actual dirt pick-up behaviour over years [225].

Polymers on the market

For the formulation of crack-bridging coatings, the following dispersions are encountered in the market:

- acrylic/styrene copolymers; and
- straight acrylics.

The Ak/S copolymers are often plasticizer-free binders with very good long-term extensibility, and a T_g in the range -10°C to -45°C. Increasingly, crosslinked Ak/S dispersions of low T_g's and reduced dirt pick-up are becoming available. More recent binder generations for these applications are distinguished, furthermore, by lower odour. They often also utilise a number of crosslinking systems, and so may be characterized as follows:

- free of ammonia;
- T_g approx. -30°C;
- solid content approx. 60%;
- greater extensibility than earlier products, with virtually the same tensile strength;
- low water absorption; and
- low dirt pick-up.

For this field of application, the straight acrylics encountered in the market are not necessarily the products of first choice, since their greater hydrophilicity, relative to that of Ak/S copolymers, carries with it performance disadvantages. Moreover, increased dirt pick-up potential has been found with the straight acrylics encountered to date.

As an example of how a crack-bridging coating can be formulated with a standard commercial binder, a typical formulation is given in Table 4.26.

Table 4.26: Starting formulation for a crack-bridging coating

Component	wt.-%
Water	101
Dispersing agent, e.g., "Pigment disperser" MD 20[1]	10
Preservative	2
Defoamer, e.g., "Agitan" 731[2]	5
Binder, e.g., "Acronal" S 321[1]	200
Titanium dioxide, rutile, e.g., "Kronos" 2056[3]	205
Filler, "Omyacarb" 5 GU[4]	128
Filler, "Naintsch" KL 30[5]	77
Filler, "Plastorit" 00[5]	51
Disperse at high speed...	
Binder, e.g., "Acronal" S 321[1]	170
Defoamer, e.g., "Agitan" 731[2]	3
"Collacral" PU 751, diluted 1:4 with propylene glycol	48
Add in the stated sequence	
Total	**1000**
Solids content: approx. 68%, PVC 40%	

Use: for bridging existing cracks up to 0.3mm; primer: formulation diluted 1:4 with water topcoat: (400g m^{-2})

[1] BASF SE, [2] Münzing, [3] Kronos, [4] Omya, [5] Naintsch

Summary: The decisive criteria for binders for crack-bridging systems are as follows:

- appropriate mechanical properties even at low temperatures;
- high dirt pick-up resistance; and
- effective moisture protection for the façades.

The mechanical properties are met by low-T_g binders crosslinked with metal salts. Light-induced surface crosslinking improves the inherently low dirt pick-up resistance without adversely affecting the mechanical properties.

4.6.5 Synthetic resin plasters and exterior insulation and finish systems (EIFS)

Plasters are a coating for interior and exterior walls, and are usually applied in several layers and in high layer thickness (several millimetres). The purpose of plastering is to smooth a surface or enhance its appearance and, in the case of exterior applications (e.g. stucco), to protect the substrate against moisture.

DIN 18 550, Part 1 (1985) distinguishes between plasters containing mineral binders and plasters containing organic binders. The standard goes on to reveal that the latter are actually coating materials comprising organic binders – with predominantly coarse fillers – which lead to a coating having the appearance of plaster.

The synthetic resin plasters are described in more detail in DIN 18 558 (1985).

The technique of stucco is very old. In the earliest times, loam mortars were used. Loam-plastered buildings can still be found in southern countries. Lime was the first mineral binder to be used for weathering-resistant exterior plasters. Lime plasters from antiquity can still be seen today, e.g. in Greek temples.

Synthetic resin plasters are a class of product which is just 50 years old. They consist of a polymer dispersion binder and usually coarse fillers (up to a maximum of 15mm). They serve exclusively as decorative plaster finishes, both outside and inside. They dry physically, with the binder bonding the various ingredients and producing adhesion to the substrate.

The advantages of synthetic resin plasters are

- good adhesion to many substrates;
- low susceptibility to cracking;
- imperviousness to driving rain coupled with water vapour permeability;
- ability to be processed as a finished product in one operation, without subsequent operations; and
- great scope for surface design and colouring.

Synthetic resin plasters were developed from emulsion paints on one hand and mineral plasters on the other. Emulsion paints admixed with fairly coarse fillers and applied by brush or roller were referred to as "spreadable plasters". Mineral plasters to which a dispersion or dispersion powder was added to improve moisture protection, adhesion, and elasticity were considered "dispersion-enhanced mineral plasters". Both approaches led ultimately to the simple synthetic resin plaster.

Formulated correctly, synthetic resin plasters are highly durable. In virtually all climatic zones, it is now possible to see damage-free buildings aged 30 years or more. Despite the decorative effect of plasters, however, it should not be forgotten that they are a wear layer intended to protect the stuccoed wall.

In composition, the conventional synthetic resin plasters offered on the market are very similar to one another. The product diversity arises from the scope for styling surfaces with synthetic resin plaster, and from the different processing techniques.

In addition to "conventional" synthetic resin plasters, there are a number of plasters that possess special properties:

- Silicone resin plasters (see Chapter 4.6.3); and
- Silicate plasters (see Chapter 4.6.2) [226].

Further specialty plasters, such as those for soundproofing, require no special binder or binder combinations and are therefore, as far as the binder is concerned, classified as normal synthetic resin plasters.

4.6.5.1 Classification of synthetic resin plasters and technical requirements

There are a number of standards relating to plasters:

- DIN 18 550, Section 1: Plaster; terminology and requirements
- DIN 18 550, Section 2: Plasters made of mortars containing mineral binders; application
- DIN 18 550, Section 3: Rendering; rendering systems for thermal insulation purposes made of mortar consisting of mineral binders and expanded polystyrene (EPS) as aggregate
- DIN 18 555: Testing of mortars containing mineral binders
- DIN 18 557: Factory mortar - Production, control and delivery
- DIN 18 558: Synthetic resin plasters
- DIN 18 556: Testing of coating materials for synthetic resin plasters
- DIN 4108, Section 3: Heat insulation in buildings; protection against moisture caused by the climate; requirements and directions for planning and construction
- DIN 4102, Section 4: Fire behaviour of building materials and building components; synopsis and application of classified building materials, components and special components [227]

According to DIN 18 550 Part 1 and DIN 18 558, there are two classes of synthetic resin plaster:

- P Org. 1 – Interior and exterior plasters
- P Org. 2 – Interior plasters

Table 4.27: Minimum binder amount for diverse plaster types

Type	Max. particle-size [mm]	Binder fraction [solid based on total solid content, wt.-%]
Exterior and interior plaster:		
P Org. 1	≤ 1	8
P Org. 1	≥ 1	7
Interior plaster only:		
P Org. 2	≤1	5.5
P Org.2	≥1	4.5

The technical requirements (e.g., weatherability, moisture protection) are much higher for the first class than for the second.

Suitable binders for synthetic resin plasters are all types of binder available on the paints market, including Ak/S copolymers, straight acrylics, VAc copolymers, and high-pressure polymers containing ethylene. Of course, the choice of formulation ingredients (binder, additives, and fillers) influences the behaviour of the material, e.g., the WA of the plaster.

Synthetic resin renders must contain minimum quantities of binder, the precise levels of which vary with the type of filler and its maximum coarseness, as shown in Table 4.27 (adapted from DIN 18 558).

For long-term durability in the order of many years, the particle size of the predominant fraction of fillers in the plaster should be greater than 0.125mm.

The following properties are normally tested (in accordance with DIN):

- on the wet plaster: crack-free drying, ease of processing at 5°C and storage without separation
- on the dried plaster: WA, water vapour permeability, frost resistance on exposure to moisture and cold (-15°C) and resistance to hydrolysis

Requirements imposed on the binder

The most important requirements imposed on the binder are:

- high binding capacity (for coarse fillers)
- low water absorption as binder, as well as in the formulation/good moisture protection
- low thermoplasticity (unless expressly required for specialty applications)
- high alkali resistance
- non-flammability for EIFS
- good processing properties

Binding power

The lower limits on the quantity of binder used are set by the respective standards. Ak/S dispersions feature very high binding power, although the binding power of the other types of binder is also entirely adequate for use in plasters. Unlike the case for paint, stating a PVC for a plaster is fairly meaningless owing to the use of coarse fillers of a relatively low specific surface area. For test methods and their description, see Chapter 4.6.1.3.

Water absorption and moisture protection

In order to provide effective moisture protection for the coated wall, a plaster must on one hand be highly impervious to rain and on the other be as open as possible to permit diffusion of water vapour. These two requirements ensure that the wall is not penetrated by too much liquid water and yet that water which has penetrated evaporates again as rapidly as possible. Both requirements are described in Chapter 4.6.4.1. DIN 18 558, a standard on plasters, puts the maximum value for the product of W s_d at $\leq$ 0.2kg m $h^{-0.5}$. The values shown in Table 4.28 have been found for commercial products.

These values are within the permitted range for moisture protection. Comparison with standard building materials (concrete: s_d = 7m, masonry: s_d = 2.4m) shows that synthetic resin plasters are sufficiently watertight and water-vapour permeable. The WA and WVP of the binder on their own do not provide conclusive evidence of the ready-formulated plaster's suitability. Nor could anything different be expected for a coating that has a very low binder fraction.

The values listed above apply only to the plaster alone. If, as is often the case, it is provided with a coloured coating, it is necessary to consider the whole system comprising plaster + coating. A coating applied in a thick layer can have a positive effect on WA and a negative effect on WVP [226].

Thermoplasticity, alkali resistance, and flammability

Apart from binders for special flexible plasters, the most suitable binders for plasters are those which have an MFFT of 15 to 20°C. Higher MFFTs lead to brittle plasters, while lower MFFTs increase the sensitivity to dirt pick-up.

Table 4.28: Water absorption of typical commercial products

	W [$k_g/(m^2\ h^{0.5})$]	s_d [m]
Scraped plaster	0.05 to 0.3	0.5 to 1.0
Grooved plaster	0.08 to 0.1	0.3 to 0.6
Coloured stone plaster	0.03	0.1 to 0.12

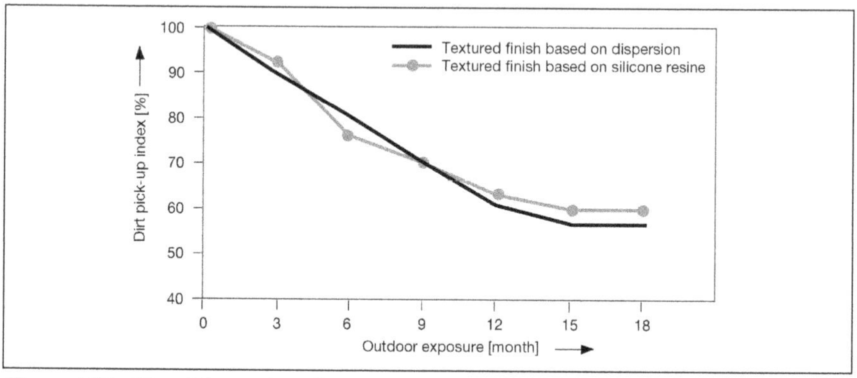

Figure 4.28: Dirt pick-up of synthetic resin and silicone resin plasters (in accordance with ASTM D 3719 – 95) [228]

Advance statements as to the dirt pick-up tendency of a plaster are extremely difficult to make with any accuracy. Factors determining the dirt pick-up tendency of plasters include the site (industrial zone, heavy traffic roads, town, and country), projecting roofs, building height, plaster structure, and substrate, together with weathering effects. Experience shows that dirt pick-up by plasters cannot generally be assessed until after 9 to 15 months of outdoor weathering on a weather stand. Dirt pick-up generally peaks after a total of 18 to 24 months' outdoor weathering (with samples stored at 45°), see Figure 4.28. The greater the binder content of a plaster, the more it becomes soiled.

Normal synthetic resin plasters are thermoplastic enough to provide reliable bridging of "dead" plaster cracks (shrinkage cracks and wind cracks). The bridging of larger cracks requires specially adapted flexible plasters, or repair work prior to plastering.

Alkali resistance is a requirement imposed on exterior plasters, especially those on concrete substrates. In this respect, Ak/S copolymers have advantages over the other types of binder (Chapter 4.6.1).

The requirement for non-flammability (Class B1) based on DIN 4102 applies not to the dispersion itself, but only to the ready-processed plaster system when applied to a solid mineral substrate. Correctly formulated synthetic resin plasters meet this requirement and can therefore be used on buildings up to a height of 20m.

Processing properties

The processing properties of a plaster – and the assessment of the worker applying it – determine the success or failure of the product far more than the technical properties. Despite the fact that the bulk of plasters are now applied by machine, and so the plasters must be suitable for machine processing, smoothing and surface treatment of the plaster are still largely manual operations. The following properties are important in this respect:

- surface working with minimal application of force;
- no smearing or sagging of the plaster; and
- sufficiently long open time to avoid areas of premature setting.

The processing properties can be influenced primarily through auxiliaries, such as those for modifying rheological properties. The binder has only a minor role to play in this respect.

Surface structures

The diversity of the surface effects attainable with synthetic resin plasters is governed by the particle composition on one hand and by the application technique on the other.

The following types of plaster exist:

- spreadable plaster;
- rolling plaster;
- sprayable plaster;
- scraped plaster;
- float plaster;
- grooved plaster;
- modelling plaster; and
- coloured stone/natural stone plaster.

Brushable, rolling and sprayable plasters take their names from the respective application technique and do not contain coarse fillers. Where subsequent working of the plaster surface is appropriate, modelling plaster is used.

Scraped plasters contain equal fractions of aggregates of very different particle sizes. On drying, the large particles project markedly (as a result of volume shrinkage). The result is a surface similar to that of a scraped mineral plaster.

Float plasters and grooved plasters contain a small fraction of particles of substantially greater size, which on smoothing leave grooves in the surface. Depending on the type of smoothing, a large number of rounded or linear structures may be obtained.

Coloured and natural stone plasters contain exclusively coloured sand or stones with a particle size of 1 to 4mm. The binder turns transparent when it dries.

Since there are no fine fillers, the grains of coloured sand become visible. Usually, the binder is exposed directly to sunlight. It is therefore necessary to use pure acrylics for such plasters; Ak/S copolymers are unsuitable because they turn yellow and are degraded by UV light.

4.6.5.2 Exterior insulation and finish systems (EIFS)

A substantial proportion of synthetic resin plasters nowadays goes into EIFS. These systems were developed in the 1960s and experienced a boom following the energy crisis of 1973, although this led to many compensation claims, because lack of practical experience meant that a great deal of trial and error was involved. The situation has since changed. There are now a very large number of trialled systems which meet the technical requirements. Thermal insulation is applied not only to old walls, but is also provided for in new buildings at the planning stage. In the erection of prefabricated houses, as well, there is almost always provision for a thermal insulation system, although in this case the critical factor tends to be the crack-bridging properties of the EIFS. A decisive point is that the system used should be a complete system from just one manufacturer, and its individual components must not be swapped for others. The requirements profile, moreover, always relates to the system as a whole.

In accordance with Figure 4.29, an EIFS consists at least of:

- an adhesive layer;
- a thermal insulation material;
- a filling compound with embedded reinforcement; and
- a finish coat.

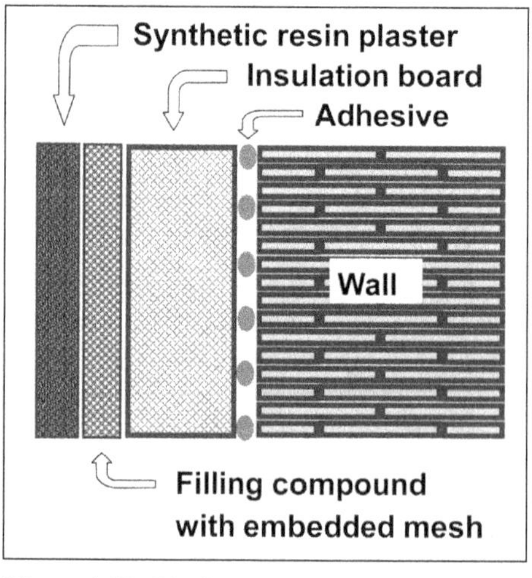

Figure 4.29: EIFS structure

The adhesive must ensure contact between the thermal insulation and the substrate, and therefore must adhere well to both components. The thermal insulation (usually mineral fibreboards or "Styropor"[1] boards) forms the actual core of the EIFS. The boards are required to have a low thermal conductivity and adequate tensile strength, and to be rot resistant. On critical substrates, they are additionally fastened with specialty dowels.

The binder used for the filling compound is almost always the adhesive.

[1] BASF SE

The reinforcement is embedded into the compound, and equalizes the mechanical and thermal stresses between the substrate and/or insulation boards and the finish coat. This layer in particular is required to have a high tensile strength with low extensibility. The mesh embedded in it is usually made of glassfibre; plastic meshes are unsuitable owing to their thermoplasticity. The finish coat should have a certain degree of flexibility, although in principle a normal synthetic resin plaster is suitable for this purpose. The standard requirements apply to processing of the plaster.

An important property of an EIFS is the non-flammability of the **entire** system. Here, it needs to be remembered that the fire behaviour is determined not only by the flammability properties of the insulation board but also by the organic components in the adhesive, reinforcement, and finish coat. The significant factor is the polymer class used to produce these components. The various classes of polymer exhibit differences in fire behaviour, given equal proportions in the respective components.

In order that the suitability of the various classes of polymer for EIFS may be assessed, the modular composition of the assembly must be analysed. Table 4.29 shows the modules in an EIFS. They cannot all be combined with each other at random. For example, a silicate plaster cannot be applied over a mineral insulation board as it is sensitive to water. This type of insulation material requires a topcoat of a polymer-modified plaster. If the insulation board is based on "Styropor"[1], a silicate plaster can be applied over it.

The fire behaviour of different classes of polymer in the fire shaft or "Brandschacht" [231, 232] are described in the literature. From these results, it is clear that Class B1 is hard to achieve with a purely "organic" composition. Such systems can only attain a Class B1 fire rating if a binder of low calorimetric value is used throughout the system. Polyvinyl chloride-containing binders meet the fire shaft test in every respect, even without flame retardant (such as aluminium hydroxide). Topcoat plasters often contain combinations with Ak/S dispersions in practice due to better weatherability and higher PBP.

Table 4.29: Modular structure of an EIFS

Inorg.	Adhesive	Insulation material	Reinforcement	Topcoat
	cementitious powder, polymer modified	mineral	mineral	cementitious powder, polymer modified
	as paste + 30% cement	poly styrene	poly styrene	textured finish on silicate basis
Org.	as paste, without cement	(hard) foam	(hard) foam	synthetic resin plaster

[1] BASF SE

Such a topcoat plaster must be formulated with a fire retardant (aluminium hydroxide) if it is to achieve a Class B1 rating.

The greater the number of inorganic fractions in the whole system, the lower the influence of the binder on fire behaviour.

Table 4.30: General guidelines for the formulation of a synthetic resin plaster

Component	wt.-%	Action
Polymer dispersion, e.g., "Acronal" 290 D[1]	132	introduced initially; amount meets standard; the high inherent viscosity of polymer dispersions such as "Acronal" 290 D reduces the amount of thickener required
Sodium polyphosphate, 5%	8	wetting agent and dispersing agent, important for storage stability; amount depends on binder and size distribution of the fillers; optimum amount to be determined by experiment
Preservative	3	usually a formaldehyde donor; preserves formulation during storage
Defoamer	3	foam may impact adversely on the processing properties of the plaster; adequate defoaming is therefore important!
"Latekoll" D[1] 8% in ammonia	8	thickener; influences consistency, levelling, processing and structure; amount must be limited owing to water sensitivity
White spirit (180 to 210°C)	10	film-forming agent; ensures crack-free drying even at the minimum processing temperature as per standard (5°C)
Butyl diglycol	10	retards setting; max. addition 2%, since also raises water sensitivity
"Basophob" WDS[1]	6	water repellent; improves early rain resistance and prolongs open time
Titanium dioxide, rutile	28	as with paints, the pigments and fillers should be added in the order of increasing particle size
"Omyacarb" 40 GU[2]	395	filler
"Omyacarb" 130 GU[2]	255	filler
"Plastorit" 0.5[3]	65	filler
Quartz shingle	45	particles with oversize for achieving groove effect on smoothing
Water	32	
Total	**1000**	
Solids content: 86%		
Binder fraction (dry) 7.5%		

[1] BASF SE, [2] Omya, [5] Naintsch

4.6.5.3 Formulation scheme for synthetic resin plasters

The general guidelines for formulating a synthetic resin plaster will be illustrated with a float plaster. As Table 4.30 shows, this product is based on a commercial Ak/S dispersion.

The choice of thickener is very critical for the processing properties of the plaster. A carboxy-functional polymer, such as "Latekoll" D[1], ensures adequate wet adhesion and permits absolutely smear-free smoothing.

Cellulose derivatives can also influence the properties of the plaster. Methyl or hydroxyethyl cellulose allow good wet adhesion, although in this case smearing of the plaster may well occur. Methyl hydroxypropyl cellulose permits good wet adhesion without smearing. The choice of defoamer is likewise very important. In the case of curved stone plasters, indeed, a certain quantity of foam may even be desirable for obtaining a plaster of smooth consistency which is easy to process.

Plasters of this kind are nowadays usually produced in horizontal forced mixers. Since the goal is both a very high solid content and high viscosity, the dispersion must be added first. All of the additives are then added to it, followed finally by the fillers.

4.6.5.4 Typical binders for synthetic resin plasters

Typical products for synthetic resin plasters are dispersions of VAc copolymers, Ak/S and straight acrylic types. The customary VAc copolymers are notable for their relatively good fire performance. Ak/S copolymers have a high binding capacity and are widely used for polymer-modified plasters due to their cost/benefit ratio. In many cases, these products are also available as ammonia-free versions. This allows the formulation of odourless plasters. The MFFT is usually in the range of 0 to about 20°C. Binders of lower MFFT are suitable for producing emission-free and solvent-free plasters if no ammonia is present. Such products can be rendered particularly low in odour by means of chemical and physical deodorisation.

When plasters are formulated with such binders, however, caution is needed if the polymer is thermoplastic, as this can have a deleterious effect on the dirt pick-up capacity of the plaster.

Straight acrylics confer particularly good weatherability, this fact indicating that they could be suitable for coloured stone plasters. Nevertheless, the weatherability of these systems still needs to be demonstrated in preliminary tests under extremely high, direct sunlight in southern climate zones.

To ensure non-flammability, synthetic resin plasters for EIFS are formulated with vinyl chloride-containing binders, which serve as sole binder or in combination with Ak/S dispersions.

[1] BASF SE

4.6.6 Adhesives

In the applications described so far, the substrate has always been treated with a coating which adhered through the agency of an intermediate layer. Where coatings are directly applied to the substrate, adhesion phenomena play a very important role.

In the case of adhesives, these phenomena are pivotal, as the adhesives need to create a permanent link between both materials. Adhesives are non-metallic substances which are capable of joining materials through surface bonding (adhesion), while the join itself exhibits inner force (cohesion) [233]. They are usually composed of adhesives raw materials, primarily binders which define the level of adhesion and its inner force (cohesion), and other auxiliaries, which regulate the desired end and processing properties [234]. In many cases, bonded joints are as effective as other joining techniques, such as riveting, welding, soldering and bolting. In some cases, adhesives complement these techniques and show numerous advantages. Naturally, process parameters and material properties which are linked to the chemical structure can limit the use of adhesives. Adhesive bonding has the advantages that the bond between two pieces can be realized efficiently, is often very quick and has an excellent cost/benefit ratio [235]. A common important feature of bonded joints is their highly uniform force distribution over the entire adhesion area, especially when compared with rivet and bolting points. The uniform distribution of stress permits better use of the materials' properties. Therefore, slimmer and, as a consequence, lighter material joints can be achieved. This leads to materials savings and, as in the automotive, aircraft, capital goods and furniture industries, to new structural elements (sandwich elements) [233].

As all plastic materials are derived from polymers, the properties of adhesives are markedly dependent on the temperature. The physical properties of adhesion can only be maintained within a relative limited temperature range. Adhesives which are thermally crosslinked and therefore function practically independent of temperature, can withstand temperatures of 150°C [236].

4.6.6.1 Theories of adhesion

The atoms or molecules in every solid are held in place by forces of cohesion. These inter-atomic and inter-molecular forces are conservative, i.e. they do not depend on the treatment of the material [237]. When the solid is broken, the two pieces cannot be easily re-joined to restore their original state. Reactivation of the cohesion forces is only possible if the original distances between the atoms and molecules are restored. The same applies to the joining of two different materials; only when the pieces approach each other across their full faces to the extent that molecular interaction can take place will adhesion occur in the same order of magnitude as cohesion [238]. In practice, this is not possible due to three reasons: first, the surface roughness of the substrate to be bonded; second, the reduction in effective surface area needed for interface reactions due to the surface tension of the adhesive; and, third, the formation of the so-called weak interface layer [239, 240]. This is a region near the interface layer in the adhesion film, where the reactive groups of molecules are aligned in the direction of the substrate and therefore cannot contribute to the hardening process. As a consequence, the weak interface layer is always a weak point in a bonded joint. The task of the adhesive in the bonding zone is to achieve ideal wetting of the substrate and as a consequence to reduce the effects of surface roughness.

Over the years, different theories about adhesive action and the mechanism of adhesion have been advanced. These range from a simple mechanical theory through the electrostatic and adsorption theories to the diffusion theory [241].

Mechanical theory of adhesion

This is the oldest theory of adhesion. It is based on mechanical anchorage of the adhesive in the pores and the surface roughness of the substrate and is principally adduced in respect of wood and other porous materials [242] (see Chapter 4.6.1.2 and 4.6.1.5).

Electrostatic theory of adhesion

In the electrostatic theory [243] the forces of adhesion between the adhesive and the substrate are built up by contact and transfer potentials. These transfer potentials lead to the build-up of an electrical double layer in the interface region between the adhesive and the substrate and corresponding *Coulombic* attraction forces between the two components. The existence of transfer potentials can be ruled out, as electrostatic discharges during the destruction of bonded joints prove. The potential usefulness of the forces of attraction in connection with these discharges is still being debated [233, 237, 244].

Adsorption theory

As the name indicates, this adhesion theory uses surface forces to explain the observed phenomena [245]. It looks on adhesion as a special property of phase interfaces. The forces believed to be responsible for this process are called secondary valence or *van der Waals* forces. They have three components: *Keesom's* dipole effect, *Debye's* induced dipole effect and *London's* dispersion effect [246 - 248]. In order that that these forces may be rendered active, the distance between the molecules of adhesive and the molecules of the substrate must be at the molecular scale. Good adhesion can be expected if the adhesive (in liquid form or in a liquid) wets the substrate [242].

Diffusion theory

The diffusion theory of adhesion lends itself especially to the bonding of large polymer structures. In this theory, adhesion is obtained by intermingling of the substrate and the adhesive [249, 250]. The mobility is based on the basic properties of large polymer structures, namely their chainlike structure and resulting mobility, with the possibility that the chains are subjected to *Brownian* motion. On account of their greater mobility, adhesive molecules play a greater part in the diffusion process. However, it must be remembered that when the adhesive is in a liquid state (which usually is the case) and when the substrate exhibits low solubility in the liquid medium, substrate molecules or parts of it can diffuse into the adhesive. A diffusion band is characterised by the disappearance of a clear boundary between the two phases and a gradual transition between the two phases. The adhesion mechanism therefore develops as a three dimensional process and is not limited to the interface.

The inter-diffusion of the polymer molecules of the adhesive and the substrate depends on numerous parameters, such as pressure, time, temperature, and molecule size. Naturally, as described, it also depends on mutual solubility, as shown by the correlation between polymer compatibility and adhesion quality [251].

None of the theories alone can completely describe every aspect of adhesion, and there is still no sign of a unified theory. Nevertheless, the performance of polymer dispersions in many applications, such as in emulsion paints (Chapter 4.6.1), silicate systems (Chapter 4.6.2), silicone resin systems (Chapter 4.6.3), elastic coatings (Chapter 4.6.4), plasters (Chapter 4.6.5) and others, is still subject to the above-mentioned theories. This aspect is developed in the introduction to the following section on dispersions as adhesives.

4.6.6.2 Polymer dispersions as adhesives

Applications of acrylic dispersions in adhesives range from pressure sensitive adhesives (PSA) for tapes, labels, decals, lamination, contact adhesives for assembly in construction and products, packaging adhesives, carpet adhesives and others. A good description of acrylate adhesion technology, including some formulations, is provided by *Gehman* [253]. *Satas* [254] treats PSAs in detail.

Transparency, non-yellowing, resistance to hydrolysis and oxygen, and versatile adhesion performance through monomer selection and ease of formulation are important attributes that acrylic dispersions bring to adhesives. However, where these effects are not desired, their higher costs relative to olefin, diene and styrene adhesives prove to be a drag and limit their applications.

In particular, P-EHA, P-EA and P-BA, along with EA and BA are important raw materials for adhesives applications, as they have good tack. The attainment of adequate elastic modulus, as shown in Figure 4.30, in combination with good peel and shear strength can be influenced by the molecular weight, as shown in Figure 4.31 [256].

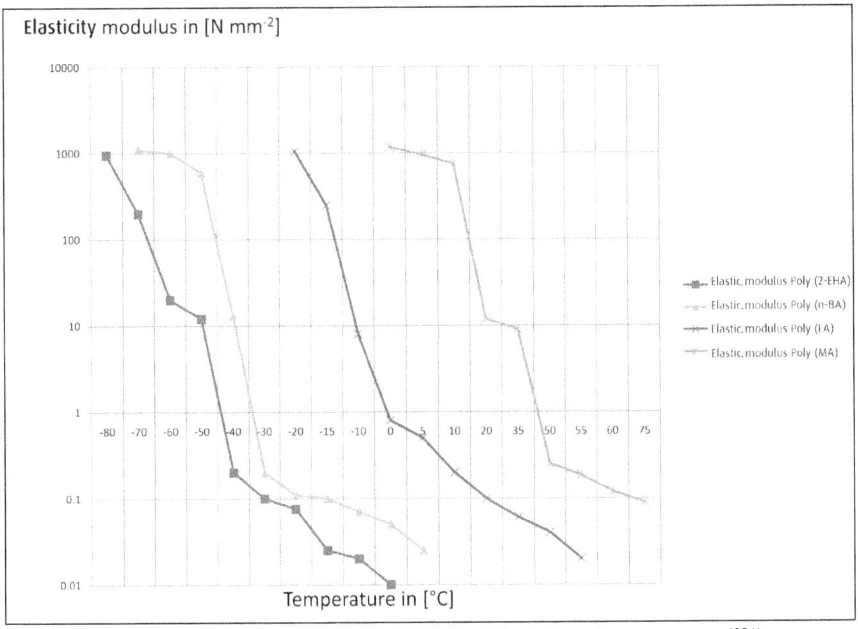

Figure 4.30: Elasticity modulus of some PSAs as a function of temperature [254]

Their copolymers with different esters containing films of tailor-made hardness or additional functional groups (such as carboxylic, amide, amine, methylol or hydroxyl groups) serve as adhesive raw materials for improving adhesion properties and crosslinking the adhesive layer to a certain extent (see also Chapter 4.6.4 and 4.6.1.4).

Methacrylates often serve as comonomers for improving the hardness of the adhesion layer. Ak/S copolymer dispersions are often used to bond paper layers [233].

Furthermore, acrylate polymer dispersions are also employed in combination with other adhesive raw materials, such as starch and casein, be it as thickener (see Chapter 4.6.1 "Acrylate thickeners") or for improving specific properties, such as water resistance.

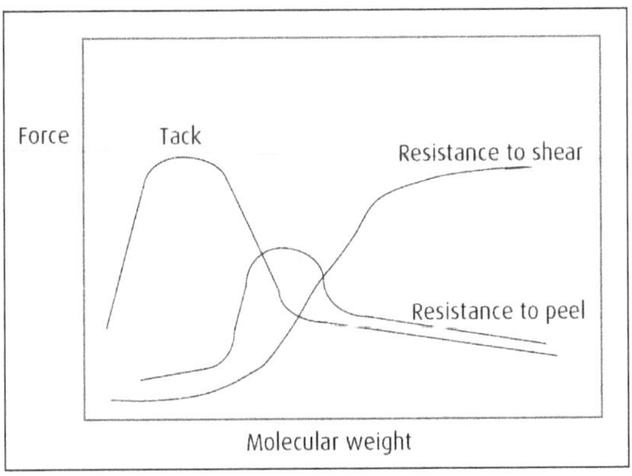

Figure 4.31: PSA properties as a function of molecular weight (adapted from Satas [255]*)*

Acrylic dispersions acting as adhesives raw materials release their water content during the bonding process to form a film (see also Chapter 4.1.5) [233, 252]. The addition of plasticizers, solvents or other additives can improve adhesion, lower the MFFT and enhance tack. The addition of polyvinyl alcohol, cellulose ether or other hydrocolloidal additives improves the open time. Polyvinyl alcohol also improves tack in the wet stage.

4.6.7 Construction chemicals

Acrylic dispersions are also employed in new constructions and renovations. The addition of acrylic dispersions to construction materials such as sand, cement and other hydraulic binders, leads to clear improvements, as indirectly described in the sections above. Therefore, only the effects will be discussed in this chapter.

The main applications of acrylic dispersions in construction chemicals are:

- cementitious applications;
- flooring adhesives;
- sealers;
- new construction materials;
- PVC modifiers;
- ready to use; and
- additives.

These are treated below.

Acrylates in cementitious systems

Cement is a relatively heavy building material which is burdened with high emissions due to its production process, and is relatively expensive to the user. This explains the interest in using additives in cementitious systems which confer better long-term properties on the final concrete.

As cement is mixed with water and other additives to ultimately yield concrete, an acrylic dispersion must withstand such a highly alkaline milieu. As shown in Chapter 4.6.2 in respect of silicate paint, an alkaline system, straight acrylic dispersions and Ak/S dispersions are normally sufficiently alkali-stable to be used in cementitious systems.

The addition of such dispersions must be factored into the water/cement ratio because the flow properties of cementitious systems are changed dramatically by this addition. Usually, the addition of an acrylic dispersion to the system also changes the curing time (hardening time) as a function of the polymer composition and the auxiliaries. Therefore, besides the normal hardening time, curing should also be checked after 7d and 30d curing. In normal cases, the addition of acrylic dispersions to cementitious systems confers better end hardness and better water content through improved water resistance.

Flooring adhesives

These adhesives, based on acrylic dispersions, are used to affix flexible flooring materials, such as textiles, PVC, rubber and parquet. As these binders are mostly used to make bonded joints on the construction site at normal outside temperatures, they should have a T_g range below 10°C. Such dispersions enable low-odour and low-emissions pasty adhesives and primers to be formulated which can also act as construction adhesives, as additives to hydraulic setting systems, such as fillers and putties, and for flexible sealants.

These polymers can also be used for other flooring adhesives, such as parquet, and as blending components for improving the cohesion of adhesives for textile flooring. Again, the emphasis here is on the bonding action.

Sealants

Cavities generated during construction or renovation must be sealed, e.g. to avoid the penetration of rain water. Acrylic dispersions are prime candidates here due to their excellent water resistance and high WVTR (see Chapter 4.7.3).

New construction materials

For the production of new construction materials, acrylic dispersions are the products of choice due to their excellent weatherability. They are also used as latent heat reservoirs and for modifying loam.

In their role as latent heat reservoirs, the dispersions raise the heat capacity of the building material, thereby improving temperature behaviour indoors, eliminating temperature peaks and lowering energy consumption.

The addition of acrylic dispersions to clay-containing preparations bestows greater resistance to water and weathering on natural construction materials [257].

PVC modifiers

These are acrylic copolymers, which are used for modifying PVC moulding compounds. The addition of acrylic dispersions to PVC leads to improvements in the processing properties, in weatherability and mechanical profile [258].

Ready to use

Acrylic dispersions are also used in construction chemistry to formulate specialty coatings. Typical applications are:

- flexible roof coatings for the formulation of weatherable horizontal roof seals;
- water-vapour barriers for formulating water-repellent coatings of damp indoor rooms;
- pasty tile adhesives for the production of adhesives for laying ceramic flooring; and
- primers, especially those based on acrylate styrene dispersions (see Chapter 4.6.1.1).

Additives

In construction chemistry, acrylates are prized additives on account of their excellent performance, especially the alkali resistance of acrylate thickeners; see also [66].

4.6.8 Fibre bonding/non-wovens

One of the first commercial applications of acrylic dispersions was in the field of textile finishing, where even today large volumes are used. They are employed in textiles and non-wovens because of their contribution to dimensional stability and to abrasion resistance.

In fibre bonding and processing of fibres to non-wovens, shaped parts and other products, acrylic dispersions play a "bonding" role. Application forms there include

- polymer dispersions or water-borne polymer solutions for fibre bonding;
- non-wovens; and
- additives.

Acrylic dispersions for fibre bonding

In industry, many different fibre materials are used for the production of often intricate articles. Examples are wooden or natural-fibre mouldings for automotive indoor panels, cork agglomerates, leather and cellulose fibres, mineral fibre boards for noise and heat insulation, abrasives, cellulose filters, and decorative films. In order that these various articles may retain their shape and functionality, the binder must transfer its properties to the system. On account of their excellent water resistance in the dry state, acrylic dispersions are the products of choice for these applications. There are numerous descriptions in the literature the good shape performance of acrylate-based articles [259 – 261].

Acrylic dispersions for non-wovens

The quality and functionality of a non-woven are influenced by the fibre material, the non-woven process and the manner in which it solidifies. A wide range of binders and additives is available for chemically bonding the various types of non-wovens. Typical applications are:

- supports for non-wovens based on PET and glassfibres;
- filter non-wovens;
- non-wovens for use in hygiene and medicine; and
- insert and non-woven wadding.

For such applications, the shape must also be retained in humid conditions, a fact which makes acrylic dispersions the first-choice products due to their water resistance, as described in Chapter 4.6.1.1.

A large number of acrylic dispersions for textile and non-woven applications are crosslinked or are hardened after film formation to improve water and solvent resistance, and durability during washing and cleaning. A quite common self-crosslinker contains the N-methyololamide function. Melamine formaldehyde condensates or urea formaldehyde condensates, which also contain the N-methyololamide function, can be blended to improve the crosslinking density. Figure 4.32 shows how this hardening is activated by heat or acid. Typical crosslinking conditions are 1 to 5min in a temperature range of 125 to 160°C.

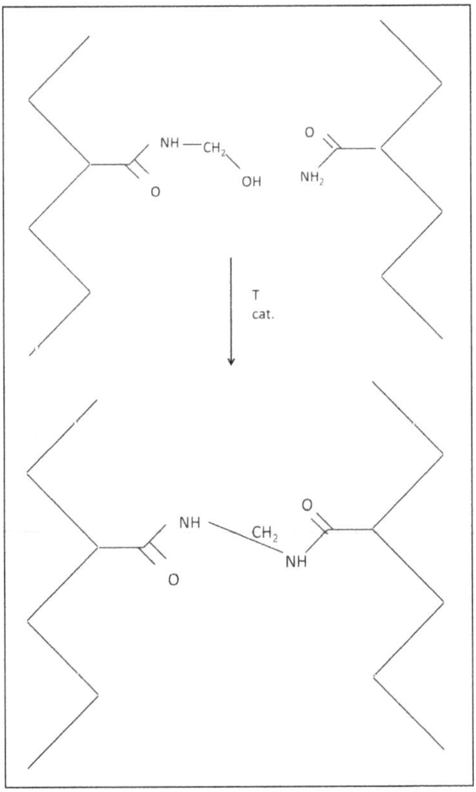

Figure 4.32: Hardening mechanism of the N-methylolamide functionality [256]

Acrylic dispersions as additives

For the fibre bonding and non-wovens which must retain their shape, additives based on water-resistant acrylate thickeners are used; see also [66].

4.6.9 Floor polishes

Floor polishes are used to improve aesthetics and to prolong the life of floors against abrasion and damage through the formation of a protective layer. The polishes must be easy to remove, which is normally realized with paint strippers on alkali or amine basis, before being re-applied. The acrylic dispersions for this application normally contain large quantities of acid groups of relatively low molecular weight and an MFFT in the range of 40 to 80°C. In the formulations, the dispersion particles coalesce to form ideal films through the addition of solvents and plasticizers. The main monomers are BA, MMA and styrene, whereby styrene contributes its gloss and water resistance and MMA its hardness. The addition of zinc salts which form complexes with the acid groups renders the polishes resistant to cleaners, but can be dissolved by paint strippers based on ammonia [262]. These cleaning agents have a long life-time and can be over-painted. They are still an important technology [256].

4.7 Comparison of acrylate primary dispersions with other binders

There are numerous types of primary dispersions, such as Ak/S, and VAc copolymers, in addition to acrylates, especially straight acrylates. The results of direct comparisons are discussed below.

4.7.1 Whitestone test

It is widely known that artificial weathering methods have difficulty in simulating outdoor exposure [119]. This is due to differences in the interaction between temperature, humidity and ultra-violet radiation. Nevertheless, it is still possible to gain meaningful results even from accelerated natural weathering tests if certain principles are observed.

There is a clear difference between the UV resistance of pigmented and unpigmented formulations: pigments and fillers act as UV absorbers, reducing the destructive effects on the binder. This is particularly true for masonry paints with PVC of 40 to 50%. In order to eliminate these effects, binders should be tested in unpigmented formulations. However, on what substrate should such a test be carried out?

Unpigmented films can be tested in two ways:

- as free films, attached to a carrier and
- as a coating applied to a substrate.

Neither method is without its problems: polymer films can exhibit different T_g's; their consistency can range from hard, almost brittle – so that film formation at 20°C is only possible with the aid of coalescing solvents – to soft, almost cold flowing. Securing free films to a carrier for extended periods is therefore problematic. This is in complete contrast to a method referred to as the "whitestone test".

Table 4.31: Formulation for a whitestone test

Component	wt.-%
Polymer dispersion, 50%	195
Hydroxyethyl cellulose, 4% soln.	36
Butyloxyethyl acetate	15
Preservative	2
White marble chippings, 2mm	750
Defoamer	2
Total	**1000**

The formulation used for the "whitestone test" is similar to that for a coloured aggregate, except white marble chippings are used instead of coloured particles. The marble aggregate coating can be formulated with most polymer-dispersion coating binders. For testing purposes, the formulation presented in Table 4.31 can be used [120].

The solids content and viscosity of the coating must be taken into consideration, of course. To obtain a coating of optimum working viscosity, it may be necessary to add either a thickener or extra water. Although the quantity of thickener (hydroxyethyl cellulose) required is ultimately determined by the viscosity of the polymer dispersion, the quantity added should be kept to a minimum.

The point at which the polymer film starts to decompose can be made more noticeable by tinting the binder with a small quantity of phthalocyanine blue (0.03% of a pigment preparation – "Luconyl" Blue 6900[1] – expressed in terms of the whole formulation is sufficient). The thickener solution and the pigment preparation are premixed together for more accurate metering. The blue pigment greatly improves the contrast between the film and the marble, making it easier to distinguish the two materials on a photograph.

To achieve efficient film formation, a suitable coalescing solvent must be added to all polymer dispersions with an MFFT above 5°C. To avoid different drying times in a test series, butyloxyethyl acetate can be added to each formulation. A preservative is unnecessary if the coating material is applied within a few days of its being prepared. The quantity of defoamer should be such that not all the foam is eliminated: some foam is desirable as it prevents the marble aggregate from settling, thus making it easier to achieve a consistent film.

The whitestone coating is applied to primed fibre-cement board in a thickness of 4mm, twice the particle size (2mm) of the marble aggregate. The jaggedness of the marble chips leads to considerable variation in the thickness of the polymer film on their surface – well below 100µm in some places. To make it easier to assess the effects of weathering, it is recommended that samples be exposed at a 45° angle and in a vertical position. Experience shows that a surface placed at 45° weathers twice as fast as a vertical one.

[1] BASF SE

When exposed to the weather, white-marble coatings become dirty – the extent of this depends on the type of binder. Moreover, chlorinated binders (e.g. vinylidene chloride (VDC) or vinyl chloride (VC)) are prone to yellowing. There is also some yellowing of styrenic polymers, but it is only noticeable when compared against weathered coatings containing straight acrylic binders. Dirt pick-up and yellowing will become noticeable after 4 to 8 weeks, but it will take several months for the degradation of VC and VDC binders to become evident (the precise duration will depend on the VC and VDC content). As the binder degrades and the film weathers away to expose the underlying marble, the coating will appear much brighter.

4.7.1.1 Comparison of artificial weathering methods

Naturally weathered samples were not only assessed, but also compared with artificially weathered samples of the "whitestone plasters". Three different commercial polymer dispersions as shown in Table 4.32 were used to formulate blue-tinted whitestone coatings, which were then weathered artificially in Suntest and Xenotest 1200 weathering machines [120].

As mentioned earlier, accelerated and natural weathering tests on unpigmented binders cannot be expected to yield comparable results, particularly if the accelerated test involves no exposure to humidity. This is the case with the Suntest weathering machine, which was used only with a light-source exposure device. Here, a power density of 1000W m^{-2} from light with wavelengths in the range 270 to 800nm only caused the VCE 1 binder to become yellow – no destruction of the binder film on the marble chips occurred.

A Xenotest 1200 machine was used for a further test in which the coatings were exposed to both light and humidity. The test panels, irradiated permanently, were subjected to three minutes of water spray every 20 minutes.

Surprisingly, in the Xenotest, the formulation based on dispersion Ak/S 1, an Ak/S polymer, was found to degrade much more than the VCE 1 coating – the opposite is true outdoors. This is a typical example of how easy it is to make an unsound assessment about a coating's performance from the results of an accelerated test. How else can it be that Ak/S copolymer dispersions are regar-

Table 4.32: Commercial polymer dispersions used for artificial weathering

Binder	Monomers	MFFT [°C]	Water absorption after 24 h [%]
Dispersion Ak3	MMA, n-BA	13	12
Dispersion Ak/S1	Styrene, n-BA	20	8
Dispersion VCE1	VAc, VC, ethylene	12	27

ded (mistakenly) as only being suitable for interior coatings? Investigations by Schwartz and Kossmann [109, 263] highlight the importance of natural weathering trials for assessing the practical performance of coatings.

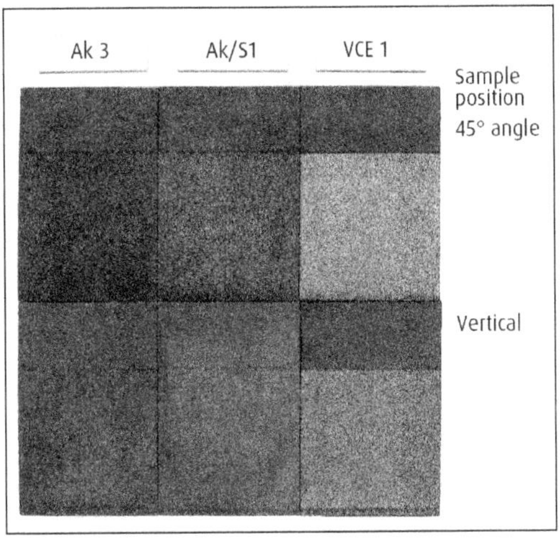

Figure 4.33: Outdoor weathering 45°/vertical comparison [120]

If these results are compared with those of 22 months' natural weathering (45° and vertical, facing south (on the northern hemisphere)), it is apparent that the whitestone test produces more realistic results. Whereas there is no difference in appearance between the vertical specimens coated with dispersion Ak 3 and Ak/S 1, destruction of the surface of the VCE 1 dispersion is clearly noticeable. At 45° exposure, the entire marble surface has been stripped of VCE 1 binder, whereas, in the case of the AS 1 dispersion, the white marble is just starting to show through (Figure 4.33). These comparative tests thus prove that artificial and natural weathering can produce very different results [264].

4.7.1.2 Comparison of binders

How do the performances of the different dispersion binders compare? Although the findings of the whitestone test are paramount, other results from the test series are also of interest.

For comparison purposes, it is useful to group different dispersion binders according to their chemical composition; this should exhibit similar behaviour during outdoor exposure:

Group A	=	Acrylic copolymers (straight acrylics)
Group B	=	Acrylic/styrene copolymers
Group C	=	Vinyl ester dispersions (homopolymers and copolymers)
Group D	=	Copolymers with VC and/or VDC

The polymers within each group may well have different T_g's or exhibit some distinct property such as a tendency to self-crosslinking. This is particularly the case with dispersions in groups A and B. Group C comprises copolymers of VAc and

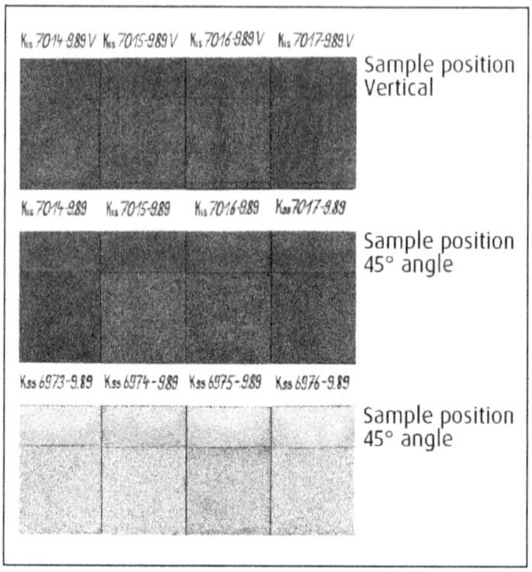

Figure 4.34: Straight acrylic coatings after 30 months' natural weathering [120]

vinyl propionate, but not vinyl chloride. VC is a comonomer in combination with some vinyl ester and acrylic copolymers; VDC on the other hand is usually copolymerized with acrylates and VC. Group D comprises chlorinated binders, whose chlorine contents can vary widely.

Let's first look at Group A dispersions. These are so-called pure (or straight) acrylic polymers built from monomeric esters of (M)AA. The T_g's vary to some extent, by an amount depending on the proportion of the individual monomers. A further variant is the self-crosslinking polymer dispersions. Formulations based on dispersion Ak 3 were found to have the highest photochemical stability (Figure 4.34, panel Kss 7014). However, slight decomposition of the polymer film was observed with the self-crosslinking dispersions (panels Kss 7015 and 7016).

Within Group B, the Ak/S copolymers, comparison of Kss 7031 (dispersion Ak/S 1), Kss 7032 and Kss 7033 (Figure 4.35), all of which are composed of n-BA and styrene, makes for interesting reading. Kss 7032, owing to its higher proportion of nBA, yields a softer film, which makes it more prone to dirt pick-up, but less susceptible to attack, due to the protective film of dirt. Kss 7033 has the same monomer composition as Kss 7031, but is slightly modified to make it self-crosslinking. As with the straight acrylic self-crosslinking formulations, its photochemical stability is somewhat lower.

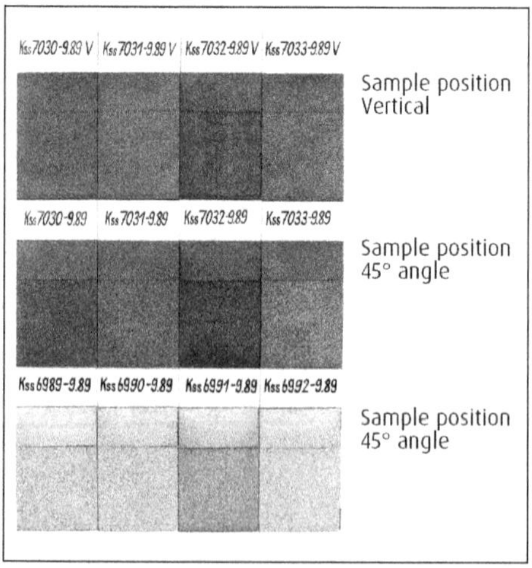

Figure 4.35: Acrylic/styrene coatings after 30 months' natural weathering [120]

Astonishingly, the photochemical stability of the vinyl ester copolymers without VC approaches that of the straight acrylates. The formulations Kss 7044 and Kss 7045 consist of VAc/ethylene copolymers. Kss 7043, a Group D polymer, contains

VC as a hard component which is responsible for the film's rapid decomposition during natural weathering. However, it should always be remembered that, apart from UV stability, the main reasons for preferring acrylic and Ak/S copolymers to vinyl ester binders are better alkali resistance and lower WA.

Summary

When different polymers are subjected to natural weathering, various degrees of change are observed after a certain duration of exposure – there is zero change in the case of straight acrylics; but in other cases destruction of the film until exposure of individual marble chips occurs; complete erosion of the top layer of binder; or even penetrative destruction of the coating so that the marble chips become loose.

Although not a substitute for tests on fully formulated pigmented systems, the "whitestone test" does provide a quick method – four to five times faster than usual – for assessing and comparing the photochemical stability of polymer dispersion binders in exterior coatings under natural weathering conditions. Initial film degradation can be rendered more distinguishable by slightly tinting the binder with phthalocyanine blue pigment.

The whitestone test yields information on possible applications for different types of binders. A binder that shows visible signs of degradation after only six months should not be used in binder-rich systems, such as gloss paints, or unpigmented (i.e. colourless) coatings. Since paint formulations usually differ in their pigmentation, their outdoor performance should also be tested [120, 264].

4.7.2 Comparison of acrylic dispersions with acrylate/styrene dispersions

Precisely for application as masonry paint, the principal uses for the binder groups straight acrylics and Ak/S dispersions overlap, as Figure 4.5 shows. It is in the field of masonry paint applications that the acrylic-versus-Ak/S question arises continually. To answer this question, the differences and the similarities between both types of acrylic dispersion are discussed in the following.

Two typical copolymer dispersions – one a straight acrylic (Ak), the other an A/S (Ak/S) – were used in the studies (see Table 4.33 [109]).

Table 4.33: Characteristics of the tested dispersions

Product	Monomers	MFFT [°C]	H_2O absorption after 24 h [%]
Dispersion Ak	MMA, nBA	13	12
Dispersion Ak/S	S, nBA	20	8

Table 4.34: Characteristics of the pigments used

Pigment	Density [g cm^{-3}]	Oil number [g (100 g)$^{-1}$]
Rutile titanium dioxide	4.1	18
Yellow iron oxide	4.1	65
Red iron oxide	5.0	26

Tests were carried out on films formed from the pure dispersion and from various masonry paints in which the dispersion was present as film former. The paints were applied to a substrate and allowed to weather naturally; they were examined for evidence of chalking and colour change at specific intervals.

Naturally, laboratory experiments cannot tell the whole story; but they can be used to support long-term weathering tests.

The following variables were taken into account:

- type of dispersion (acrylic versus acrylic/styrene);
- type of pigment (titanium dioxide versus iron oxide);
- pigment volume concentration; and
- type of pigment extender.

The paints used in the tests were produced in two colours: the first a TiO_2-containing formulation tinted with yellow iron oxide, the second a mixture of red and yellow iron oxide only. The most important data of the pigments are shown in Table 4.34 [109].

A 83:17 blend of calcite (calcium carbonate) and talc served as extender in those tests where the type of pigment extender was not the variable (i.e. a. to c. above) [109].

Fibre-cement board was chosen as the alkaline substrate on which to test the paint. The board was first primed with a water-borne emulsion primer (to even out any differences in absorbency and to prevent efflorescence due to leaching of alkali salts from the substrate), followed by two coats of the test paint in a total coat weight of 300 g m^{-2}. The panels were placed in a south-facing (on the northern hemisphere) exposure rack inclined at 45° (experience has shown that this is equivalent to the weathering of coatings on vertical exterior surfaces, but is 2 to 2½ times faster). At set intervals, the coatings were photographed and examined for evidence of chalking (DIN 53 159) and colour change (DIN 6174).

4.7.2.1 Effect of the binder type and pigment volume concentration

The effect of binder type and PVC in matt masonry paints was investigated.

A useful measure of weatherability is the extent of chalking. Weathering is the process whereby UV radiation and humidity destroys the binder at the surface of the coating, thus exposing the particles of pigment and filler. Chalking was assessed on a scale of 1 (heavy chalking) to 10 (no chalking). As a rule, no colour change occurs at ratings between 6 and 10.

Paints were formulated with straight acrylic and Ak/S binders and PVCs of 35%, 45% and 55%. Coated test panels were kept outdoors for a nine-year period.

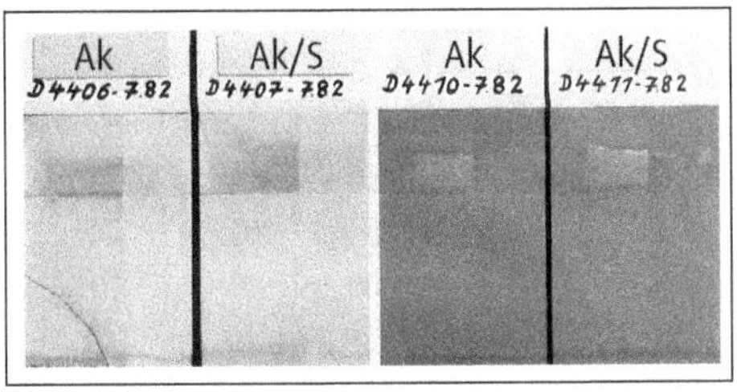

Figure 4.36: Ak and Ak/S masonry paints with 45% PVC after 9 years' weathering at 45°, facing south (The upper third of the panels was covered, the right half was washed with water) [109]

At PVC of 45%, the effect of the binder type is small: after 2 years' exposure, the chalking was rated at just 8, i.e. there was no noticeable colour difference between the acrylic and Ak/S formulations (Figure 4.36).

The PVC has a greater influence than the type of binder. When chalking of paints with different PVCs is compared, a difference is seen only after two years (the change is roughly the same for each binder type). After this time, chalking falls to a rating of 9 for PVC of 35% and to 7 for PVC of 55% (Table 4.35). Chalking, and with it colour fade, becomes more pronounced after 9 years. Ratings for each value of PVC are 5 or 6; the pigment loading therefore makes little difference. Again, the effect of the binder is small. Similar results were observed for the red-brown, full-shade paints.

A paint with a super-critical PVC applied to a vertical substrate (a wall, essentially) tends to chalk and therefore fade earlier than when applied to a south-facing inclined surface like the exposure rack. The reason is that the inclined surface collects air-borne dirt more easily. And because the dirt protects the binder from UV radiation, the coating exhibits less chalking in the initial stages of weathering. However, this effect evens out after many years of weathering.

4.7.2.2 Effect of pigment/extender ratio

Weathering tests were carried out on paints with the same binders and colour shades, but the chosen PVC was 45% and 55%. The difference relative to the aforementioned paints was the use of different ratios of pigment and extender.

Type of binder	Straight acrylic			Acrylic/styrene		
Years	1	2	9	1	2	9
PVC 35%	9	9	6	9	9	6
PVC 45%	9	8	5	9	9	6
PVC 55%	9	7	5	9	7	5

Table 4.35: Chalking of cream-coloured masonry paint due to weathering (DIN 53159) [109]

What is the point of such tests? Coloured pigments

Table 4.36: Total colour difference Δ E in cream-coloured paints after 5 years' weathering*[109]

Binder type	Acrylic		Acrylic/styrene	
	PVC 45%	PVC 55%	PVC 45%	PVC 55%
Pigment/extender				
10:90	6.6	5.7	7.0	6.0
30:70	3.0	2.7	2.5	1.9
50:50	1.6	2.8	2.1	2.0

with good hiding power (e.g. iron oxides) are often mixed with a high proportion of pigment extender to lower the cost of the formulation. However, pigment extenders can also add to paint's hiding power by protruding from the surface of the dry film, where they scatter incident light. The reason for the increased hiding – known as dry hiding – is the larger difference in the refractive indices of air and extender compared with that for air and binder. The phenomenon of dry hiding is put to good effect in highly extended interior wall paints (see Chapter 4.6.1.3).

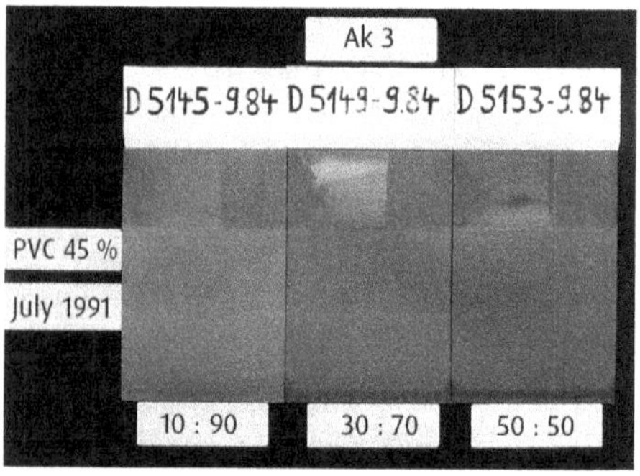

Figure 4.37: Ak red–brown masonry paint with 45% PVC and various pigment/extender ratios after 7 years' weathering [109]

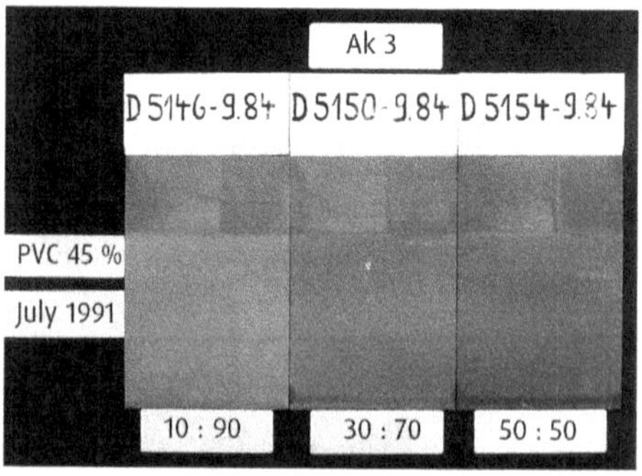

Figure 4.38: Ak/S red–brown masonry paint with 45% PVC and various pigment/extender ratios after 7 years' weathering[109]

Masonry paints that contain a lot of extender pigment are prone to premature fading. This, however, does not necessarily mean that much chalking occurs in the initial stages of the binder's photochemical disintegration: the exposed particles of pigment and extender may still be firmly anchored in the binder. Because of this, only fading was investigated; this was done by photometrically determining the total colour difference Δ E* after 5 years' weathering. The results are given in Table 4.36.

Paints with the higher proportion of extender (90%) faded the most. After 7 years' weathering, most test samples had been photographed. The Figures 4.37 and 4.38 confirm what *Kresse* [265] suggested: that the extender can cause fading in formulations containing less than 30% pigment. Once again, the findings showed no dependence on the type of binder.

Table 4.37: Types of filler and their properties [109]

	Silica SiO_2 [%]	Specific surface area [m^2 g^{-1}]	Oil number [g $(100g)^{-1}$]	pH	Refractive index
Calcite	< 0.1	3.3	18	9.0	1.59
Dolomite	< 0.6	2.3	15	10.0	1.62
Barytes	6.0		11	8.0	1.64
Mica	44.7	8.4	51	8.4	1.58
Quartz	> 99.0	3.5			
Talcum/dolomite	38.9	6.6	40	9.5	1.57
Talcum	61.0	16.0	43	9.5	1.57
China clay	55.0	8.0	50	7.0	

4.7.2.3 Effect of the extender type

Most masonry paints contain pigment extenders. These substances not only make the formulation more economic to produce, they also impart desirable properties – for example, they improve rheology, increase abrasion resistance, optimize the spacing of TiO_2 particles, produce certain surface texture, and, as already mentioned, impart dry hiding.

Most extender pigments are mineral substances with a wide variety of chemical compositions. Table 4.37 lists some common minerals used as extenders.

In this test, the only variable was the type of extender: the same binders, PVC and colour shades were adopted from the previous trials; a pigment/extender ratio of 30:70 was chosen. The results of the weathering tests were very informative. The results after 3 years are summarised in Table 4.38 and in Figures 4.39 to 4.40. Except for calcite, dolomite and barytes, which caused little chalking (ratings of 8 and 9), the rest of the extenders produced significant chalking in most cases. There was also a slight difference in the chalking of the cream and red-brown paints, the extent of which depended on the type of extender used: the red-brown paints, which contained no TiO_2, exhibited less chalking when extended with mica, quartz or talc; there was no difference with china clay. Weathering of

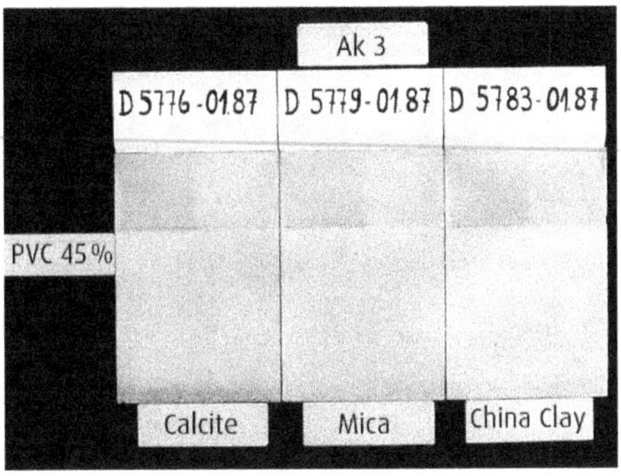

Figures 4.39: Ak cream masonry paint with 45% PVC and various pigment/extender ratios after 3½ years' weathering [109]

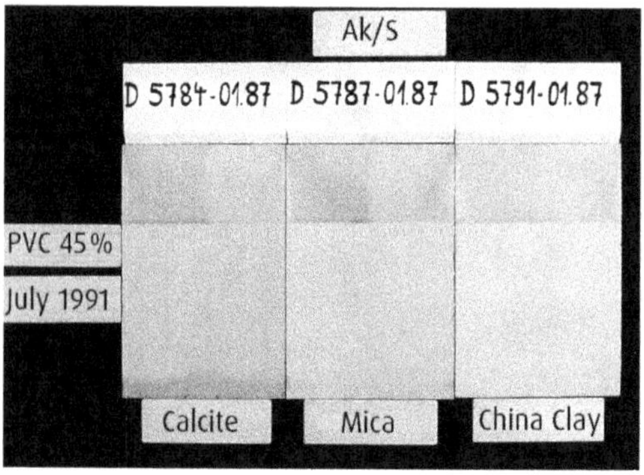

Figures 4.40: Ak/S cream masonry paint with 45% PVC and various pigment/extender ratios after 3½ years' weathering [109]

paint films for 4½ years on the 45° exposure rack – equivalent to about 10 years on a vertical wall – confirms earlier results.

An important point to note is that extenders high in silica lead to pronounced chalking; no such effect occurs in those containing no silica – calcite and dolomite. Such a comparison is certainly not without its difficulties. For example, the extenders have very different specific surface areas despite having the same particle size. Moreover, paints containing only mica or talc are uncommon in practice; on the other hand, coatings highly loaded with quartz sand were more or less standard.

Summary

As weathering tests proved, there is no difference in the outdoor performance of acrylic and Ak/S latex paints with PVCs in the range 35 to 55%. What is important, however, is the use of an optimal PVC and pigment/extender ratio (a ratio of 30:70 produces only slight fading). The choice of extender is also important: calcite, dolomite and barytes cause no fading. Testing formulations with supercritical PVCs reveals problems on inclined substrates due to the protective effect of atmospheric dirt that collects in the pores of the coating. The critical PVC of straight acrylics is slightly lower than that of Ak/S copolymers.

The type of binder used is unimportant, provided the paints are properly formulated [109].

Table 4.38: Effect of various extender pigments on chalking properties (DIN 53 159; 1 = heavy chalking, 10 = no chalking) [109]

	Masonry paint			
	cream		red-brown	
Binder system	**Ak**	**Ak/S**	**Ak**	**Ak/S**
Calcite	8	9	8	8
Dolomite	9	9	8	8
Barytes	9	8	8	8
Mica	4	4	7	5
Quartz	3	4	6	6
Talcum/dolomite	4	4	4	6
Talcum	5	5	5	7
China clay	3	4	3	3

4.7.3 Comparison of acrylic dispersions with vinyl ester dispersions

Comparison of the binders based on poly(meth)acrylates with those based on PVAc show that the coatings formulated with the acrylates are more hydrophobic, more water resistant (see Chapter 4.6.1.1), more stable to hydrolysis (see Chapter 4.6.1.1) and therefore more weatherable. On account of their higher refractive index and normally finer particles, dispersions based on acrylates permit higher gloss grades in coatings than those based on polyvinyl esters.

As a consequence, simple PVAc copolymers and VAc-ethylene pressure polymers are preferentially used in highly pigmented interior coatings (see Chapter 4.6.1.3), where the polymer character does not tend to dominate the coatings properties and weatherability is not a basic requirement. Vinyl ester binders only meet the necessary requirements for outdoor applications when copolymerized with significant quantities of the expensive and sterically demanding monomers, such as versatic acid vinyl esters. However, this lowers the cost advantage of vinyl esters over acrylates.

An in-depth comparison of the outdoor behaviour of vinyl ester dispersions and acrylic dispersions is presented in Chapter 4.7.1.2.

4.7.4 Comparison of acrylic dispersions with polyolefin dispersions

Traditionally, polyolefins have been supplied as thermoplastic pellets which can be transformed by conventional thermoplastics processes such as extrusion, injection moulding and blow moulding. The lack of a feasible EP method prevented the production of water-borne dispersions of these polymers. Such dispersions could serve as coatings, binders and other applications, where normally emulsion polymers are used. Through new processes in which dispersion is achieved mechanically, a large number of highly concentrated polyolefin dispersions with

Table 4.39: Adhesion properties of formulations with polyolefin dispersions on polypropylene carpets in comparison with other binders

Dispersion or latex	Filler	Force needed for delamination [N]	Tuftlock [N]
Polyolefin dispersion A	none	41	42
Polyolefin dispersion A	$CaCO_3$ (200 pph)	73	36
Vinyl acetate-ethylene (VAE) latex	none	41	22
Nitrile latex	none	20	17
Ethylene-acrylic acid latex	none	12	42

an average particle size of around 1µm have become accessible. These dispersions are applied to all kinds of substrates, where they demonstrate their outstanding properties in respect of resistance to weathering and chemical substances, adhesion on polyolefin substrates and low-temperature flexibility [266 - 269].

A comparison of polyolefin dispersion with other binders is presented in Table 4.39. Such good bonding to polyolefin substrates is hard to achieve with acrylic dispersions [270].

4.7.5 Comparison of acrylic dispersions with styrene/butadiene dispersions

Styrene/butadiene dispersions which, as a consequence of their main monomers styrene and butadiene, are readily available and inexpensive, have an economic advantage over acrylic dispersions. For applications where highly filled or highly pigmented formulations are used and no outdoor weathering is needed, styrene/butadiene dispersions are an excellent choice, as applications on fibres (e.g. carpet and paper coatings) and the like confirm.

As the concentration of filler and pigment decreases in the formulation, the disadvantages of the aromatic system styrene, discussed in Chapter 4.7.3, come to the fore; hence, for such applications, acrylates have their advantages.

In the coatings industry, styrene/butadiene dispersions serve as primers, where they show excellent corrosion inhibition [271].

4.8 Outlook

In summary, the development of acrylic dispersions and their coatings as well as their applications can be broken down into quite distinct time periods, which can be termed generations. The advent of a new generation always brought fairly important improvements in water-borne coatings or spawned new applications, such as thermal insulation, one-pot silicate paints and gloss emulsion paints. This is especially true of the 3rd generation (see Chapter 4.2.2).

However, there is no doubt that the future has already begun. Consider, for example, polyurethane dispersions with their vast array of raw materials, or polymer dispersions, where improvements can surely be made in crosslinking and core-shell polymers. A key task that remains is the replacement of solvent-borne paints and coatings by less harmful ones; here, besides powder coatings and radiation-curable systems (see Chapter 5), water-borne coatings will play an important part. Acrylate and vinyl acetate copolymers will be the most important classes of binder, with acrylates dominating gloss emulsion paints and façade systems, and vinyl acetates finding greater use in indoor paints. The future will bring plenty of surprises, plus the advent of new generations and applications.

4.9 Literature

[1] H. Warson, The application of synthetic resin emulsions, London: Benn Publishers, London 1972

[2] F. Hölscher, Dispersionen synthetischer Hochpolymerer, part 1, Eigenschaften, Herstellung, Prüfung, Berlin: Springer-Verlag, 1969

[3] H. Reinhard, Dispersionen synthetischer Hochpolymerer, part 2, Anwendung, Berlin: Springer-Verlag, 1969

[4] a) D. Distler, Wässrige Polymerdispersionen, Synthese, Eigenschaften, Anwendungen, Wiley-VCH, Weinheim 1999, b) S. Theisinger, Synthese von funktionellen Hybridnanopartikeln und Verkapselung von Aktiv- und Wirkstoffen mittels Miniemulsionspolymerisation, Ph.D. thesis, University of Ulm, 2008

[5] V. Verkholantsev, Colloid chemistry, part II: stability of dispersions and emulsions, European Coatings J., (1997) 614–622

[6] F. Walker, Polymerization processes: III, J. Coatings Technology 72, No. 903 (2000) 27–32

[7] B. Schlarb, S. Haremza, W. Heckmann, B. Morrison, R. Müller-Mall, M. Rau, Hydroresin dispersions: Tailoring morphology of latex particles and films, Progress in Organic Coatings 29 (1996) 201–208

[8] a) D. Stoye, W. Freitag, Lackharze, Carl Hanser Verlag, München (1996), Section 7.1.5.7, Polyurethandispersionen, p. 200–203, b) G. Hakim, Surface coatings, Vol. 1, Chapman & Hall, London, 3rd edition (1993), Section 10, Waterborne urethane resins, p. 173–178

[9] Ullmann: Encyclopedia of industrial chemistry, A 21, 5. edition, p. 317ff, VCH Verlagsgesellschaft mbH, Weinheim (1985)

[10] Houben-Weyl: Methoden der Organischen Chemie, E 20, 4. edition, p. 1 ff, G. Thieme Verlag, Stuttgart (1987)

[11] H. Elias: Makromoleküle, 4. edition, Hütig & Wepf Verlag, Basel (1981)

[12] J. Brandrup und E. Immergut: Polymer Handbook, 2nd edition, Wiley, New York (1975)

[13] B. Vollmert: Grundrisse der Makromolekularen Chemie, Bd. 1. E. Vol., Imert Verlag, Karlsruhe (1982)

[14] C. Bamford, W. Barb, A. Jenkins, P. Onyon: Vinyl polymerisation by radical mechanism, Butterworths Sci. Publ., London (1958)

[15] I. Bewington: Radical polymerisation, Academic Press, London (1961)

[16] A. North: The kinetics of free radical polymerisation. Pergamon Press& Oxford (1965)

[17] D. C. Blackley, High polymer latices, Vol. I, II, London: Maclaren & Sons Publishers, London 1966

[18] I. Piirma, Emulsion polymerisation, New York: Acadamic Press Inc., 1982

[19] D. Blackley, Emulsion polymerisation, theory and practice, London: Applied Science Publisher, 1975

[20] R. Buscall, T. Corner, J. F. Stagemann, Polymer colloids, London: Elsevier Applied Science Publishers, 1985

[21] R. Athey, Emulsion polymer technology, New York: Marcel Dekker, 1991

[22] G. Poehlein, Encyclopedia of polymer science and engineering; Volume 6, Emulsion Polymerisation, New York: J. Wiley, 1986

[23] P. A. Lovell, M. S. El-Asser, Emulsion polymerisation and emulsion polymers, New York: J. Wiley, 1997
[24] J. M. Asua, Polymeric dispersions: principles and applications, (NATO ASI Series E: Appl. Science - Vol. 335), Dordrecht, Klewer Academic Publishers, 1997
[25] A. Schmidt, H. Lagemann, G. Ley, Houben Weyl, Bd. E 20 (1987) 218–313
[26] F. Walker, Polymerization processes: II, J. Coatings Technology 72, No. 902 (2000) 79–86
[27] F. Bovery, I. Kolthoff, A. Medalia, E. Meehan, Emulsion polymerization in high polymers, Vol. IX, Interscience, New York 1955
[28] R. G. Gilbert, Emulsion polymerization, a mechanistic approach, London Academic Press, 1995
[29] H. Gerrens, Kinetik der Emulsionspolymerisation, Adv. Polym. Science 1 (1958) 234
[30] J. Vanderhoff, Recent advances in the preparation of latexes, Chem. Engineering Science 48, No. 2 (1993) 203–217
[31] W. Harkins, J. Am. Chem. Soc., 69 (1947) 1428
[32] W. Smith, R. Ewart, J. Chem. Phys., 16 (1948) 592
[33] R. Fitch, C. Tsai, Polymer Colloids, R. Fitch ed., New York 1971, p. 73
[34] F. Hansen, J. Ugelstad, J. Polym. Sci., Polym. Chem. 16 (1978) 1953 ff
[35] H. Kast, Polymerdispersionen – Synthese, Morphologie, Filmbildung, 12. Kolloquium Chemie und Technologie makromolekularer Stoffe, FH Aachen, Fachbereich Ingenieurwesen am 08.05.1987, p. 5–17
[36] Ullmann's encyclopedia of industrial chemistry, Vol. A1, E. Penzel, Polyacrylates 157–178
[37] M. Horn, Acrylic resins, Reinhold Publ., New York 1960
[38] H. Rauch-Puntigam, T. Völker, Acryl- und Methacrylverbindungen, in Chemie, Physik und Technologie der Kunststoffe, Vol. 14, Springer Verlag, Berlin 1967
[39] H. Kittel, Lehrbuch der Lacke und Beschichtungen, Vol 2, Bindemittel für lösemittelhaltige und lösemittelfreie Systeme, ed. W. Krauß, 2nd edition, S. Hirzel Verlag Stuttgart, Leipzig 1998, Section 2.3.2, Polyacrylate und Polymethacrylate, p. 355–436
[40] Lackharze, ed. D. Stoye, W. Freitag, Lackharze, Chemie Eigenschaften und Anwendungen, Carl Hauser Verlag München, Wien, 1996; P. Denkinger, Section 8, Polymerisate, p. 281–336
[41] a) P. Swaraj, Water-borne acrylic emulsion paints, Progress in Organic Coatings 5 (1977) 79–96, b) P. Swaraj, Surface coatings, science & technology, 2. edition, John Wiley and Sons (1996), p. 312–334
[42] T. Yanagihara,The recent progress of acrylic emulsion for coating industries, Progress in Organic Coatings 11 (1983) 205–218
[43] C. Hare, Anatomy of paint, J. of Protective Coatings & Linings (1993) 69–79
[44] H. Warson, Developments in emulsion polymerization (mainly vinyl esters and acrylics), Part 2 – Acrylic & styrene polymers and copolymers, Polymers Paint Colour J. 180, No. 4265 (1990) 507–510
[45] J. Padget, Polymers for water-based coatings – a systematic overview, J. of Coatings Technology, 66, No. 839 (1994) 89–105

[46] CD Römpp Chemie Lexikon – Version 1.0, Stuttgart/ New York: Georg Thieme Verlag 1995
[47] H. Kittel, Lehrbuch der Lacke und Beschichtungen; Bd. 1, part 3, Grundlagen, Bindemittel, 1974, Verlag, W.A. Colomb in der H. Heenemann GmbH, Berlin – Oberschwandorf, p. 920–1001
[48] S. Eckersley, A. Rudin, Mechanism of film formation from polymer latexes, J. of Coatings technology, 62, No. 780 (1990), 89–100
[49] M. Joanicot, K. Wong, B. Cabane, Structure des films de latex, Double Liaison 452–453 (1993) 17–25
[50] A. Zosel, G. Ley, Film formation from polymer latices, Progr. Colloid, Polym. Sci. 101 (1996) 86–92
[51] M. Winnik, Latex film formation, Curr. Opin. in Colloid and Interface Sci., Materials Aspects 2 (1997) 192–199
[52] A. Toussaint, M. De Wilde, A comprehensive model of sintering and coalescence of unpigmented latexes, Progress in Organic Coatings 30 (1997) 113–126
[53] J. Mulvihill, A. Toussaint, M. De Wilde, Onset, follow up and assessment of coalescence, Progress in Organic Coatings 30 (1997) 127–139
[54] K. Hoy, Coalescence and film formation from latexes, J. Coatings Technology, 68, No. 853 (1996) 33–39
[55] J. Feng, M. Winnik, Effect of water on polymer diffusion in latex films, Macromolecules 30 (1997) 4324–4331
[56] V. Verkholantsev, Colloid Chemistry, Part IX: Film formation of dispersions, European Coatings J. (1998) 60–66
[57] Ch.-L. Zhao, W. Heckmann, Mechanical strength and morphology of polymer latex films, 27th International Water-Borne, High Solids and Powder Coatings Symposium, March 1-3, New Orleans, USA
[58] Ch. Kan, Role of particle size on latex deformation during film formation, J. Coatings Technology, 71, No. 896 (1999) 89–97
[59] D. Jensen, L. Morgan, Particle size as it relates to the minimum film formation temperature of latices, J. of Appl. Polym. Science, 42 (1991) 2845–2849
[60] Wässrige Polymerdispersionen, Synthese, Eigenschaften, Anwendungen, Ed. D. Distler, Section 4, H. Wiese, Eigenschaften von Polymerdispersionen und Messmethoden, Wiley-VCH, Weinheim 1999, p. 31–66
[61] R. Hoffman, Factors affecting the viscosity of unimodal and multimodal colloidal dispersions, J. Rheol. 36, 5 (1992) 947–965
[62] F. Buckmann, F. Bakker, Fast drying high solids latices, European Coatings J. (1995) 922–929
[63] R. Becker, A. Hashemzadeh, H. Zecha, High solids polymer dispersions, Macromol. Symp. 151 (2000) 567–57
[64] RAL-UZ12a: Schadstoffarme Lacke, Ausgabe Jan. 1997, Vergabegrundlagen in http://www. blauer-engel.de/Produkte/uz/012a-ef.htm (07.11.2000)
[65] G. Mehos, D. Quick, Removal of residual monomers from polymer emulsions by steam stripping, Separation Science and Technology, 29, 14 (1994) 1841–1856
[66] M. Schwartz, R. Baumstark, Water-based Acrylates, Ed. U. Zorll, Vincentz Verlag, Hannover/Germany, 2001

[67] H. Kossmann, Kunstharzgebundene Putze – heute. Eigenschaften und Anwendungen, Die Mappe 8 (1974), p. 527–533

[68] H. Lehmann, Die Bedeutung der Kunstharzdispersionen in der heutigen Anstrichtechnik, Farbe + Lack 11 (1963), p. 845–846

[69] H. Kossmann, Fassadenbeschichtungen auf der Basis von Acryl-Copolymerdispersionen und Silikonemulsionen, Farbe + Lack 5 (1991), p. 412–415

[70] H. Kossmann, G. Florus, Zur Frage der Haftfestigkeit von Dispersions-Glanzfarben auf oxidativ getrockneten Anstrichen, FATIPEC 1978, Budapest, Congress paper, p. 337–343

[71] H. Rinno, Bindemittel für emissionsarme Beschichtungstoffe, Farbe + Lack 8 (1993), p. 697–704

[72] H. Kossmann, Polymerdispersionen und wässerige Anstrichmittel, APPLICA 7/2003, p. 4–12

[73] P. Swaraj, Water-borne acrylic emulsion paints, Progress in Organic Coatings 5 (1977) 79–96

[74] P. Swaraj, Surface coatings, science & technology, 2. edition, John Wiley and Sons (1996), p. 312–334

[75] C. Hare, Anatomy of paint, J. of Protective Coatings & Linings (1993) 69–79

[76] a) Additives for coatings, Ed. J. Bieleman, Wiley-VCH, Weinheim 2000, Section 6.2, K. Dören, Coalescing agents, p. 180–199, b) E. Bagda, VdL-Richtlinie zur Deklaration von Inhaltsstoffen in Bautenlacken und -farben sowie verwandte Produkte, Farbe + Lack 102, 7 (1996) 110–111, c) VdL-RL01; Richtlinie zur Deklaration von Inhaltsstoffen in Bautenlacken, Bautenfarben und verwandten Produkten, Revidierte Ausgabe August 1997, Verband der Lackindustrie e.V. Frankfurt, d) E. Bagda, Th. Brenner, Emissionen aus Bautenlacken und Bautenfarben, Farbe + Lack 102, 10 (1996) 109–118, e) A. Toussaint, M. De Wilde, A method to predict the distribution coefficient of coalescing agents between latex particles and the water phase, Progress in Organic Coatings 30 (1997) 173–177, f) B. Van Leeuwen, Zum Filmbildungsmechanismus von Polymerdispersionen, Farbe + Lack 101, 7 (1995) 606–609, g) O. Lückert, Karsten Lackrohstoff-Tabellen, 10th edition, Vincentz Verlag 2000, h) H. Rinno, Kunststoffdispersionen als Bindemittel für emissionsfreie Innenfarben; XIX FATIPEC Congress, Aachen 1988, Vol. 1, p. 85–104, i) H. Rinno, Bindemittel für emissionsarme Beschichtungsstoffe, Farbe + Lack 99, 8 (1993) 697–704

[77] D. Feustel, Emulsifier systems for modern surface coatings, Surface Coatings Australia 31, 1/2 (1994) 6-17

[78] H. Stache, K. Kosswig, Tensid-Taschenbuch, 3rd edition, Carl Hanser-Verlag, München (1990)

[79] Y. Vandenberghe, Alkylphenol ethoxylates: a threat to human health and the environment?, The Coatings Agenda Europe (1998) p. 106–107

[80] Ch. Arz, Gibt es Alternativen zu Alkylphenolethoxylaten?, Farbe + Lack 104, 1 (1998) 49–53

[81] K. Holmberg, Novel surfactants for paints, Färg Och Lack Scandinavia 5 (1993) 125–133

[82] A.-Ch. Hellgren, P. Weissenborn, K. Holmberg, Surfactants in water-borne paints, Progress in Organic Coatings, 35 (1999) 79–87

[83] H. Kittel, Lehrbuch der Lacke und Beschichtungen; Bd. 1, part 3, Grundlagen, Bindemittel, 1974, Verlag, W.A. Colomb in der H. Heenemann GmbH, Berlin-Oberschwandorf, p. 920–1001
[84] V. Eliseeva, Morphology and phase structure of latex particles, their influence on the properties of latices and films, Progress in Organic Coatings 13 (1985) 195–219
[85] J. Daniel, Latex de particules structures, Makromol. Chem., Suppl. 10/11 (1985) 359–390
[86] L. Morgan, Multi-feed emulsion polymers: the effects of monomer feed sequence and the use of seed emulsion polymers, J. of Applied Polymer Sci. 27 (1982) 2033–2042
[87] M. Devon, J. Gardon, G. Roberts, Effects of core-shell latex morphology on film forming behavior, J. of Applied Polym. Sci. 39 (1990) 2119–2128
[88] R. Arnoldus, R. Adolphs, W. Zom, Progress in acrylic emulsion developments, Polymers Paint Colour Journal, Vol. 181, No. 4287 (1991), 405- 409, 418
[89] S. Lee, A. Rudin, Control of core-shell latex morphology, ed. E. Daniels, E. Sudol, M. El-Aasser, ACS Symposium Series 492 (1992), Section 15, p. 234–254
[90] M. Heurts, R. le Febre, J. van Hilst, G. Overbeek, Influence of morphology on film formation of acrylic dispersions; ACS Symposium Series 648 (1996), Section 8, p. 271–285
[91] D. Lee, The control of structure in emulsion polymerization, Macromol. Chem., Macromol. Symp. 33 (1990) 117–131
[92] A. Rudin, Practical methods to control morphology of heterogeneous polymer particles, Macromol. Symp. 92 (1995) 53–70
[93] R. Baumstark, S. Kirsch, B. Schuler, A. Pfau, Mehrphasige Polymerpartikel für lösemittelfreie Dispersionslacke, Farbe + Lack 106, 11 (2000) 125–132, 145–147
[94] a) Encyclopedia of polymer science and technology, plastics, resins, rubbers, fibers, Vol. 5, J. Wiley and Sons, New York (1966), p. 847, b) K. Nagai, Y. Ohishi, K. Ishiyama, N. Kuramoto, J. Appl. Polym. Sci. 1989, 38, 2183, c) C. H. M. Caris, L. P. M. V. Elven, A. M. V. Herk, A. L. German, Br. Polym. J. 1989, 21, 133, d) Y. Haga, T. Watanabe, R. Yosomiya, Angew. Makromolek. Chemie 1991, 189, 23, e) Frost & Sullivan cited in PRNewswire "Tightening Building Energy Efficiency Regulations to promote the Growth of the European Thermal Insulation Markets", internet access on February 2009 (http://www.prnewswire.co.uk/cgi/news/release?id=163432)
[95] G.A. Vandezande, O.W. Smith, D.R. Bassett, in P.A. Lovell and M.S. El-Aasser (eds.), Emulsion Polymerization and Emulsion Polymers, John Wiley and Sons, Chichester, 1997, Ch. 16
[96] D.C. Blackley, in P.A. Lovell and M.S. El-Aasser (eds.), Emulsion Polymerization and Emulsion Polymers, John Wiley and Sons, Chichester, 1997, Ch. 15
[97] Verband der Lackindustrie e.V., Jahresbericht 2009/2010, Frankfurt
[98] D. Stoye, W. Freitag, Paints, coatings and solvents, 2nd edition, Wiley-VCH Verlag GmbH, Weinheim (1998), Section 3.3 Waterborne paints, p. 109–115
[99] V. Verkholantsev, Colloid chemistry, part VII: Latex paints, European Coatings J. (1997) 1036–1043
[100] a) H. Dörr, F. Holzinger, KRONOS Titanium Dioxide in Emulsion Paints, Herausgeber KRONOS International Inc., Leverkusen, Deutschland 1990, b) W. Könner,

Optimum formulation of emulsion paints, II. International Conference of Advances in Coatings Technology, 7-9.10.1996, Katovice (Polen), Paper 25, 32 Seiten
[101] J. Becker Jr., D. Howell, CPVC concentration of emulsion binders, Official Digest (1956) 775–790
[102] G. Bierwagen, D. Rich, The critical pigment volume concentration in latex coatings, Progress in Organic Coatings 11 (1983) 339–352
[103] J. Weaver, CPVC, critical pigment volume concentration – an overview, J. of Coatings Technology 64, No. 806 (1992) 45–46
[104] W. Asbeck, A critical look at CPVC performance and application properties, J. of Coatings Technology 64, No. 806 (1992) 47–58
[105] F. Floyd, R. Holsworth, CPVC as point of phase inversion in latex paints, J. of Coatings Technology 64, No. 806 (1992) 65–69
[106] G. Bierwagen, Critical pigment volume concentration (CPVC) as transition point in the properties of coatings, J. of Coatings Technology 64, No.806 (1992) 71–75
[107] E. Schaller, Critical pigment volume concentration of emulsion based paints, J. of Paint Technology 40, No. 525 (1968) 433–438
[108] a)G. del Rio, A. Rudin, Latex particle size and CPVC, Progress in Organic Coatings 28 (1996) 259–270, b) M. Khorassani et al., Estimation of Critical Pigment Volume Concentration in Latex Paint Systems Using Gas Permeation, Iranian Polymer Journal 14 (11), 2005, 1000–1007
[109] M. Schwartz, H. Kossmann, Acrylat- und Acrylat/Styrol-Copolymere, Farbe + Lack 103, 10 (1997), 109–122
[110] O. Wagner, Polymerdispersionen in Silikatsystemen, Farbe + Lack 97, 2 (1991) 109–113
[111] J. Lamprecht, Verhalten von unpigmentierten Kunststoffdispersionsfilmen gegenüber Wassereinwirkung, Farbe + Lack 79, 3 (1973) 209–213
[112] R. Baumstark, C. Costa, M. Schwartz, Farben auf Basis von Acrylat-Dispersionen, Farbe + Lack 106, 10 (2000) 103–110, 123–124.
[113] J. Snuparek, Some aspects of water absorption in free films from non-pigmented copolymer latex binders, XXIII FATIPEC Congress, Brüssel, 10.–14.06.1996, Vol. B, B232–B244
[114] a) H. Koßmann, The formulation of aqueous anticorrosion paints, AFTPV Congress in Nice, 20th September 1989, b) M. Morcillo, Soluble salts: their effect on premature degradation of anticorrosive paints, Progress in Organic Coatings, Volume 36, Issue 3, July 1999, Pages 137–147
[115] a) H. Künzel, Beurteilung des Regenschutzes von Außenbeschichtungen, Institut für Bauphysik der Fraunhofer-Gesellschaft, Mitteilung 18, 1978, b) E. Carretti, L. Dei, Physicochemical characterization of acrylic polymeric resins coating porous materials of artistic interest, Progress in Organic Coatings, Volume 49, Issue 3, April 2004, Pages 282–289
[116] P.M. Lesko and P.R. Sperry, in P.A.Lovell and M.S. El-Aasser (eds.), Emulsion Polymerization and emulsion polymers, John Wiley and Sons, Chichester, 1997, Ch. 18
[117] A. Guez, F. Huguet, M. Arami, Use of ultrafine latex in primers for architectural paints, Piturre e Vernici Europe - Coatings 6 (1998) 21–27

[118] a) J. Mondt, Umweltfreundliche Bindemittel mit guten Penetrationseigenschaften auf Basis von feinteiligen Kunststoffdispersionen, Farbe + Lack 83, 1 (1977) 13–17, b) G. Effinger, Fassadenanstriche für mineralische Untergründe, APPLICA 12/2004, Seiten 10–12. Schweizerischer Maler- und Gipsermeister- Verband, CH-8304 Wallisellen
[119] Nissler, Klug und Rody, Klassifizieren kaum möglich, Farbe+Lack 101, 11/1995, Seite 1034–1040
[120] H. Kossmann, M. Schwartz, Freibewitterung als Kurzprüfung, Farbe + Lack 105, 04/1999, p. 217–224
[121] R. Baumstark, C. Costa, M. Schwartz, Farben auf Basis von Acrylat-Dispersionen (part 1), Farbe + Lack 105, 02/1999, p. 30–37
[122] A. Zosel, Lack- und Polymerfilme, Viskoelastische Qualitätsmerkmale, Vincentz-Verlag, Hrsg. U. Zorll, 1996, Section 5.8, p. 100–109
[123] M. Bradac, R.Novak, Einfluss von Wasser auf die mechanischen Eigenschaften von Anstrichfilmen aus wässrigen Polymerdispersionen, Coating 25, 8/1992, p. 278–280
[124] S. Kawaguchi, E. Odrobina, M. Winnik, Non-ionic surfactant effects on polymer diffusion in poly(butyl methacrylate) latex films, Macromol. Rapid Commun. 16, 1995, p. 861–868
[125] S. Lam, A. Hellgren, M. Sjörberg, K. Holmberg, H. Schoonbrood, M. Unzue, J. Asua, K. Tauer, D. Sherrington, Surfactants in heterophase polymerization; A study of film formation using AFM, J. Appl. Polym. Sci. 66, 1997, p. 187–198
[126] V. Stevens, The effect of styrene-acrylic latexes on paint yellowing and chalking: I. A correlation of accelerated and field exposures, Waterborne, Higher Solids and Powder Coatings Symposium, February 9-10 1994, New Orleans, p. 577–593
[127] M. Schwartz, Ch. Zhao, Acrylat-Copolymerdispersionen - Bindemittel für Fassadenfarben, Farbe + Lack 105, 09/1999, p. 61–70
[128] E. Reck, J. Wilford, Benetzung und Verträglichkeit von Titandioxid, Farbe + Lack 104, 01/98, p. 40–48
[129] Th. Rentschler, Determination of hydroxyl groups on the surface of TiO_2 pigments, European Coatings Journal 10/97, p. 939–941
[130] H. Kossmann, M. Schwartz, Freibewitterung unpigmentierter Polymerfilme, applica 1-2/99, 106, p. 6–11, Schweizerischer Maler- und Gipsermeister- Verband, CH-8304 Wallisellen
[131] M. Schwartz, Acrylat-Copolymerdispersionen – Bindemittel für Fassadenfarben (part 1), APPLICA 5/2000, 107, p. 8–17, Schweizerischer Maler- und Gipsermeister- Verband, CH-8304 Wallisellen
[132] F. Holzinger, XI. FATIPEC Congress paper, 143 (1972)
[133] H. Dörr, F. Holzinger, KRONOS Titandioxid in Dispersionsfarben, Herausgeber KRONOS International Inc., 1990, Leverkusen, Germany
[134] H. Dörr, F. Holzinger, KRONOS Titandioxid in Dispersionsfarben, p. 73, Herausgeber KRONOS International Inc., 1990, Leverkusen, Germany
[135] H. Dörr, F. Holzinger, KRONOS Titandioxid in Dispersionsfarben, p. 51, Herausgeber KRONOS International Inc., 1990, Leverkusen, Germany
[136] H. Dörr, F. Holzinger, KRONOS Titandioxid in Dispersionsfarben, p. 79, Herausgeber KRONOS International Inc., 1990, Leverkusen, Germany

[137] K. Büppelmann, Vergleich der Messverfahren zur Bestimmung der Wasch- und Scheuerbeständigkeit von Innenwandfarben, XXIV Fatipec-Buch 1998, Volumen D, Seite 173–188

[138] Th. Brettner, R. Kuropka, H. Petri, Recipe developments of low VOC interior paints, XXIV Fatipec-Buch 1998, Volumen A, pp. 259–267

[139] R. Rinno, Bindemittel für emissionsarme Beschichtungsstoffe", Farbe und Lack 99 (1993) 697–704

[140] E. Reck, J. Holmes, DIY emulsion paints in Northern Europe, European Coatings J. (1998) 366–369

[141] J. Ulyatt, Solvent-free wall paints for interior use, Polymers Paint Colour J. 183 (1993); No 4335, 412–413

[142] L. Simpson, Factors controlling gloss of paint films, Progress in Organic Coatings, 6 (1978) 1–30

[143] a) F. Fensterseifer; Glanzmessung zur objektiven Oberflächenprüfung, Coating 5 (1999) 207–209, b) Byk Gardner, Gloss and reflection haze, Pitture e Vernicci, Paints and Varnishes (1996) 7–16

[144] J. Braun, Gloss of paint films and the mechanism of pigment involvement, J. Coatings Technology 63, No. 799 (1991) 43–51

[145] C. Rodriguez, J. Weathers, B. Corujo, P. Peterson, Formulating water-based systems with propylene-oxide-based glycol ethers, J. Coatings Technology 72, No. 905 (2000) 67–72

[146] RAL-UZ12a: Schadstoffarme Lacke, Ausgabe Jan. 1997, Vergabegrundlagen in http://www. blauer-engel.de/Produkte/uz/012a-ef.htm (07.11.2000)

[147] a) B. Schuler, R. Baumstark, S. Kirsch, A. Pfau, M. Sandor, A. Zosel, Structure and properties of multiphase particles and their impact on the performance of architectural coatings, 25. International Conference in Organic Coatings, Vouliagmeni (Athen), 1999, p. 203–218, b) R. Baumstark, S. Kirsch, B. Schuler, A. Pfau, Mehrphasige Polymerpartikel für lösemittelfreie Dispersionslacke, Farbe + Lack 106, 11 (2000) 125–132, 145–147

[148] M. Winnik, J. Feng, Latex blends: an approach to zero VOC coatings; J. Coatings Technology, 68, No. 852 (1996) 39–50

[149] H.-Ch. Krempels; Polymer-Blends als Lackbindemittel, Farbe + Lack 100, 1 (1994) 13–18

[150] S. Eckersley, B. Helmer, Mechanistic considerations of particle size effects on film properties of hard/soft latex blends, J. Coatings Technology 69, No. 864 (1997) 97–107

[151] K. Zimmerschied, Mehr Umweltschutz mit Reinacrylat-Dispersionen. part III: Einfluss von pH-Wert und Neutralisationsmitteln bei der Herstellung von Dispersionslacken, Kunstharz-Nachrichten 26 (1990) 7–13

[152] J. Bieleman, Additives for Coatings, Section 4.1, Wetting and dispersing agents, p. 67–99, Wiley-VCH, Weinheim 2000

[153] J. Braun, D. Fields, Gloss of paint films: II. Effects of pigment size, J. of Coatings Technology 66, No. 828 (1994) 93–98

[154] B. Emelie, U. Schuster, S. Eckersley; Interaction between styrene-butylacrylate latex and water soluble associative thickener for coalescent free wall paints, 22. International Conference in Organic Coatings, Vouliagmeni (Athen), 1997, p. 107-125

[155] T. Svanholm, F. Molenaar, A. Toussaint, Associative thickeners: their adsorption behaviour onto latexes and the rheology of their solutions, Progress in Organic Coatings 30 (1997) 159–165
[156] P. Howard, E. Leasure, S. Rosier, E. Schaller; A systems approach to rheology control, Proc. ACS Div. Polym. Mater. Sci. Eng., 61 (1989) 619 ff
[157] B. Richey, A. Kirk, K. Eisenhart, S. Fitzwater, J. Hook, Interactions of associative thickeners with paint components as studied by the use of a fluorecently labeled model thickener, J. Coatings Technology 63, No. 798 (1991) 31–40
[158] C. Glancy, D. Bassett, Effect of latex properties on the behaviour of nonionic associative thickeners in paints, Proc. ACS Div. Poly. Mat. Sci. Eng., 51 (1984) 348
[159] A. Schwartz, A formulator's guide to water-borne wood finishing, American Paint & Coatings J. (1995) 43–50
[160] G. Fonnum, J. Bakke, F. Hansen, Colloid Polym. Sci. 271 (1993) 380 ff
[161] J. Glass (Ed.), Polymers in aqueous media, Advances in Chemistry Series 223, ACS, Washington DC, 1989
[162] Deutsche Gesellschaft für Holzforschung e.V., Merkheft No. 11, Oberflächenbehandlung von Holz im Außenbereich, München 1991
[163] J. Hein, Holzschutz, ROTO Fachbibliothek, Bd. 2, 1998, Wegra Verlag
[164] J. Sell, T. Zimmermann, Werkstoff Holz, Spektrum der Wissenschaft, No. 4 (1997) 86–89
[165] Deutsche Bauchemie e.V., Frankfurt, Schutz von Holz im Bauwesen, Druckschrift für Architekten und Planer, Druck Zechnersche Buchdruckerei GmbH, Speyer, 2. Ausgabe 1998
[166] J. Hein, Holzinhaltsstoffe sicher blockieren, Die Mappe 6 (1999) 53–56
[167] Institut für Fenstertechnik e.V. (Hrsg). (1991); Merkblatt Anstrichsysteme für Holzfenster, Fenster und Fassade No. 2, Rosenheim
[168] Braunschweiger Richtlinien; Anhang 2 zum Abschlussbericht des operativen Eigenforschungsprojekts, Wilhelm Klauditz Institut, Braunschweig 1999
[169] a) G. Hora, A. Belz, Moderne Beschichtungssysteme für Holzfenster, Holz-Zentralblatt vom 12.02.1999, 258–259, b) J.E.P. Custódio, M.I. Eusébio, Waterborne acrylic varnishes durability on wood surfaces for exterior exposure, Progress in Organic Coatings, Volume 56, Issue 1, 1 May 2006, Pages 59-67
[170] L. Matthäi, Erfahrungen mit Holzschutzlasuren auf Basis wässriger Polyacrylatdispersionen; Farbe + Lack 89, 5 (1983) 339–345
[171] H. Kastien, Influence of the binder and additives on the weathering stability of colourless, aqueous wood varnishes, Polymers Paint Colour J., 181, No. 4286 (1991) 366–368, 371
[172] A. Broek, An acrylic emulsion paint for wood: An adequate alternative?, XIX. FATIPEC-Congress Aachen 1988, Vol. II, p. 361–380
[173] G. Rodsrud, J. Sutcliffe, Alkyd emulsion properties and application, Surface Coatings International 77 (1994) 7–16
[174] M. de Meijer, H. Militz, Adhesion of low-VOC coatings on wood: a quantitative analysis, Verfkroniek (1999) 25–30
[175] M. de Meijer, K. Thurich, H. Militz, Comparative study on penetration characteristics of modern wood coatings, Wood Science and Technology 32 (1998) 347–365

[176] a) E. Schmid, Holzaußenanstriche und Glasumwandlungs-Temperatur, Farbe + Lack 93, 12 (1987) 980–983, b) E. Schmidt, Glasumwandlungstemperatur und Wasseraufnahme von bewitterten Holzlasuren, Farbe + Lack 98, 5 (1992) 330–333

[177] E. Schmid, Außenbewitterung von Holzlasuren, Applica 105, No. 3 (1998) 10–17

[178]R. Arnoldus, R. Adolphs, W. Zom; Progress in acrylic emulsion developments, Polymers Paint Colour J. 181, No. 4287, (1991) 405–409, 418

[179] a) U. Desor, S. Krieger, G. Apitz, R. Kuropka, Water-borne acrylic dispersions for industrial wood coatings, Surface Coatings International (1999) 488–496, b) A. Overbeek, Polymer heterogeneity in waterborne coatings, J. Coat. Technol. Res., DOI 10.1007/s11998-009-9201-5, published online 23 September 2009

[180] M.-C. von Trentini, V. Carlson, P. Gerosa, R. Kuhn, K. Wood, Advanced technologies for formaldehyde-free waterborne acrylic wood coatings, 3. Nürnberg Kongreß, 1995, Paper 6

[181] M. Grimberg, Neuartige wässrige, glänzende und halbglänzende Anstrichmittel für Holz, Farbe + Lack 72, 5 (1966) 437–441

[182] R. Kreis, A. Sherman, Development of a ureido functional monomer for promoting wet adhesion in latex paints; Water-Borne and Higher-Solids Coating Symp. 02/1988; New Orleans, 222–243

[183] B. Singh, L. Chang, R. diLeone, D. Siesel, Novel wet adhesion monomers for use in latex paints; 23. International Conference in Organic Coatings, Vouliagmeni, Athen 1997, p. 427–441

[184] P. Ahola, H. Derbyshire, G. Hora, M. de Meijer, Water protection of wooden window joinery painted with low organic solvent content paints with known composition; Holz als Roh- und Werkstoff 57 (1999) 45–50

[185] A. Hofland, The use of alkyd-emulsion and alkyd/acrylic hybrid systems in high-gloss architectural coatings, FATIPEC-Kongreß 1992, Buch 2, 207

[186] P. Weissenborn, A. Motiejanskaite, Drying of alkyd emulsion paints, J. Coatings Technology 72, No. 906 (2000) 65-74

[187] T. Nabuurs, R. Baijards, A. German , Alkyd-acrylic hybrid systems for use as binders in waterborne paints, Progress in Organic Coatings 27 (1996) 163–172

[188] E. van Hamersv, F. Cuperus, J. van Es, Oil-acrylic emulsions as binders for waterborne coatings, 24th FATIPEC-Congress, Interlaken, 8.-11.06.1998, D 247-D 258

[189] H. Pecina, O. Paprzycki, Lack auf Holz, Vincentz Verlag, Hannover 1995

[190] U. Desor, Wässrige Lacke und Beschichtungen für Holz auf Basis von Polymerdispersionen; WKI-Bericht No. 31, Umweltfreundliche und emissionsarme Möbel, Seminar vom 09.-10.11.1995 in Braunschweig; ed. T. Salthammer, R. Marutzky, P. Böttcher, p. 57–81

[191] E. Daniels, A. Klein, Development of cohesive strength in polymer films from latices, Progress in Organic Coatings 19 (1991) 359–378

[192] M. Oaka, H. Ozawa, Recent developments in crosslinking technology for coating resins, Progress in Organic Coatings 23 (1994) 325–338

[193] R. Athey, Additives for waterborne coatings, Part 7: Curatives, European Coatings J. 11 (1996) 569–571

[194] J. Feng, H. Pham, P. Macdonald, M. Winnik, J. Geurts, H. Zirkzee, S. van Es, A. German, J. Coatings Technology 70, No. 881 (1998) 57–68

[195] W. Reich, K. Menzel, W. Schrof, Wasser im UV-Lack, ein Widerspruch?, Farbe + Lack 104, 12 (1998) 73–80
[196] Ch. Hegedus, K. Kloiber, Aqueous acrylic-polyurethane hybrid dispersions and their use in industrial coatings, J. Coatings Technology 68, No. 860 (1996) 39–48
[197] U. Desor, B. Leon, R. Kuropka, Comparison of water-borne acrylic and polyurethane dispersions and mixtures of both in 1-component water-based clear varnishes for wood, 25th FATIPEC-Congress, 19.–22.09.2000, Turin, Vol. 2, p. 117–134
[198] H.-J. Luthardt, R. Spohnholz, R. Simon, Wässrige Dispersionssysteme für Holzlacke, Farbe + Lack 104, 11 (1998) 30–38
[199] H.-H. Bankowsky, A. Eichfelder, E. Gulbins, Selbstvernetzende Polymerdispersionen für chemikalienfeste Holzlacke, Asian Pacific Coatings Show, Kuala Lumpur 1996
[200] O. Wagner, Wässrige Polymerdispersionen als Bindemittel in Silikatfarben, p. 233–261, in W. Schultze et al., Dispersions-Silikatsysteme, expert-Verlag, Renningen-Malmsheim, 1995
[201] B.R. Vijayendran, Effect of carboxylic monomers on acid distribution in carboxylated polystyrene latices, J. Appl. Polym. Sci., 23, 893–901 (1979)
[202] S. Muroi, Some physicochemical properties of poly(ethyl acrylate) emulsions containing carboxyl groups, J. Appl. Polym. Sci., 10, 713–729 (1966)
[203] S. Muroi et al., Alkali solubility of carboxylated polymer emulsions, J. Appl. Polym. Sci., 11, 1963–1978 (1967)
[204] E. Engler, Lösliche Silikate, Seifen-Öle-Wachse 100, p. 165-170, 207-212, 269-271, 298–300 (1974)
[205] E. Bagda, Beschichtungen im Wetter, Farbe und Lack 91, 07/1985, p. 595–599
[206] D. Y. Perera, Hygric aspects of coated porous building materials, Prog. Org. Coat. 8, p. 183–206 (1980)
[207] H. Künzel, Anforderungen an Außenanstriche und Beschichtungen aus Kunstharzdispersionen, Kunststoffe Bau Heft 12, p. 26–32 (1968)
[208] H. Meyer, H. Weber, Anstriche als Beschichtungen für mineralische Fassadenbaustoffe, Vol 3, Expert Verlag, Ehningen (1991)
[209] W. Sittenthaler, Siliconharzfarbe – die Mineralfarbe der Zukunft, XIX. Fatipec-Congress, Vol II, p. 329 (1988)
[210] K. H. Schmidt, Mineralische Anstrichstoffe, Definition und Abgrenzung zu Siliconharzfarben, p. 7–10, APPLICA 16 (1989)
[211] E. Bagda, Echte und weniger echte Siliconharzanstriche, XX. Fatipec- Congress, p. 471 (1990)
[212] E. Bagda, Siliconharzfarben und ihre positiven Eigenschaften, Die Mappe, 23 (1990) No. 3
[213] O. Wagner, Polymerdispersionen als Bindemittel in Siliconharzsystemen, p. 113–164, in W. Schultze, Wässrige Siliconharz-Beschichtungssysteme für Fassaden, Expert Verlag, 1997
[214] T. Helmen et al., Ermittlung der KVPK an Dispersionsanstrichen durch Messung der inneren Spannung, Farbe+Lack 96, 10/90, p. 769–772 (1990)
[215] P.E. Pierce, R.M. Holsworth, Determination of CPVC by measure of density of dry paint films, Off. Dig. Fed. Soc. Paint Technol., 37, p. 272–283 (1965)

[216] R.J. Cole, Determination of CPVC in dry surface coating films, J.O.C.C.A., 45, p. 776–780 (1962)
[217] R. Bachmann, Ein Beitrag zur Berechnung von Anstrichmittel-Rezepturen, Fette, Seifen, Anstrichmittel, 60. Jahrgang, 12, p. 1177–1186 (1958)
[218] H. Dörr, F. Holzinger, Kronos Titandioxid in Dispersionsfarben, Kronos Titan GmbH, Leverkusen, Germany (1989)
[219] A. Smith, O. Wagner, Factors affecting dirt pickup in latex coatings, J. Coatings Technology, Vol. 68, No. 862, Nov. 1996, 37–41
[220] P. Pföhler, A. Zosel, R. Baumstark, Elastic coating systems for the renovation of facades, Surf. Coat. Austr. 1995, p. 18–23
[221] P. Gieler, Dissertation "Überlegungen und Versuche zur Rissüberbrückungsfähigkeit spezieller Beschichtungssysteme an Fassaden", University of Dortmund, 1989
[222] P. Peyser, Glass transition temperatures of polymers, VI/S. 209–277 in J. Brandrup, E.H. Immergut, Polymer Hand Book, Interscience Publ., New York, 3rd ed., 1989
[223] H. Kossmann, M. Schwartz, Using natural weathering to assess polymer dispersion performance, Paint and Coatings Industry Magazine, März 1999, p. 48–56, Troy, Michigan (USA)
[224] A. Zosel, Lack- und Polymerfilme, Viskoelastische Qualitätsmerkmale, Hrsg. U. Zorll, Vincentz Verlag, Hannover (1996)
[225] Carlos Costa, personal note, 2001
[226] Peter Pföhler, personal note, 2001
[227] H. G. Meier, Der mineralische Fassadenputz, in Fassadenschutz und Bausanierung, expert-Verlag, Renningen-Malmsheim, 1994
[228] M. Güntert, Siliconharzputze in der Praxis, p. 460 ff., in W. Schultze, Wässrige Siliconharz-Beschichtungssysteme für Fassaden, expert Verlag, 1997
[229] H. Zeh, Weiß, weißer, reinweiß – Zur Wirkungsweise von Rezepturbestandteilen auf die Verschmutzungsresistenz von Kunstharzputzen in der Freibewitterung und im Laborexperiment, Vortrag auf der Jahresversammlung der Fachgemeinschaft Kunstharzputz e.V. am 29.04.1999 in Dresden
[230] H. Zeh, personal note, 1999
[231] H. Zeh, F. Jodlbauer, Brandverhalten von Wärmedämmverbundsystemen und brandtechnische Anforderungen, Vortrag auf dem 2.Seminar Beschichtungen und Bauchemie am 07.10.1998 in Kassel
[232] Testing of FMPA, Stuttgart
[233] G. Gierenz, W. Karmann (eds.), Adhesives and Adhesive Tapes, Wiley-VCH Verlag, Weinheim 2001
[234] Neue polymere Werkstoffe 1969–1974, Hanser Verlag, München – Wien, 1975
[235] M. Michel, Adhäsion und Klebtechnik, Hanser Verlag, München, 1969
[236] National Materials Advisory Board, National Research Council: Treatise of Adhesion and Adhesives („Structural Adhesives with Emphasis on Aerospace Applications"), Vol. 4, Marcel Dekker, New York – Basel, 1976
[237] J.J. Bikerman, The Science of Adhesive Joints, 2nd ed., Academic Press, New York – London, 1968
[238] C. Heitler in D.J. Alner (ed.), Aspects of Adhesion, Vol. 4, University of London Press, London – Liverpool – Prescot, 1968
[239] G. Habenicht, Kleben, 3rd edition, Springer, Berlin, 1997

[240] W. Brockmann, J. Adh. 37 (1992) 173–179
[241] K.W. Allen in D.J. Alner (ed.), Aspects of Adhesion, Vol. 5, University of London Press, London – Liverpool – Prescot, 1969
[242] W.C. Wake, Polymer, 19 (1978) no.3, 291
[243] B.V. Deryagin, N.A. Krotova, V.P. Smilga, Adhesion of Solids, Consultants Bureau, New York – London, 1978
[244] W.A. Zisman, Ind. Eng. Chem., 55 (1963) 18
[245] L.H. Shorpe, H. Schonborn, Adv. Chem. Ser. 43 (1964) 189
[246] R.J. Good in R.L. Patrick (ed.), Treatise on Adhesion and Adhesives, Vol. 1, Marcel Dekker, New York, 1967
[247] D.H. Kaelble, Physical Chemistry of Adhesion, Wiley-Interscience, New York 1971
[248] R.H. Ottewill in D.J. Alner (ed.), Aspects of Adhesion, Vol. 2, University of London Press, London – Liverpool – Prescot, 1966
[249] S.S. Voyutskii, Autohesion and Adhesion of High Polymers, Interscience, New York – London – Sydney, 1963
[250] R.M. Vasenin, Adhesion, Fundamentals and Practice, Ministry of Technolgy, MacLaren, London, 1969
[251] J.R. Huntsberger in R.L. Patrick (ed.), Treatise on Adhesion and Adhesives, Vol. 1, Marcel Dekker, New York, 1967, p. 136–137
[252] R.L. Patrick (ed.), Adhesion and Adhesives, Vol. 2: Materials, Marcel Dekker, New York, 1969
[253] D.R. Gehman, Acrylic adhesives, in Handbook of Adhesives, 3rd ed., I. Skeist (ed.), Van Nostrand Reinhold, New York, 1990, Ch. 25
[254] D. Satas, Acrylic adhesives, in handbook of Pressure Sensitive Adhesvive Technology, 2nd ed., D. Satas (ed.), Van Nostrand Reinhold, New York, 1989, Ch. 15
[255] D. Satas, Adhesives Age, 15 (10), 19 (1972)
[256] P.M. Lesko and P.R. Sperry, in P.A. Lovell and M.S. El-Aasser (eds.), Emulsion Polymerization and emulsion polymers, John Wiley and Sons, Chichester, 1997, Ch. 18
[257] M. Sandor et al., US Patent Application 2004/0087702 on 03.09.2001
[258] W. Siol et al., EP Patent Application 0321878 on 16.12.1988
[259] B. Dotzauer et al., US Patent Application 5,145,625 on 08.09.1992
[260] B. Dotzauer et al., Deutsche Offenlegungsschrift 4102344 vom 26.01.1991
[261] W. Beckerle et al., Canadian Patent Application 2059626 on 20.01.1992
[262] H.W. Tiggemann, Twenty years of floor polish latex technology, in Polymer Technology, Chemical Specialties Manufacturers association, Washington, DC, 1974
[263] H. Kossmann, Aufhellung von Fassadenbeschichtungen auf Basis wässriger Polymerdispersionen, XXI. FATIPEC-Congress 1992, Vol. II Seite 341–359
[264] H. Kossmann, M. Schwartz, Photoinduzierter Abbau von Polymerdispersionen in der Freibewitterung, XXIV Fatipec-Congress 1998, Volumen A, p. 223–241
[265] P. Kresse, Nachbehandelte anorganische Pigmente, Farbe + Lack 97, 5/1991, p. 399–404
[266] Noblesse oblige, Bericht zum 31st Münchner Klebstoff- und Veredelungssymposium, part 2, access www.mkvs.de/img/content/102/Coating-12-2006-Teil2.pdf (20.02.2010)

[267] Ethylene Acrylic Acid Copolymer – High Performance EAA Copolymer Dispersions for Applications in Coatings by Michelman, access www.azom.com/details.asp?ArticleID=4865 (20.02.2010)
[268] T. Hirose et al., POLYOLEFIN RESIN DISPERSION COMPOSITION AND PROCESS FOR PRODUCING THE SAME, Canadian Patent Application CA000002672498 (19.06.2008)
[269] R. Weavers et al., AQUEOUS POLYOLEFIN DISPERSIONS FOR TEXTILE IMPREGNATION, US Patent Application US020090253321 (08.10.2009)
[270] Ch. F. Diehl et al., Waterborne Mechanical Dispersions of Polyolefins, The Dow Chemical Company, access http://www.dow.com/PublishedLiterature/dh_007a/0901b8038007a9c1.pdf?filepath=dowpod/pdfs/noreg/789-00801.pdf&fromPage=GetDoc (20.02.2010)
[271] H. Kossmann, personal communication, 2002

5 Acrylate resins for radiation curable coatings

Reinhold Schwalm

5.1 Introduction and definitions

In the preceding chapters, acrylate resins are described which are produced by radical polymerization in a reaction vessel either in organic solvent or aqueous dispersion. In contrast to these resins, this chapter is concerned with acrylate resins where the unsaturated double bond of the acrylate is still present. These low molecular weight acrylates are synthesized by poly-addition or poly-condensation reactions. Formulations containing these resins are applied onto a substrate and the polymerization then occurs in the film by a radiation-induced radical cross-linking reaction (Figure 5.1).

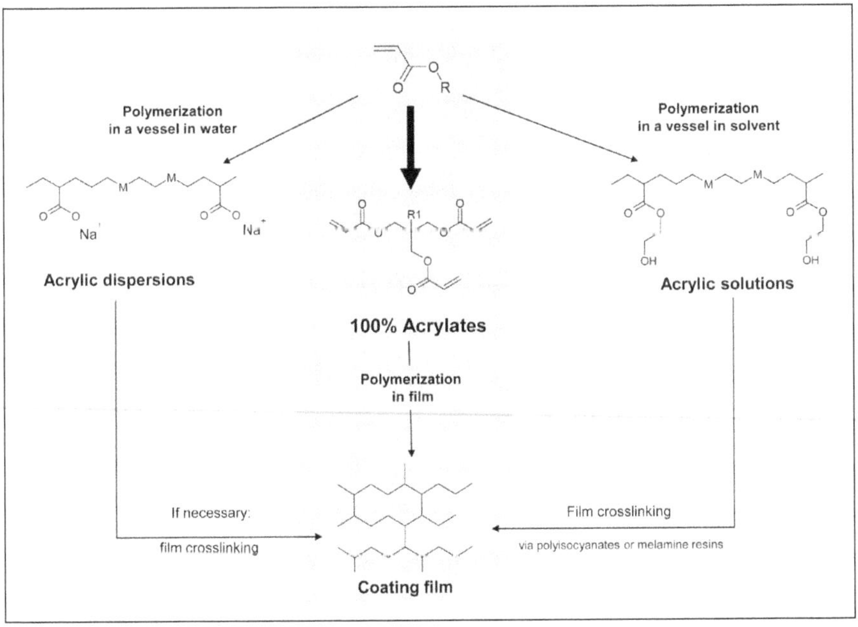

Figure 5.1: Different approaches to acrylate and acrylic coatings

Poth/Schwalm/Schwartz: Acrylic Resins

ISBN: 978-3-86630-857-2

The use of acrylic ester terminated coating raw materials in radiation curable systems has been established over the past 30 years and developed into a promising eco-efficient technology. The radiation curing technology and the available acrylate coating raw materials as such have already been widely described [1].

In the early days of the technology, the term "radiation curing" led to confusion and misunderstandings, since it was associated with the use of dangerous radiation (e.g., X-rays).

In order to avoid the term "radiation" the descriptive "energy curing" has been created. However, this term also comprises the classical thermal curing and therefore does not characterize the technology exclusively. Radiation curing describes the crosslinking of coatings by reactions initiated by radiation rather than by heat. In the meantime, the term "radiation curing" is well established and includes all types of radiations used, ranging from electron beams to ultraviolet (UV-C to UV-A) and visible.

The radiation is used to generate initiators at room temperature to start a polymerization reaction. More than 80% of the radiation curable systems are based on radical polymerizing acrylate terminated oligomers and/or (multifunctional) monomers. The radiation induces the formation of free radicals, either directly through bond cleavage by electron beam radiation, or indirectly through the use of photoinitiators. Depending on the absorption spectrum of the photoinitiators, radiation ranging from UV-C to UV-A is applied to generate the radicals. The main light sources are based on mercury lamps or the recently developed LED (light emitting diode) technology.

The radiation curable formulations are typically solvent-free and are considered as low VOC (volatile organic compounds) coatings. These systems do not have to be dried, such as solvent or water-based formulations. If occasionally the term "UV drying" is used, it refers to the fast conversion of the liquid UV coatings into a solid network by crosslinking, accompanied by an increase in glass transition temperature. However, if it is required by the application or the property spectrum, one can also modify these coating systems in order to be diluted by water or solvents.

5.2 History

The curing of coatings by means of light and air was already known in ancient times. Thus, mummies in ancient Egypt were wrapped with Syrian asphalt, which contained significant amounts of unsaturated hydrocarbons, and cured under the action of sunlight. Around 1826, the inventor of photography, *Joseph Niepce* also took advantage of the UV curable properties of Syrian asphalt by transferring images onto stone by exposing lithography templates to light. A little later, around 1877, the polymerization of methyl methacrylate to polymethyl methacrylate was discovered.

However, it took up to the 1960s until the first acrylic ester compounds were used in commercial radiation curable systems. Since the classical radiation curing technology is executed at room temperature, the first applications were on temperature-sensitive substrates, like paper and wood. For example, printing inks based on highly functional monomers such as pentaerythritol triacrylate, trimethylolpropane triacrylate and hexanediol diacrylate are described as early radiation-curable systems [2].

Most of the radiation curable wood coatings used at that time were based on unsaturated polyesters diluted with styrene as a monomeric reactive diluent. Back then, the radiation curable technology did not gain considerable market share because of the low polymerization rate of the polyester systems, the then available low lamp power and the oxygen inhibition occurring at the surface. With the development of more powerful lamps, reactive acrylate resins and better photoinitiator systems in the 1970's, the breakthrough of the radiation curing technology occurred. The growth of the radiation-curable systems has been accelerated with the development of less irritating acrylate monomers and increasing legal requirements for low-emission paints. In the course of time, successively other substrates were coated with radiation curable formulations, besides wood and paper. Such new applications were on plastics (PVC, linoleum, automotive plastic parts, audio and video media, CD, DVD, or PSAs) or metal substrates (steel pipe, steel furniture). These are already established or in development (coil coating). Furthermore, the use of radiation for the production of relief structures using masks, or laser direct-write (e.g. printing plates, photoresists, and stereolithography process) progressed constantly.

Before 1980, in Europe, less than 2500 tons of radiation curable coatings were consumed per year. This market has since grown to more than 120,000 tons in 2009. However, a significant market share over other technologies is still only noticeable in the wood and paper coating industry.

The global market for UV/EB resins is valued at about 296,000 tons in 2008, whereof the consumption in Asia is the largest with about 134,000 tons and exhibits the largest growth rates.

5.3 Basics of the radiation curing technology

The setup of the whole coating process is largely determined by the desired properties of the coating layer on the chosen substrate. Examples are, for instance, a scratch-resistant coating on a finished parquet floor or a high gloss lacquer on a brochure or a cosmetic packaging container. The desired function of the coating, for example, the protection of the item against scratches, corrosion, weathering or exposure to liquids (red wine, mustard, coffee, gasoline, etc.) determines the type of properties required and the thickness of the coating.

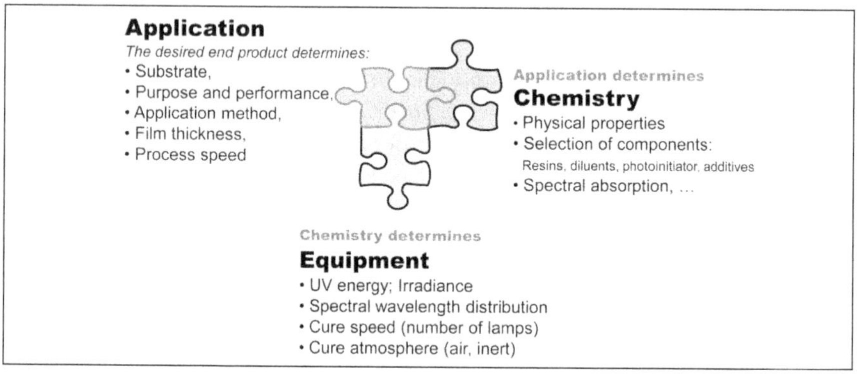

Figure 5.2: Interaction of the radiation curing process parameters

The property requirements and the desired application method (rolling, casting, spraying) dictate the selection of components (resin, thinner, photoinitiator, additives) used in the formulation and the composition of the formulation. The choice of components (e.g., photoinitiator, pigments) then influences the selection or settings of the exposure systems, such as the lamp spectrum, light intensity, and exposure time (Figure 5.2).

Advantages of radiation curing technology

The advantages of the radiation curing technology have been summarized by RadTech, the European Association for the promotion of UV/EB curing, under the "e5 emblem" (Efficient, Energy saving, Enabling, Economical, Environmentally friendly) [3]:

- **Environmentally friendly**, because the traditional UV curable varnish almost exclusively contains components that are incorporated into the paint and thus contain low or no volatile organic compounds (VOC)
- **Energy saving**, because UV curing usually proceeds at room temperature and thus consumes considerable less energy compared to traditional baking ovens
- **Efficient**, since the curing reaction is completed within seconds and the cured parts can be immediately packaged or further processed
- **Economical**, because material recycling of the UV varnish can be easily established and often the eco-efficiency on cost basis is also beneficial to alternative technologies
- **Enabling**, because UV curing can be used within all existing technologies, e.g., 100% liquid coatings, water-based UV coatings, UV powder coatings and dual cure systems (UV and thermal curing).

In order to explain the advantages of the UV curing process in more detail, a typical coating process for coating of parquet panels is shown in Figure 5.3.

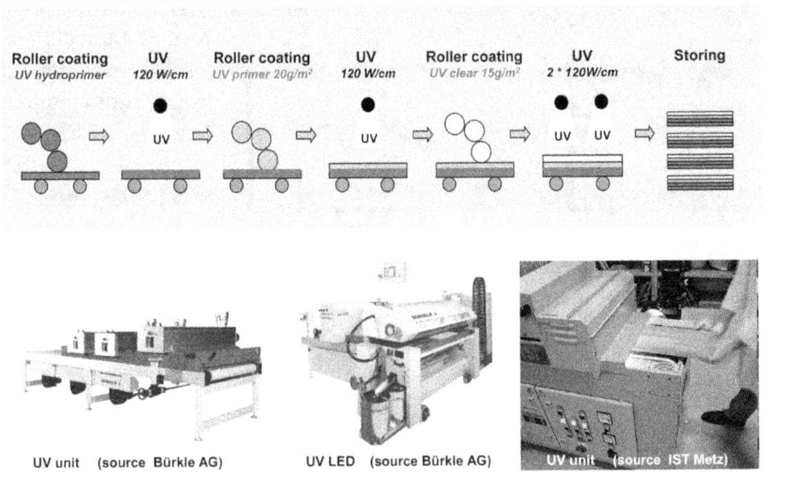

Figure 5.3: Fabrication of a ready-to-use parquet with a multiple layer structure by an UV coating process

The scheme clearly demonstrates the benefits of the UV curing process. Since the UV induced polymerization is completed within seconds, the complete assembly of a multilayer coating can be done in line on a conveyor system. The hydroprimer responsible for the adhesion to the substrate, the primer to ensure the abrasion resistance and the clear coat for high gloss, scratch and chemical resistance, are applied and cured consecutively in a row. Optionally, also a sanding unit can be installed after each curing unit. After passing through the different steps, the finished boards can be packed immediately.

Acrylate based formulations for radical UV curing

Due to several advantages, discussed in the following, UV curing formulations used in industrial coatings are predominantly radical curing systems based on acrylate resins. The chemical process of the UV induced free radical polymerization is shown schematically in Figure 5.4.

The formulations are based on three essential components

- **binders** containing unsaturated, usually polyfunctional acrylate compounds,
- mono-or difunctional or even multifunctional acrylates as **reactive diluents**, which take over the function of the solvent to adjust viscosity and
- **photoinitiators**, which produce radicals upon radiation induced exposure.

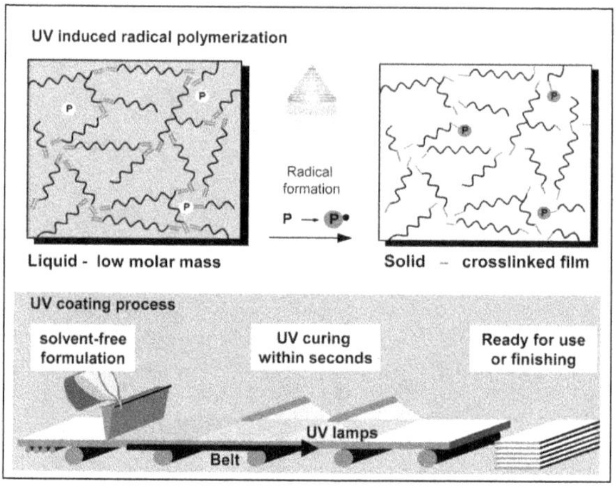

Figure 5.4: Scheme of the UV induced polymerization of acrylate resins

The photochemically formed radicals initiate a spontaneous polymerization at room temperature, leading within seconds to curing and solidification of the entire coating. The typical UV curable formulations are 100% systems requiring no solvents and are therefore an environmentally friendly alternative to conventional paint systems.

The photo-induced polymerization proceeds analogously to the classical free-radical polymerization. However, the starting radicals are produced by a radiation induced reaction rather than by heat. Whereas electron beam exposure forms unspecific radicals by bond cleavage, the exposure with UV light requires the use of photoinitiators, which form free radicals as a result of rearrangements of excited states. These radicals add to a monomer unit to form a new radical. By multiple additions of monomers, long polymer chains are generated. Difunctional or multifunctional acrylate monomers act as crosslinkers and form networks quickly. The growth of the chain is interrupted only when the active radical centre is deactivated in a termination reaction. The thereby occurring basic reactions are shown in Figure 5.5. In this figure, only the main reactions are depicted, however, several other side reactions can occur (see also Chapter 2.1). For example, the initiating radical can be generated by either a hydrogen abstraction or a cleavage reaction depending on the photoinitiator type used. Furthermore, in the propagation step the monomer addition competes with a hydrogen atom abstraction from another unit within the network. Here, the polymer chain is terminated, however, the kinetic radical reaction progresses, since the hydrogen atom abstraction leaves another radical at the other unit within the network.

The termination reaction can proceed by recombination of chain radicals, by reaction of a chain radical with primary radicals or by elimination reactions. Due to the high proportion of photoinitiators applied, the number of radicals generated is comparatively high and therefore the chain length of the polymer forming reaction is relatively short. To get a feeling about the chain length, within the exposure time of a few seconds, up to 1,000 acrylate molecules may add to a radical. When using multifunctional acrylates, high molecular weight networks are formed.

Typical formulations of radiation curable coating systems contain 25 to 95% oligomeric resins, 0 to 60% of reactive monomeric diluents, 1 to 5% photoinitiator and 0 to 3% of other additives (such as levelling agents, wetting additives, defoamers). The acrylic resins and reactive diluents used are based on the value chain of acrylic acid (Figure 5.6).

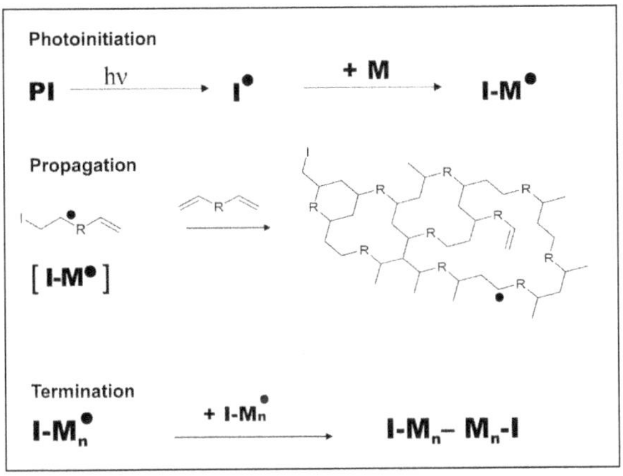

Figure 5.5: Reaction scheme of photopolymerization

The **resin** classes contain almost entirely terminal acrylate groups and are differentiated by the "polymer backbone". They are based on the acrylic acid value chain, as exemplified in Figure 5.6. All reactive diluents are synthesized by the etherification of alcohols, diols or polyols with acrylic acid. Alkoxylated polyols, which can be used as reactive diluents or resins, are produced in the same manner. Epoxy acrylates are prepared by the addition of acrylic acid to epoxides (EA). Urethane acrylates (UA) are the addition products of hydroxyalkyl acrylates to polyisocyanates. The resin classes differ in their typical properties.

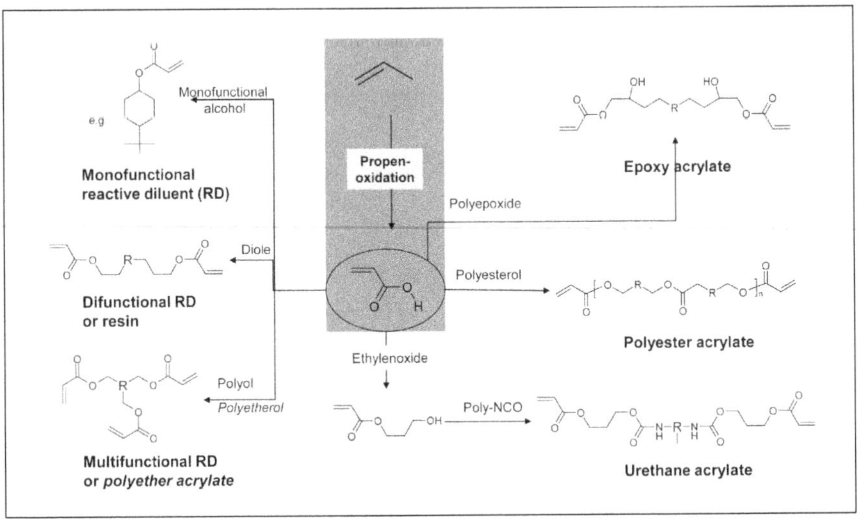

Figure 5.6: Synthesis of UV curable acrylates based on the acrylic acid value chain

While the largest class of epoxy acrylates, based on bisphenol-A-diglycidether, exhibit high hardness and chemical resistance, typical urethane acrylates are highly flexible and abrasion resistant. Due to their low viscosity, polyether acrylates are mainly used as reactive diluents, but, can also be used as sole resins. Polyester acrylates have a balanced range of properties. In many formulations, however, combinations of different resins are utilised.

The synthesis of compounds and structure-property relationships are described in the following sections in more detail.

The applied coating process in a manufacturing line is often determined by the already existing set-up and equipment. However, if one considers starting the whole coating process from scratch, UV curing systems are often superior, not only in terms of ecological aspects, but also in terms of economic criteria. This holistic approach, in which criteria such as energy consumption, raw material costs, emissions and risk potential are evaluated by eligible coating alternatives for a defined customer benefit (e.g., in square meter of coated surface) has been established under the term eco-efficiency analysis. An example of such an eco-efficiency analysis is shown in Figure 5.7 for the coating of 1000 wooden door fronts [4].

Figure 5.7 demonstrates that in the considered base case scenario the UV clear coat is in both criteria, environmental impact and in terms of total costs better than the alternatives (NC, acid curing epoxy, 2K PU, aqueous). This eco-efficiency analysis can also be used to evaluate alternative scenarios, for example, to determine the influence of recycling of the paint on costs and environmental impact.

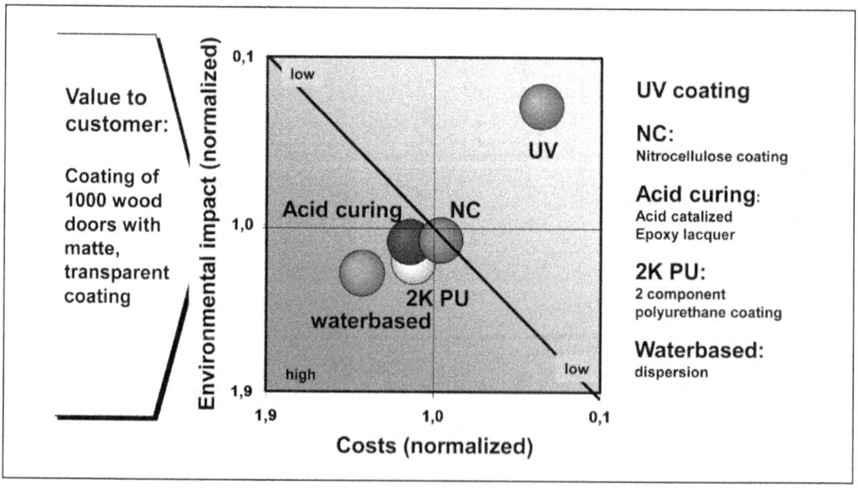

Figure 5.7: Eco-efficiency portfolio for alternatives to the door paint coating

Wood
(finished parquet, doors, furniture, ..)

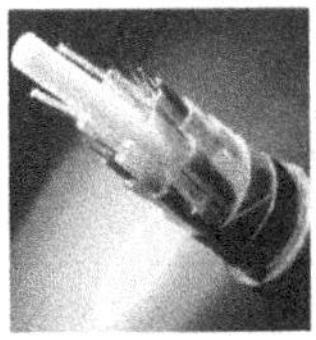

Glas fibres

Photo resists
printed circuit
boards (PCB)
chips

Paper (post cards, high gloss brochures)
Printing plates
Cosmetic packaging
Screen print inks
Flexo prining inks
Overprint varnishes

Adhesives

	2003 [k tons]			2010 [k tons]		
	conv.	UV	Sharel UV [%]	conv.	UV	Share UV [%]
Gen. industry	503	0	0	589	5	0.8
Wood	244	27	11.1	285	43	15
Anti corrosion	145	0	0	172	0	0
Auto OEM	145	0	0	172	2	1.2
Auto refinish	73	0	0	86	4	4.7
Can/coil	73	0	0	86	4	4.7
Others	220	3	1.4	257	6	2.3
Graphics	416	23	5.5	487	35	7.2
Total	1821	53	2.9	2135	89	4.6

Source: H.H. Bankowsky, RadTech US, Conf. Proceedings Charlotte, May **2004**

Figure 5.8: Traditional applications of UV curable coatings

The **classical applications** are in particular market segments where temperature-sensitive substrates are coated predominantly in industrial coating lines. Examples include parquet flooring, furniture parts, door fronts, glossy brochures, cosmetic packaging, PVC flooring and in the electronics sector, where the structural imaging can be exploited, e.g. in photoresists for IC (integrated circuits) manufacturing or photopolymer printing plates. The market share of UV coatings in the segments therefore reflects the main application areas: graphics (printing inks, overprint varnishes), wood, plastics and electronic applications (Figure 5.8). Whereas graphics is worldwide the biggest segment, the dominating segment in Asia is electronics and in Europe it is wood/plastics.

Due to the many advantages of UV curing, a lot of new applications for radiation curable coatings are emerging. Among them are **printing applications**, like UV inkjet, label printing, gravure and wide web flexo, **adhesives**, like pressure sensitive adhesives or CD bonding, **clear coats** for metalized **plastics**, exterior **coil coating** of steel and aluminum, and **automotive applications** [5]. In Asia, especially electronic applications, like photoresists and flat panel displays, are the drivers, while Asian companies are the market leaders, as well as in CD's and DVD's, where adhesives and protective coatings are UV cured [6].

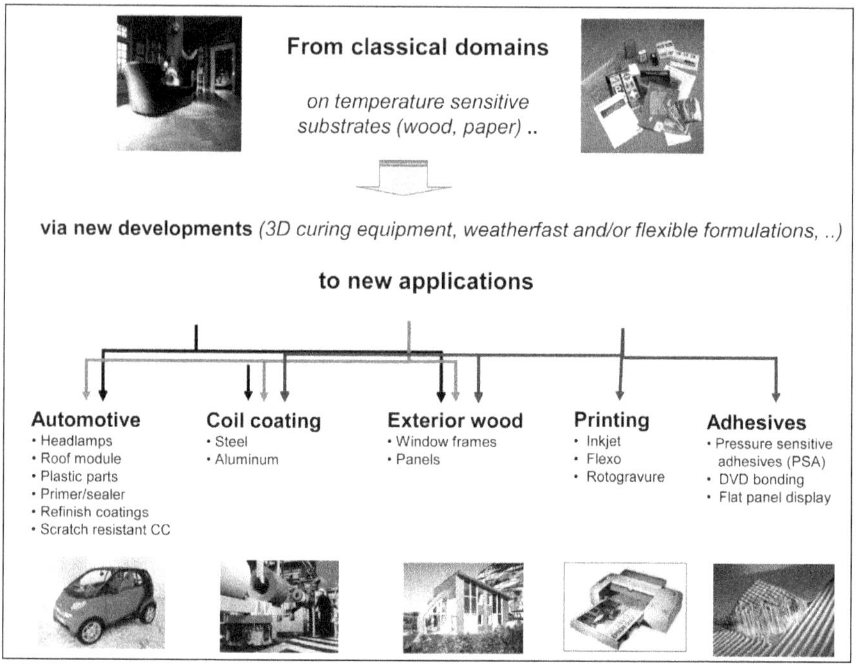

Figure 5.9: From traditional applications of UV coatings to new horizons via new developments

UV applications for 3D substrates, such as TV sets and vehicle parts, have been reported recently [7]. Especially the major advances in the equipment development for 3D applications and curing of weather resistant coatings have opened up new application fields with automotive applications as the door-opener for many other industrial coating possibilities. Besides UV applications already in use in automotive parts, like motor sealing, sensor encapsulates, electronic parts, headlamp assemblies, lens and reflector coatings, graphic identification labels, name boards, dashboard screen printings or battery labels, new automotive applications are under development (Figure 5.9).

Disadvantages of radiation curable coatings

The main reasons why radiation curable coatings have not spread as fast as expected are due to some weaknesses or disadvantages:

- Material costs are often higher than conventional paints (photoinitiator, acrylate)
- Curing of thick, pigmented coatings is not always possible (competition of photoinitiator absorption with pigment absorption)
- Curing of three-dimensional objects is difficult (lamp distance, shadow zones)

- Adhesion to smooth substrates (metal, plastic) is often problematic (shrinkage)
- Oxygen inhibition requires increased effort (energy, chemicals, inert)

Furthermore there are a lot of other factors which have contributed in the past to a retarded penetration of the technology into further market segments:

- Plant investment and material costs are often considered rather than a complete process cost evaluation
- Exposure assets for three-dimensional curing are just under development
- Raw materials have to be selected carefully (skin irritation, migration, smell)
- Exposure units necessitate high safety levels (EB, UV)
- Up to now only few exterior applications are realized (yellowing, stabilization)
- Matting of 100% liquid coatings is harder to realize compared to solvent- or waterbased coatings

Because of the excellent eco-efficiency, in particular the solvent-free formulations, low energy requirements during curing, low space requirements of the assets and good film properties, UV coatings have increased its market share in the classical applications on temperature-sensitive substrates within the last forty years considerably. In order to foster the spread of this future technology the disadvantages (oxygen inhibition, absorption interference of pigments and photoinitiators) have to be eliminated and the development of exposure equipment for three-dimensional objects has to be pushed. Considerable development activities to tackle these disadvantages are ongoing worldwide. They will contribute to a continued strong growth in radiation curing technology, particularly in view of increasingly stringent energy usage discussions and environmental (VOC) concerns.

Recommendations of related literature

- Glöckner, P., Jung, T., Struck, S., Studer, K., **Radiation Curing**, Vincentz, (2009)
- Schwalm, R., **UV Coatings**, Basics, Recent Developments and New Applications, Elsevier, (2007)
- Müller, B., Poth, U., "**Coatings Formulation**", 2nd Edition, S. 237-250 (Radiation Curing), Vincentz Network, Hannover (2011)
- Garratt, P., **Strahlenhärtung**, Vincentz, Hannover, (1996)
- Fouassier, J.P, Rabek, J.F., Radiation Curing in Polymer Science and Technolgy, Elsevier 1993
- Vol. I: Fundamentals and Methods; Vol. II: Photoinitiating Systems; Vol. III: Polymerization Mechanism; Vol. IV: Practical Aspects and Applications
- Holman, R, UV&EB Curing, Formulation for Printing Inks, Coatings and Paints, SITA, London, (1984)

5.4 Resins for radiation curing

5.4.1 Acrylates – preferred monomers

There are a number of unsaturated compounds available which have been used in radiation curable coating formulations. The range of radical curable functional groups is shown in Figure 5.10. As mentioned already in the chapter about the historical background, the acrylate-based coating raw materials were accepted as the preferred monomers. In order to get an overview about the more or less general advantages and disadvantages of different functional groups, they are described in the following section.

Unsaturated polyesters, in combination with styrene as the reactive diluent, were the earliest used UV coating resins, particularly in the wood coating industry. Their range of properties can be adjusted in a wide range, particularly through the variation of the unsaturated component (maleic acid/other diacids), the nature of the diols (difunctional, molecular weight, trifunctional), the molecular weight of the polyester and the nature of the reactive diluent (styrene, vinyl ethers, acrylates). Although many characteristics are easier to achieve with alternative resins, particularly acrylics, the unsaturated polyesters have kept a high market share for a long time, due to the low material cost. Substituting the undesirable styrene monomer with alternatives, like acrylates, however, reduces the cost benefit significantly.

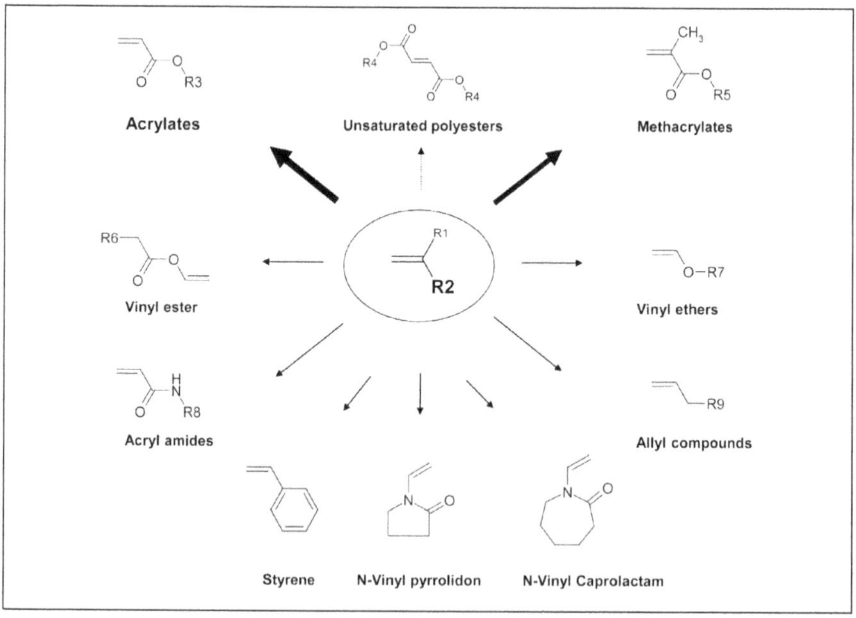

Figure 5.10: Portfolio of unsaturated radical polymerizable compounds

Table 5.1: Advantages and disadvantages of the different polymerizable groups

Unsaturated functional group		**Advantages**	**Disadvantages**
Unsaturated polyester	general	• cheap raw materials (maleic acid; fumaric acid, diols)	• slow curing speed • limited flexibility • styrene as reactive diluent
Vinyl ester	general	• reactive	• cure slower than acrylates
	vinyl acetate		• very volatile • low flash point
Vinyl ether	general	• excellent dilution power • non irritant • alternating copolymerization	• no radical homopolymerization • acid catalyzed hydrolyses (acetaldehyde formation) • unfavourable copolymerization with acrylates
N-Vinyl amide	N-vinyl-pyrrolidon	• excellent reactive diluent • good copolymerization with acrylates	• toxicity (cancerogenic suspect), odor • reactivity and resistance decrease at levels > 20%
	N-vinyl-caprolactam	• high boiling point	• solid (mp.: approx. 35°C)
Styrene	general	• good reactive diluent • increases film gloss	• no resins with terminal styrene groups available • toxicity, odor, aromatic structure
	styrene	• good copolymerization behaviour with unsaturated polyesters	• slow polymerization speed
Allyl compounds	general		• no homopolymerization • chain terminators
Methacrylates	general	• relative broad synthetic basis • higher T_g compared to acrylates	• slower polymerization speed compared to acrylatea (up to factor of 5)
Acrylates	general	• very broad synthetic basis based on the acrylic acid value chain	• high polymerization speed
	stenomeric acrylates	• reactive diluents	• most compounds are irritant
	eurymeric acrylates	• very broad spectrum of resins available	• oxygen inhibition

There are several monomers and resins available with **vinyl functional groups**, like vinyl esters, vinyl ethers, N-vinyl amides, and the vinyl aromatics (styrene, vinyl toluene).

From the **vinyl esters**, vinyl acetate is a commonly used monomer in the synthesis of polymer dispersions, however, due to the low boiling point and flash point, it is almost no longer used in radiation curable coatings. Another vinyl ester, versatic acid vinyl ester is used occasionally as a reactive diluent, but more often as a co-monomer for adhesives. In order to reduce the surface tension of radiation curable formulations, more often the reaction product of glycidyl methacrylate with versatic acid is used as a reactive diluent. New vinyl ester monomers (such as vinyl acrylate, vinyl cinnamate, vinyl maleate, divinyl fumarates) are introduced every now and then and tested in UV systems. Because of low reactivity and limited raw material basis, vinyl esters play no significant role in radiation curable formulations.

Vinyl ethers are mainly used as reactive diluents in cationic curing systems (together with epoxides), since they only homopolymerize cationically, but not radically. In the market, there is available a series of monofunctional and difunctional vinyl ethers and several vinyl ether functional urethane resins based on the reaction products of polyisocyanates with hydroxbutyl vinyl ether [8]. The main products are the divinyl ethers of triethylene glycol and cyclohexanedimethanol. The main disadvantage of vinyl ethers is their ease of hydrolysis in an acidic environment. As a consequence of the hydrolysis reaction, the formed vinyl alcohol rearranges to the toxic acetaldehyde. Thus, at least acidic components in the formulation should be avoided. Although vinyl ethers do not homopolymerize, they copolymerize radically with unsaturated polyesters in an alternating manner. The copolymerization with acrylates, however, is clearly unfavourable, and therefore, the content of vinyl ether monomers should not exceed 10% of the formulation, if a complete incorporation of the vinyl ether monomers into the acrylate network should be guaranteed [9].

Of the **N-vinyl amides**, N-vinyl pyrrolidone (NVP) was one of the most common reactive diluents in the past, but, because of toxicological concerns, smell and high price, its use decreased significantly. NVP has pretty good dilution power, however, since the homopolymer is water soluble, its proportion in the formulation is generally kept below 20%. Where applicable, acrylates are used as the alternative to NVP, and in other applications, such as plastic coatings and printing inks, NVP in particularly is replaced by N-vinyl caprolactam. Higher functional vinyl amides are not available, with the exception of the difunctional N, N´-methylene-bis-acrylamide, which however, is almost completely banned due to the high toxicity (marked T).

Significant differences in reactivity

The **vinyl aromatics**, styrene and vinyl toluene, are nowadays not anymore used widely in radiation curable formulations, due to their high toxicity and slow polymerization behaviour. But still, styrene is the preferred reactive diluent in unsaturated polyester resin systems.

Moreover, the reactivity of **allyl compounds** is some orders of magnitude lower than that of acrylates. They are therefore rarely used in radical polymerizable systems, but preferably as the ENE component in the thiol-ENE addition systems [10]. Ally compounds are also described in combination with Dual Cure systems where the allyl function can cure by a light independent oxidative curing mechanism, which allows curing in shadow areas, where the light exposure is not possible [11]. Main allyl compounds on the market are trimethylol propane triallyl ether, triallyl cyanurate and diallyl phthalate, the latter also used as a plasticizer.

Since **methacrylates** polymerize up to a factor of five times slower than acrylates, they are used mainly in specialty applications where the generally higher glass transition temperatures and lower skin irritations are beneficial, and often absolutely necessary, as in dental applications [12]. The early formulations in dental applications were based on methacrylate resins (bisphenol-A reacted with glycidyl methacrylate) and methacrylate diluents (triethylenglycol dimethacrylate) in combination with modified silica glass and photoinitiators (Camphorquinone). The main disadvantage of such composites is still their polymerization shrinkage. A considerable advantage against amalgam and other alternatives is the possibility of composites to match the tooth colour. The overall market share of methacrylates in radiation curable formulations is estimated to be well below 1%.

Generally, the raw materials are selected by the desired performance properties. This applies particularly for reactive diluents, which are especially used in order to reduce the viscosity of the formulations, but, since they are incorporated into the network, contribute significantly to the coating performance properties. Since a high reaction rate is almost always required in radiation induced curing, the reactivity is always a dominant selection criterion. The reactivity of functional groups is in general [13]:

$$\text{acrylic} > \text{methacrylic} > \text{vinyl} \gg \text{allyl}$$

The second major criterion, in addition to the reactivity, for the dominance of acrylate monomers and resins in radiation curing is the available synthetic raw material breadth, based on the acrylic acid value chain (Figure 5.6).

The **main suppliers** of coating raw materials for radiation curable formulations are: Cytec ("Ebecryl"; "Ucecoat"), CrayValley/Sartomer; Arkema, BASF SE ("Laromer"), IGM Resins, NL, ("Photomer", "Omnimer", "Omnilane"), Bayer AG ("Desmolux") Eternal, TW ("Etermer"), AGI Corporation, TW ("AgiSyn")

5.4.2 Acrylate functional reactive diluents (stenomeric acrylates)

In addition to the unsaturated acrylate resins, acrylate functional reactive diluents are an integral part of radiation-curable formulations with the following key features:

- Reduction of viscosity of the resin to the desired application viscosity
- Optimization of the property spectrum of the formulation
- Low volatility
- High reactivity and high double bond conversion
- No or low toxicological concern

Both monofunctional and multifunctional acrylates are used as reactive diluents. For reasons of classification and labeling of acrylates, the sector group "UV/EB acrylates" of CEFIC [14] has grouped acrylates into „stenomeric acrylates" (from the Greek "steno" and "meris", which means "narrow molecular weight distribution") and "eurymeric acrylates" (from the Greek "euru" and "meris", according to "broad molecular weight distribution"). The stenomeric acrylates are therefore well-defined, low molecular weight substances that can be described by an idealized structural formula. The stenomeric acrylate products offered by the various manufacturers differ only very little in the chemical composition (by-product spectrum, excipients). According to this definition, the mono- and multifunctional reactive diluents discussed in the following belong to the stenomeric acrylates. The preferred stenomeric acrylates used in radiation curable formulations are the multi-functional acrylates. The dominant products are TPGDA, with about 50 to 60% market share of the reactive diluents, followed by TMPTA (15%), HDDA and DPGDA (approx. 10%). Monoacrylates exhibit a market share of less than 10% of all the reactive diluents.

5.4.2.1 Monofunctional acrylates

Due to the requirements for low volatility, low irritation and excellent thinning behaviour, the selection of standard monomers for use in radiation curable formulations are significantly restricted (Table 5.2). The boiling points of the predominantly used acrylates are generally above 200°C (at atmospheric pressure).

According to the discussed restrictions the formulator selects the acrylate monomers with regard to the desired properties (viscosity, dilution behaviour, hardness, elasticity, adhesion contribution, reactivity, compatibility). As one selection criterion, the dilution behaviour of some standard acrylates (mono-, di-and trifunctional)

Chemical identification	Abbr.	CAS No.	M [g/mol]	Viscosity [mPas] ca.	Performance contr. positive	T_g [°C] [1]	Boiling point (extrapol. to normal P) °C @mmHg	Reaction speed Rp [mol/l*s]	Surface tension [dyn/cm] [2]	Labelling [3]	Disadvanat.
2-Ethyl-hexyl acrylate	EHA	103-11-7	184	4	soft films	-58	216	2.04		Xi R37/38; 43	odour
2-(2-ethoxyethoxy) ethylacrylate	EDGA	7328-17-8	188	5	reactivity, adhesion	-53		2.47	32.4	C R22-34-43-52/53	
Lauryl acrylate	LA	2156-97-0	240	6	hydrophobic, unpolar	-17	120°C@1 (300°C)		30.3	Xi + N R36/37/38-51/53	
Tetrahydrofurfuryl acrylate	THFA	2399-48-6	156	8	good adhesion	-20	87°C@9 (210°C)	5.3	36.1		
Isobornyl acrylate	IBOA	5888-33-5	208	8	high T_g	80	120°C@15 (240°C)	5.83	31.7	Xi + N R36/37/38-51/53	odour
4-t-Butylcyclohexyl acrylate	TB-CHA	84100-23-2	210	9	high T_g	55	85°C@1.3 (260°C)	8.47		Xi + N R36/37/38-51/53	
2-Phenoxyethyl acrylate	POEA	48145-04-6	192	9	reactivity, adhesion	7	84°C@0.2 (290°C)	4.12	39.2	Xn + N R21/36/38-43-51/53	yellowing
Isodecyl acrylate	IDA	1330-61-6	212	10	hydrophobic	-58	121°C@10 (250°C)		28.6		
Trimethylolpropan-formal-mono-acrylate	TMPF-MA	66492-51-1	200	40	low odour	40			33.1	Xi + N R36/38-43-51/53	

Table 5.2: Stenomeric monoacrylates

[1] T_g data from product brochure "AGISYN" and Ullmann Encycl. Ind. Chem., Vol. 28, Polyacrylates, p. 14; [2] Data from product brochure Sartomer; [3] no responsibility is taken for labeling data; material safety data sheets of the manufacturers are decisive. The labeling will be transferred into the globally harmonized safety system (GHS). Proposals for the new labeling can be accessed by the GHS converter of BG-Chemie (www.gischem.de).

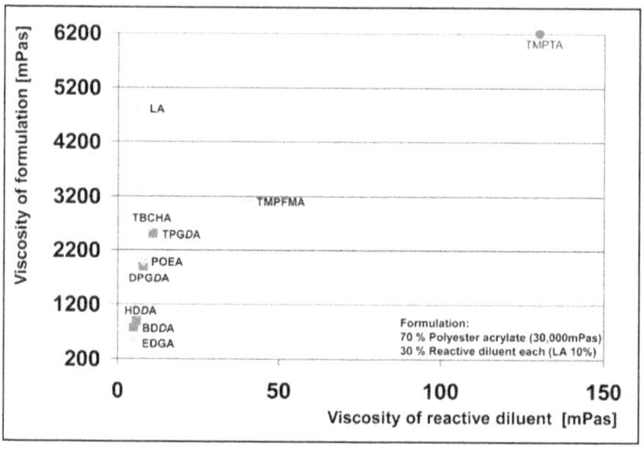

Figure 5.11: Dilution behaviour of stenomeric acrylates
Resin: "Laromer" PE 56Fwith 30% reactive diluent each (10% laury acrylate, LA). Monofunctional acrylates (rhomb), difunctional (square) and trifunctional (circle); data from brochure High lights! Basf.com/resins

in a representative resin (polyester acrylate "Laromer" PE 56F (BASF SE)) is shown in Figure 5.11.

The data of Figure 5.11 implies the general rule that the higher the viscosity of the diluent acrylate, the weaker the dilution power. Even in the low viscosity range, up to 10mPas, significant differences in the dilution effect can still be recognized. The two short-chain hydrocarbon diacrylates, HDDA and BDDA, and the ethyl diglycol monoacrylate exhibit excellent dilution power. Since DPGDA and TPGDA are highly reactive and exhibit good compatibility with most of the resins and still offer very good thinning behaviour, they are the standard diluents used in the wood finishing industry. The higher viscosity TMPFMA and TMPTA show only ordinary to moderate thinning behaviour. The thinning behaviour of lauryl acrylate (LA) is shown in figure 5.11, however, cannot be compared directly, since the 30% content led to incompatibility and therefore only 10% LA was used.

In addition to the dilution behaviour another important factor influencing the selection criteria of diluents is the reactivity. The influence of the chemical structure of the monoacrylates on the rate of homopolymerization of these monomers was studied in a systematical way by *Jansen* [15]. Jansen et al. found that two factors significantly influence the polymerization rate, see Figure 5.12:

- the **dipole momentum** of the acrylates and
- the possibility to form **hydrogen bondings**

The polar monomers with dipole moments higher than 3.5 Debye show a clear correlation of the polymerization rate of monomers with the dipole moment. However, the correlation is not that distinct in the case of monomers with dipole moments below 3.5. This scattering of such data can probably be explained with other overlapping effects, such as, T_g and/or molecular packing effects due to van-der-Waals interactions. The influence of the dipole moment on the polymerization rate has been explained with an induced increase of radical charge density within a monomer or solvent cage. By the formation of such cages, particularly the termination constant is significantly reduced. Furthermore, a pre-orientation of the monomers by the formation of hydrogen bonds results in an increase of the polymerization rate. Such an orientation of the acrylate groups should also lead to an increased fraction of isotactic polymer, which had been verified. Therefore, the reaction rate is determined by these two effects, the pre-orientation of the double bonds by the formation of hydrogen bridges and the resulting dipole moment of the acrylate monomers used. The two effects are complementary to each other, but up to now, no monomers are available that exhibit both, a very high dipole moment and the ability to form hydrogen bonds. This model can also explain the high reaction rates of the monomers studied by *Decker*[16] in the 1990's, including the monomers OXAMA and GCMA. In order to explain the high reaction rates Decker discussed the involvement of hydrogen abstraction reactions, which may additionally play a role and may be supported by the formation of the monomer cage.

The presented homopolymerization reaction rates give only general indications of the reactivity, as in almost all radiation curable formulations, the monomers are usually mixed with multi-functional resins and the reactivity of these formulations will also depend on other factors (dilution behaviour, glass transition temperature, Trommsdorff gel effect, etc.). Thus, for example, a monoacrylate per se and especially one with excellent thinning power, like EDGA, can result in very high reactivity of such a coating (Figure 5.13).

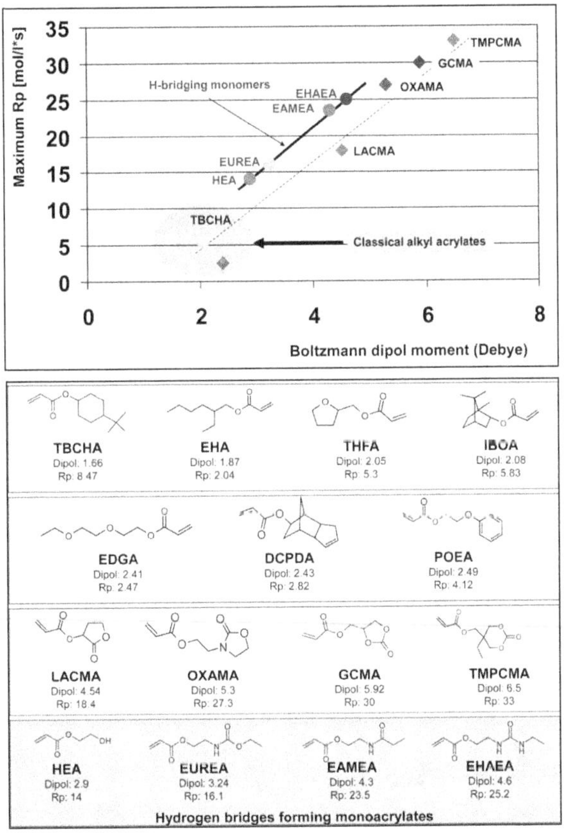

Figure 5.12: Maximum polymerization rate as a function of calculated Boltzmann dipole moment of standard monoacrylates, polar monoacrylates and monoacrylates, capable of hydrogen bond formation[15]

In addition to the previously discussed monoacrylates a number of other monoacrylates had been synthesized and introduced in the recent years. Examples are

- n-Butyl acryloyloxy ethyl carbamate (BAEC) [17]; 25mPas; Cytec Corp.
- Octyl/decyl acrylate (ODA); ca. 10mPas; AgiSyn
- Polyethylene- or polypropylene glycol mono(meth)acrylate ("Bisomer"); Cognis
- Acryloyl morpholine (ACMO); 12mPas, (non irritant), Xn; T_g ca. 88°C; as an alternative to NVP (Rahn, Mitsubishi Int. Corp)
- Glycidyl-methacrylate ("Blemmer G") and other methacrylates; NOF America Corp.
- Phenyl-acetoxy acrylate methylester and others [18],
- Dioxolan acrylate, ketales of glycerol (DOL Series) [19]: Osaka;

Additionally, there are a few hydroxyacrylates available, such as hydroxyethyl-, hydroxypropyl- and hydroxybutyl acrylate, which, however, almost exclusively are used as reaction components in the synthesis of resins, for example, for the manufacturing of urethane acrylates.

5.4.2.2 Multifunctional acrylates

The stenomeric **multifunctional** acrylate reactive diluents play a much larger role in the radiation curable market than monoacrylates. The most common ones are described in Table 5.3. They are used mainly to reduce the viscosity of the radiation-curable formulations; however, they also contribute significantly to the property profile, such as degree of crosslinking, hardness, flexibility or resistance to chemicals.

The low viscosity difunctional acrylates exhibit the best diluting properties. Due to the overall best property spectrum tripropylenglycol (TPG) diacrylate is the most widely used reactive diluent, followed by DPGDA and HDDA, while BDDA is used very rarely. Since skin irritation decreases with alkoxylation degree, TPGDA and DPGDA are the standard reactive diluents mainly in interior applications. However, ether groups are not stable to oxidation reactions and thus, the diacrylates with pure hydrocarbon backbone (HDDA) are the clear favourites for exterior applications [20].

The higher functional acrylates, with three or more acrylate groups, have a significantly decreased thinning power and are mainly used when a higher hardness or reactivity is to be achieved. Main applications for these high functional reactive diluents are therefore wood coatings (hardness), overprint varnishes and inks (curing speed).

The reactivity of the formulations as shown in Figure 5.13 (above) is an empirical value and cannot be correlated directly with the polymerization rate of the components, since the criterion for the reactivity is the obtained hardness of the coating. The reactivity is determined by passing the coating under a lamp or combina-

Table 5.3: Property spectrum of stenomeric multifunctional acrylates (MFA) used in radiation curing formulations

Chemical ident.	Abbr.	CAS Nr.	Mol [g/mol]	η [mPas]; ca.	Perform. contr. positive	Dilut. power η[1] [mPas]	Hardness pendulum [s]	Flexibility Erichsen [mm]	Labelling [> 20%][2]	Disadvantage
Butandiol diacrylate	BDDA	1070-70-8	200	5	crosslink (X-link) density	760	47	4.4	C R 21 /34 - 43	irritant
Hexandiol diacrylate	HDDA	13048-33-4	230	6	weathering stability, adhesion	880	40	4.7	Xi R36/38 - 43	irritant
Dipropylenglykol diacrylate	DPGDA	57472-68-1	240	8	standard reactive diluent	1900	43	5.4	Xi R38-41-43	weathering stability
Tripropylenglykol diacrylate	TPGDA	42978-66-5	300	11	standard reactive diluent	2500	30	6	Xi + N R36/37/38-43-51/53	weathering stability
Trimethylolpropan triacrylate	TMPTA	15625-89-5	300	110	reactivity	6200	53	2.8	Xi R36/38 - 43	
Pentaerythrit triacrylate	PETIA	3524-68-3	298	1800	reactivity, hardness	12200	113	2.2	Xi R36/38 - 43	
Pentaerythrit tetraacrylate	PT_4A	4986-89-4	352	500	reactivity, X-link density, scratch resistance	10600	113	2.2	Xi R36/38 - 43	
Di-trimethylolpropan tetraacrylate	$DTMPT_4A$	94108-97-1	467	600	reactivity, X-link density	9900	101	3.7		
Di-pentaerythrit hexaacrylate	DPHA	29570-58-9	578	7000	reactivity, X-link density, scratch resistance	17900	111	2.4	Xi R 36/37/38	

[1] Viscosity a formulation containing 70 parts “Laromer” PE 56F (resin viscosity 26Pas), 30parts MFA und 4 parts “Irgacure” 500; coatings based on these formulations resulted in the shown hardness and flexibility data

[2] no responsibility is taken for labeling data; material safety data sheets of the manufacturers are decisive. The labeling will be transferred into the globally harmonized safety system (GHS). Proposals for the new labeling can be accessed by the GHS converter of BG-Chemie (www.gischem.de).

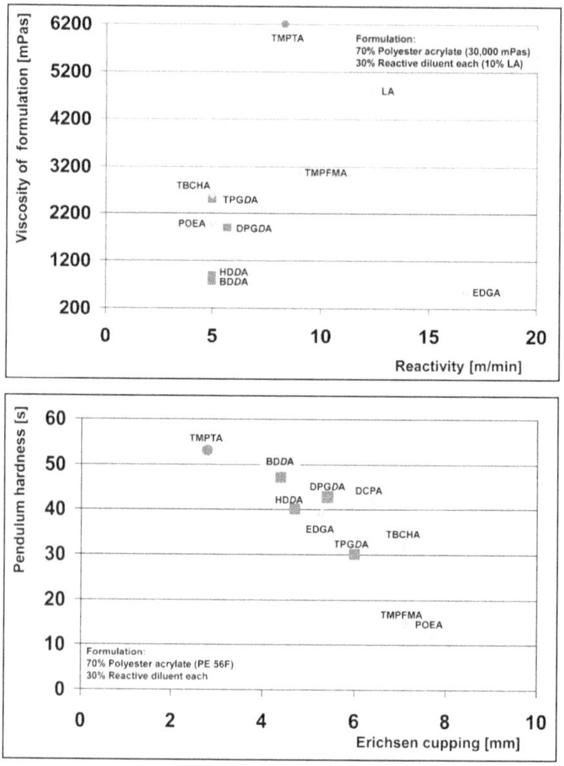

Figure 5.13 above: Effect of functionality of reactive diluents on the viscosity (of formulation) and reactivity (curing speed) and bottom: Effect of functionality of reactive diluents on the hardness (pendulum hardness) and flexibility (Erichsen cupping)

tion of lamps at certain speeds. The necessary speed to obtain a specified hardness of the coating layer (e.g., determined by a finger nail hardness) is given as the reactivity (m/min). Thus, it may be, that in such formulations a mono-functional acrylate, such as EDGA exhibits a significantly higher "reactivity" than a trifunctional (TMPTA), because the double bond conversion of the resin (e.g. Trommsdorff effect) may be much higher in this case. Thus, this viscosity-reactivity relationship is very much dependent on the formulation composition, however, often higher functional reactive diluents have higher viscosity and higher reactivity.

Provided that the curing is appropriate, it is generally observed, that with increasing functionality the hardness (e.g. pendulum hardness) of the coating increases and on the other hand the flexibility, shown as Erichsen indentation (expressed as the indentation depth of a ball-indenter, which impacts from the backside on a painted plate until the coating film cracks) decreases (see Figure 5.13 bottom).

In addition to the here shown polyfunctional acrylates, there are a number of other acrylates available in the market, which, however, are of minor importance. For example, for exterior applications, also other long-chain hydrocarbon di-acrylates are suitable, like decandiol diacrylate (DDDA) or cyclohexanedimethanol diacrylate. For good adhesion and scratch resistance sometimes the triacrylate of tris-(hydroxyethyl) isocyanurate (THEICTA) is used. In the past, also neopentylglycol diacrylate was employed, however, because of its toxicity (T labeling), it is no longer used and often replaced by ethoxylated or propoxylated NPG-diacrylates.

5.4.3 Acrylate functional resins (eurymeric acrylates)

5.4.3.1 Acrylate functionalized standard resins

The acrylate resins are also referred to as acrylic ester prepolymers or oligomers. They often exhibit higher molecular weights with a broader molecular weight distribution compared to the stenomeric acrylates and cannot be described by a simple structural formula. The exact composition is hardly ever listed in the brochures of the different manufacturers, but usually is assigned to a resin class, which is based on the chemical structure of the polymer backbone. There are four main resin classes of standard eurymeric acrylates available:

- Epoxy acrylates
- Polyether acrylates
- Polyester acrylates and
- Polyurethane acrylates

There are also a number of other specialty acrylate resins available, which are described later.

The property spectrum of the standard resins shown in Figure 5.14 describes not a structure-property relationship, but is intended to illustrate only the properties of the typical (most used) resins. The predominantly used aromatic epoxy acrylates feature high hardness and chemical resistance, and on the other hand aliphatic urethane acrylates are especially used when the coating should exhibit high flexibility and high abrasion resistance. Polyether acrylates are employed in particular because of the low viscosity as a resin or reactive diluent, while polyester acrylates

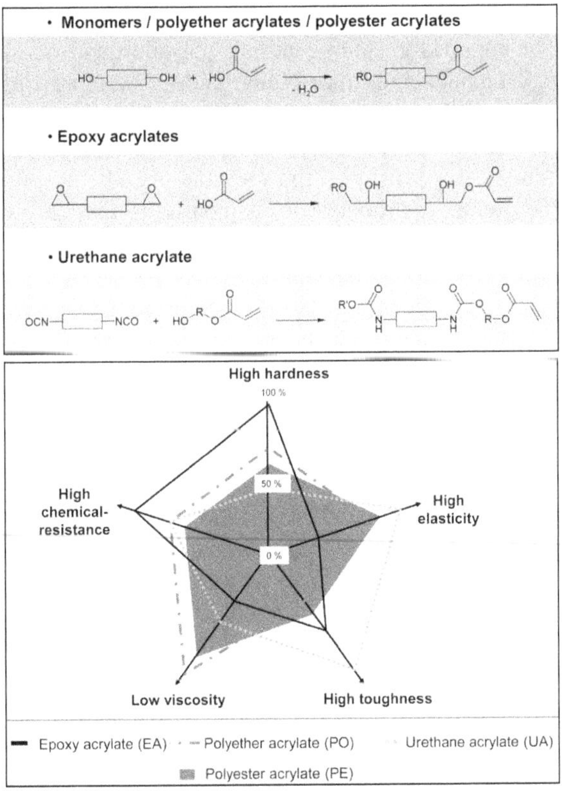

Figure 5.14: Synthetic scheme (above) and typical property spectrum of standard eurymeric acrylate resins (bottom)

predominantly offer a balanced range of properties and are often used in combination with other resins or reactive diluents.

In general, these resins, with the exception of the polyether acrylates, posses higher viscosities and need the addition of reactive diluents for processing as a solvent-free (100%) formulation. The dilution has to reach a level of about 3000 to 5000mPas for roller applications up to less than 500mPas for spray applications.

Polyether acrylates (alkoxylated acrylates)

By reacting alcohols or polyols with ethylene (EO) or propylene oxide (PO) the so-called alkoxylated alcohols or polyetherols are obtained, which are in a second step, acrylated by a classical esterification reaction with acrylic acid. This alkoxylation increases the molecular weight of the acrylates, and results in a desired reduction of their volatility, odour and skin irritation potential of the monomers. The alkoxylation is in general a polyaddition reaction which results in statistical molecular weight distribution and is therefore classified under the eurymeric acrylates. However, they are still relatively low in viscosity and therefore often used instead or in combination with the traditional reactive diluents in order to adjust the viscosity of the formulation. The advantage of propoxylation lies in the greater hydrophobicity compared to the ethoxylated products. Thus, when using the hydrophilic glycerol as a starting alcohol preferably propylene oxide is used in order to impart more hydrophobicity, and when trimethylolpropane is the starting alcohol ethoxylation is preferred. For the synthesis of the acrylate esters, the ethoxylated products are preferred, since they contain primary OH groups; which are esterified faster than the secondary OH groups, resulting from the propoxylation reaction.

The main products of polyether acrylates in the market have an average of one alkoxy group per OH group of the starting alcohol, for instance glycerol-3PO- or trimethylolpropane-3EO-triacrylate. The main advantages of the polyether acrylates are certainly the low viscosity and good dilution behaviour, but many of these products are also used as sole resins and thereby offer high reactivity and good chemical resistance. With increasing degree of alkoxylation, resin properties develop to a greater extent and result in more flexible coating characteristics, mainly due to the decreased crosslink density. With higher degrees of ethoxylation, the products are getting more hydrophilic and increasingly water soluble. Polyethylene glycol 400 diacrylate (PEG 400 DA) and glycerol or trimethylolpropane with up to 20 ethylene oxide units are water dispersible or even water soluble and are used in waterborne UV coatings and as crosslinkers in the polymerization of acrylic acid for super-absorbent polymer gels.

The synthesis of the polyether acrylates is analogous to that of the stenomeric acrylates, and is carried out almost exclusively by a catalyzed azeotropic esterification with or without the aid of an entraining agent (preferably toluene).

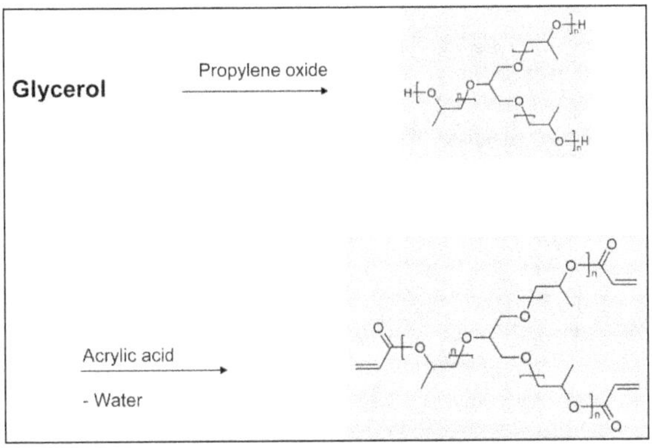

Figure 5.15: Scheme of the synthesis of a polyether acrylate

In addition to the above mentioned products based on glycerol and TMP, there are still a large number of other alkoxylated acrylates on the market, for instance the aforementioned ethoxylated and propoxylated neopentyl glycol diacrylate, which are used to provide good wetting and adhesion properties to many substrates. Furthermore, ethoxylated and propoxylated bisphenol A diacrylate need to be mentioned, which offer a good balance of hydrophobicity and hydrophilicity.

Polyester acrylates (PE)

Polyester polyols are the reaction products of di-or polycarboxylic acids with excess of di-or polyols. These polyols are well represented in the market and are mainly used for producing polyurethane foams. By esterification of the terminal OH groups with acrylic acid the described resin class of polyester acrylates results. The main components used are, on the one hand, the diols, polyols and alkoxylated alcohols or polyols, which had already been described as starting alcohols for the synthesis of stenomeric acrylates or polyether acrylates. On the other hand, the main acid component are either the polycarboxylic acids or their anhydrides, such as malonic acid, succinic acid, adipic acid, maleic acid or maleic anhydride, phthalic anhydride, pyromellitic anhydride, etc.

Polyester acrylates with functionalities from 2 to 6 are available in the product portfolio of the different manufacturers.

The synthesis of the polyester acrylates is also predominantly carried out by azeotropic esterification of the polyesterols with acrylic acid. Generally, these products exhibit broader molecular weight distributions, since the polyesterols already exhibit a molecular weight distribution and additionally, few transes-

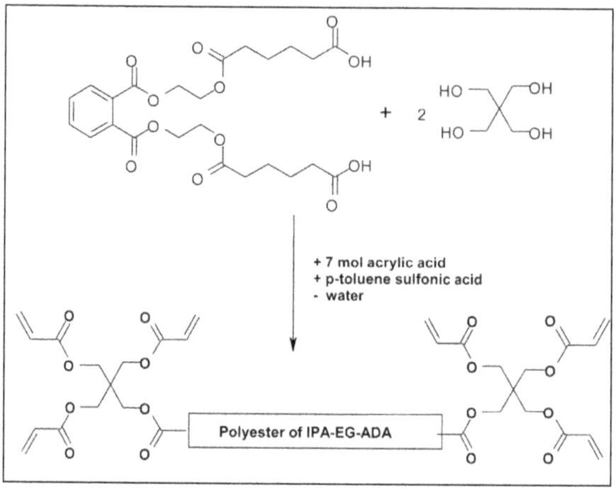

Figure 5.16: Synthetic scheme of a multi-functional polyester acrylate

terification reactions may occur during the esterification with acrylic acid. Furthermore, the purification of the final product by removing the excess acrylic acid with caustic soda is often more difficult due to higher molecular weights, higher viscosity and reduced solubility in the entrainer solvent. Thus, the phase separation is hindered and as a consequence, yield losses must be accepted.

Therefore, alternative methods, without the use of entrainers [21], alternative purification methods of the crude acrylate by consuming the excess acrylic acid with polyepoxides [22], or precipitation of the acrylic acid with bases and subsequent filtration of the precipitated salt [23], have been developed.

A procedure for the synthesis of a multi-functional polyester acrylate is described below as an example and the chemical reaction scheme is shown in Figure 5.16.

Synthesis example: synthesis of a multi-functional polyester acrylate (analog to [24])

In a double-walled glass reaction vessel (3 liters), equipped with stirrer, thermometer, nitrogen inlet, vacuum connection and a column for azeotropic distillation, 1 mol phthalic anhydride, 2 moles of diethylene glycol and 0.15 mol of toluene are heated with stirring in order to remove the reaction water. The reaction temperature was increased during 15 hours from 150°C to 180°C and one mole of water removed. After cooling the mixture to 50°C, 2 mol adipic acid, 2 mol pentaerythritol, 7 mol of acrylic acid (1 mole excess), 7 mol of toluene, 0.2 mol p-toluene sulfonic acid and 0.015 mol of copper (I) oxide were added and again heated up. When a temperature of about 85°C was reached, the elimination of water started. Within 7 hours the temperature increased to 95°C and the elimination of water was nearly complete. The reaction vessel is cooled down, diluted with further toluene (to 25% solids) and purified in successive washes with 20% aqueous NaCl solution. The organic phase is then dried over sodium sulfate and filtered. After the addition of 1000ppm hydroquinone monomethyl ether (MEHQ), the toluene is distilled off under vacuum at a temperature not exceeding 50°C. The resulting product is a viscous polyester acrylate with about 6 acrylate groups per chain and about 5% by weight of diethylene glycol diacrylate as a byproduct. The overall yield was about 94%.

Table 5.4: Portfolio of commercially available Polyester acrylates

Resin	Viscosity [mPas]	Functio-nality	Hard-ness*	Elasticity*	Reactivi-ty*	Chemical* resistance
„Laromer“						
LR 8907	1250	1.9	1	4	2	3
PE 44F	3500	3	1	3	2	3
LR 8800	6000	3	3	2	2	4
PE 56F	30000	3.1	2	3	3	3
PE 9032	20000	4	2	3	3	3
„Ebecryl“						
851	3250	2.5	3	2	2	3
885	34000	3	3	3	3	3
800	14000	4	3	2	2	4
830	50000	6	4	1	3	4

1 = low; 2 = medium; 3 = good; 4 = exellent

Based on the broad availability of polyester polyols, the commercially available polyester acrylates cover a wide range of properties. Thus, the spectrum ranges from low to high viscosities, functionalities from about 2 to 6, from low hardness and high elasticity up to high hardness and low elasticity. For the purpose of illustration of the commercially available spectrum, some products from the portfolio of BASF SE (“Laromer”) and Cytec (“Ebecryl”) are summarized in Table 5.4.

As shown already in Figure 5.14, the polyester acrylates often offer a balanced range of properties, such as good reactivity and chemical resistance at a sufficiently high hardness and high flexibility.

Epoxy acrylates (EA)

The naming of the product class as epoxy acrylates is a bit confusing, as the final products do not contain the epoxy function anymore. Polyfunctional epoxies are used merely as precursors for the synthesis of acrylic acid esters. The addition of acrylic acid to polyepoxides is synthetically straight forward and leads to β-hydroxyacrylic acid esters. This addition reaction is one of the historically longest known reactions for the production of acrylates for radiation curable systems and is also used to remove excess acrylic acid in the classical esterification

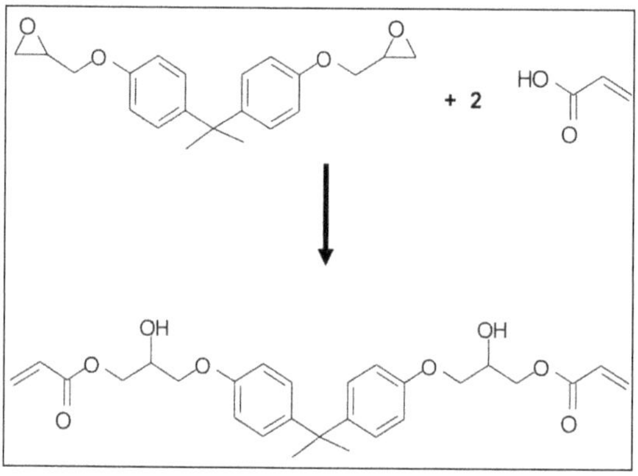

Figure 5.17: Synthetic scheme of an epoxy acryl ate

reaction in the synthesis of polyester acrylates. The availability of low cost glycidyl ethers of bisphenol A and novolacs is the reason why this class of products is still the most represented in the market of radiation-curable resins.

Synthesis example: synthesis of an epoxy acrylate (analog to [25]):

A reaction mixture of 1 mole of bisphenol-A diglycidyl ether, 2 mol of acrylic acid, 1% triethylamine and 2.5 millimoles of hydroquinone was heated to 90°C and held for 2 hours at 90°C and subsequently for about 4 hours at 100°C until the acid number fell below 5. The resulting viscous epoxy acrylate can be diluted with stenomeric acrylates to the desired viscosity.

The aromatic structure and the formed secondary hydroxyl groups are the main contributors to the very high viscosity of the predominantly used aromatic epoxy acrylates. In order to handle them during transportation they are offered almost exclusively in blends with reactive diluents (TPGDA, HDDA, GP3TA) or in solvents. These mixtures have a good cost/performance ratio by providing coating properties with high hardness and reactivity as well as good chemical resistance. Because of the aromatic character, they easily become yellow upon solar irradiation and are therefore less suitable for outdoor applications. Aliphatic epoxy acrylates, however, are resistant to yellowing, but are softer and more flexible. However, there are only a few aliphatic epoxy acrylates in the market. The most common raw materials are the bis-glycidyl ethers of butanediol, cyclohexanedimethanol and neopentyl glycol and the 3,4-epoxycyclohexylmethyl-3,4-epoxycyclohexane-carboxylate and some epoxidized unsaturated oils (e.g., epoxidized soybean oil). The short-chain aliphatic epoxy acrylates on the basis of butanediol or ethylene glycol diglycidyl ether are partially soluble in water and thus may be diluted with water. The acrylated epoxidized oils have good pigment wetting and are therefore often used in printing inks.

Furthermore, some specialty epoxy acrylates are described in the literature. By reacting glycidol with isocyanate compounds (e.g., biuret of hexamethylene diisocyanate) and subsequent reaction of the resulting trifunctional epoxy resins with acrylic acid an epoxy acrylate is formed, which should combine the properties of epoxy acrylates and urethane acrylates, such as good flexibility, good adhesion, good solvent resistance and high thermal stability [26].

Urethane acrylates (UA)

The resins of the class of urethane acrylates are obtained by the addition reaction of hydroxyalkyl acrylates to isocyanate terminated prepolymers or classical di- or polyisocyanates [27]. The range of available di- or polyisocyanate components is relatively small. For the coatings industry aromatic diisocyanates are rarely used, and if, almost exclusively, toluene diisocyanate (TDI) and hardly ever the semi-aromatic tetramethyl xylylene diisocyanate (TMXDI). Predominantly used are aliphatic di- or polyisocyanates based on hexamethylene diisocyanate (HDI) and isophorone diisocyanate (IPDI) and seldom used the hydrogenated methylene diphenyl diisocyanate (H_{12} MDI). The diisocyanates are often used in the synthesis of isocyanate terminated prepolymers. In order to reduce the volatility and increase the functionality, these diisocyanates are oligomerized to isocyanurates, biuretes or allophanates. Thus, the formally trifunctional isocyanurates of HDI or IPDI are main products in the market (Figure 5.18).

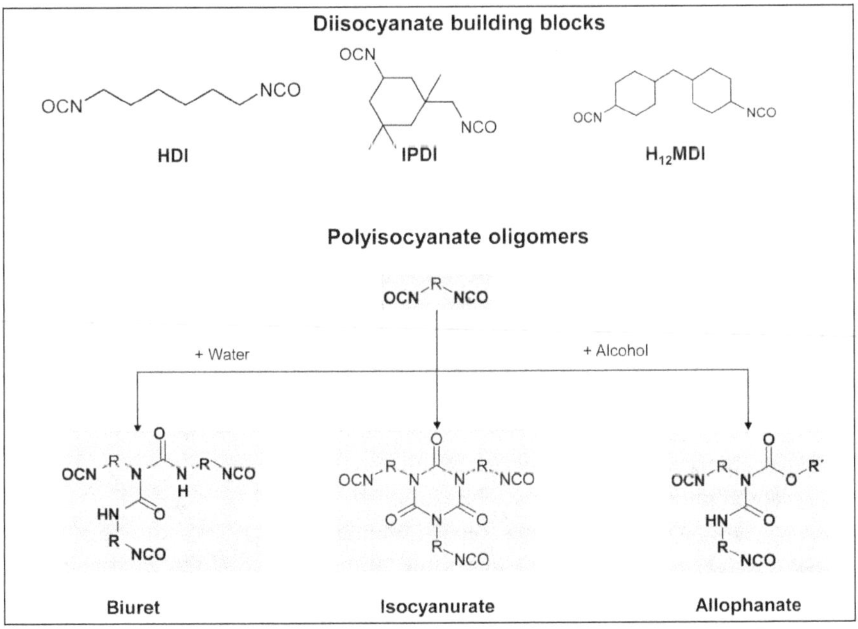

Figure 5.18: Polyisocyanate building blocks for the syntheses of urethane acrylates

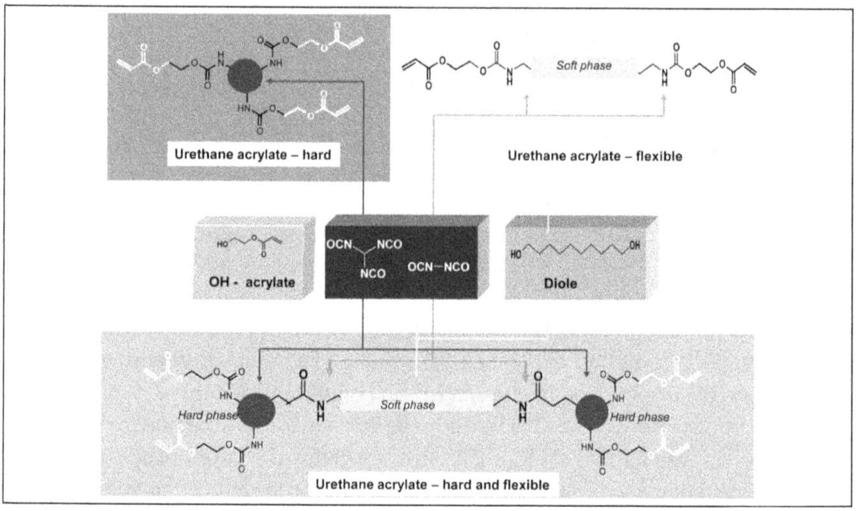

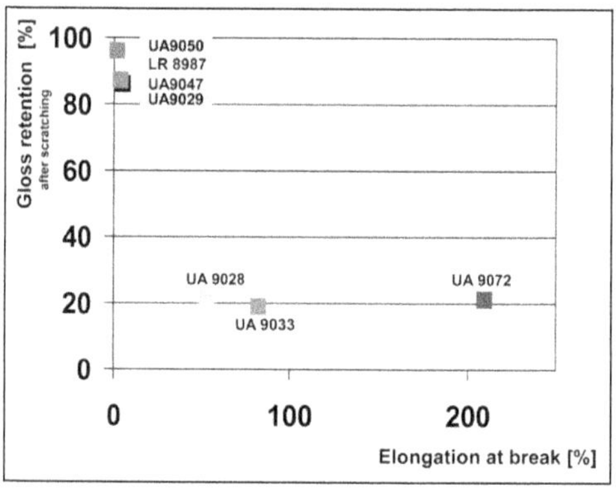

Figure 5.19 above: Schematic structures of hard and soft urethane acrylates and urethane acrylates containing both, hard and soft phases).
bottom: Property spectrum of "Laromer"-urethane acrylates in regard to scratch resistance and elongation at break
*(resin + 4% "Irgacure" 184; exposure under nitrogen at 40°C@1400mJ/cm²;stress-strain analysis: 5*2cm stripes; scratch test: gloss measurement @(20°)before and after scratching with 10 double rubs with 750g load on "ScotchBrite" fleece)*

From the building blocks of hydroxyalkyl acrylates, di-or polyisocyanates and polyols a wide range of urethane acrylate resins can be synthesized offering a broad spectrum of properties. Reacting the short chain hydroxyalkyl acrylates, like hydroxyethyl acrylate (HEA), hydroxypropyl (HPA) or hydroxybutyl acrylate (HBA) with the trifunctional isocyanurate of HDI yields urethane acrylate resins featuring hard coatings. Flexible urethane acrylate coatings will result from the termination of long chain diols with diisocyanates and subsequently adding hydroxyalkyl acrylates. By combining trifunctional polyisocyanates with long-chain diols and hydroxyalkyl acrylates, hard and soft phases in the urethane acrylate resin can be designed which should provide high flexibility and high hardness (Figure 5.19, above).

The range of the property spectrum in regard to scratch resistance and elongation at break of urethane acrylates is exemplified in Figure 5.19 (bottom).

The scratch-resistant resins UA 9050 and LR 8987 (high gloss retention after scratching with 10 double rubs of a loaded "ScotchBrite"-fleece) are relatively brittle, while on the other hand, the highly elastic resins (UA 9072 and UA 9033) are not very scratch resistant.

Despite the broad variety of synthetic design possibilities, even with urethane acrylates, not all desired properties can be optimized at the same time. Taking into account the adjustable parameters, molecular weight, glass transition temperature and crosslink density, in order to get the optimal property of one parameter results in another parameter not being optimal. Thus, choosing a high molecular weight will result in a low shrinkage values, but high viscosities. A high glass transition temperature provides high hardness, but low flexibility and high crosslinking density offers high hardness and scratch resistance, but also high shrinkage. Therefore, for each application the main requirements have to be considered and the setup of the adjustable parameters optimized in order to provide the best compromise of properties.

The outstanding property spectrum of urethane acrylates is based on the simple synthetic possibility of introducing hard and soft phases into the molecular backbone, which offer a very good hardness to flexibility ratio. Furthermore, the urethane bond provides good stability against oxidation, hydrolysis and photochemical cleavage.

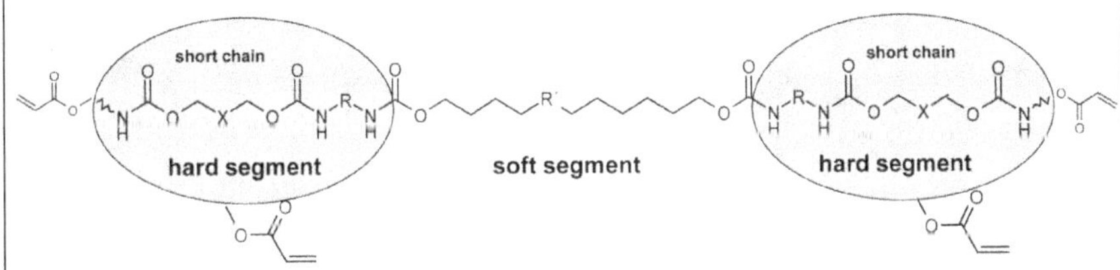

Desired property	Available by		
	Molecular weight	T_g	Crosslink density
Low shrinkage	high	low	low
Low viscosity	low		
High reactivity	high	low	high
High flexibility	high	low	low
High hardness		high	high
High chemical resistance		high	high
High scratch resistance		high	high

Figure 5.20 Principle design of urethane acrylates with hard and soft phases and impact of molecular weight, glass transition temperature (T_g) and crosslink density on the desired properties of the coating

Therefore, aliphatic urethane acrylates are the resins of choice for outdoor applications. The disadvantage of this resin class is the high price due to the use of aliphatic polyisocyanates, and the high viscosity due to hydrogen bonding interactions. The lowest viscosities are generally obtained with allophanate types, and acrylated allophanates are also available. One example is based on the isocyanato acrylate "Laromer" 9000 [46] or analogous allophanatized structures (see Chapter 5.4.3.2, Isocyanato acrylate). Since the allophanates are forming "intramolecular" hydrogen bonds, the increase in viscosity due to hydrogen bonds is considerably lower compared with reaction products of biurets or isocyanurates with hydroxyalkyl acrylates.

5.4.3.2 Acrylate functionalized specialty resins

Amino functionalized acrylates

Since acrylate polymerization in air atmosphere suffers greatly from oxygen inhibition, often amines are employed to accelerate the reaction (see for the mechanism

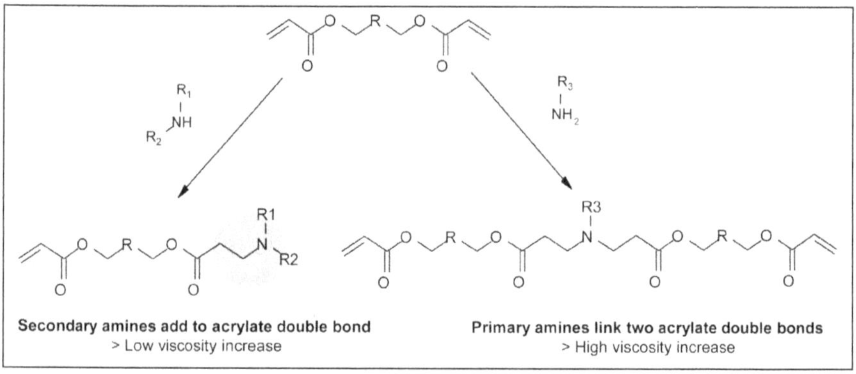

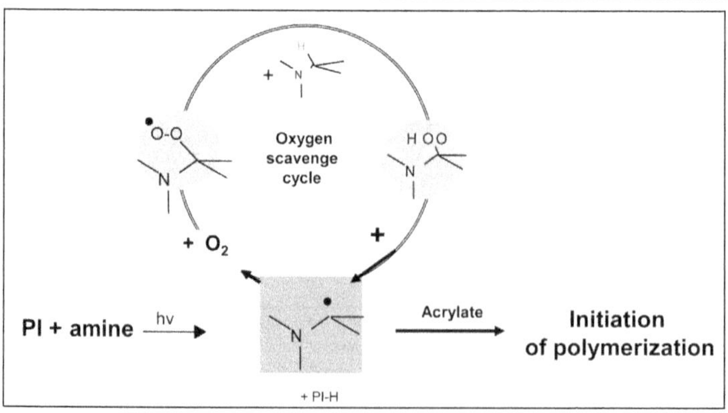

Figure 5.21 above: Scheme of amine modification of acrylate resins bottom: Mechanism of amine action to reduce the effect of oxygen inhibition

Figure 5.21 bottom). Low molecular weight added amines (e.g. methyldiethanolamine, triethanolamine, dibutyl ethanolamine) have serious drawbacks, since they may migrate and evaporate and therefore contribute to odour. Therefore, the Michael addition reaction has been utilized in order to tie the amines to the acrylate resins. Secondary amines are used preferably, since they react with a terminal acrylate group to form a tertiary amine, whereas primary amines form a secondary amine in the first step and may add to a second acrylate group resulting in a molecular weight increase, often accompanied by undesirable viscosity increase. The migration behaviour and volatility of such amine synergists as well as their ease of extractability is reduced significantly.

The range of products offered in the market covers both, almost pure synergists in which the acrylate groups are almost completely converted with amines (e.g. "Ebecryl" P115, "Laromer" LR 8956), and preferably polyether acrylates with relatively low amine levels, corresponding to amine values <50 mg KOH/g (Figure 5.21 above). These products offer relatively high reactivity under ambient air conditions. Thus, they are employed as very effective crosslinking resins preferably in the graphics industry.

Despite their beneficial action to overcome oxygen inhibition, unfortunately, the amine-modified acrylates posses some disadvantages. The most prominent is the yellowing behaviour, especially under thermal stress. Furthermore, in matt coating formulations, the matting effect is often generated by acidic matting agent, which may cause flocculation problems with the amine synergists.

Acrylated polyacrylates (APA)

Acrylated polyacrylates may be synthesized by different methods. Common methods rely on synthesizing first a polyacrylate having pendant hydroxyl or epoxy groups, which are then modified in a second step by transesterification with methyl acrylate or addition of acrylic acid to the epoxide, respectively [29]. Furthermore, copolymers containing acrylic acid may be used as starting materials, which are subsequently functionalized with glycidyl methacrylate [30] (Figure 5.22). Since glycidyl methacrylate is a fairly expensive monomer, these resins are used only in niche applications, such as adhesion promoters and some outdoor applications [31]. .

In general, the acrylated polyacrylates are up to now rarely employed, but if, mainly in the area of pressure sensitive adhesives (PSA) [32]. In order to be used in such applications the resins have to be liquid at least at the application temperatures (up to 120°C) and therefore have to be relatively low in molecular weight. Furthermore, solid acrylated polyacrylate resins have also been discussed in the applications of UV curable powder coatings (see Chapter 5.6.6.1; [112])

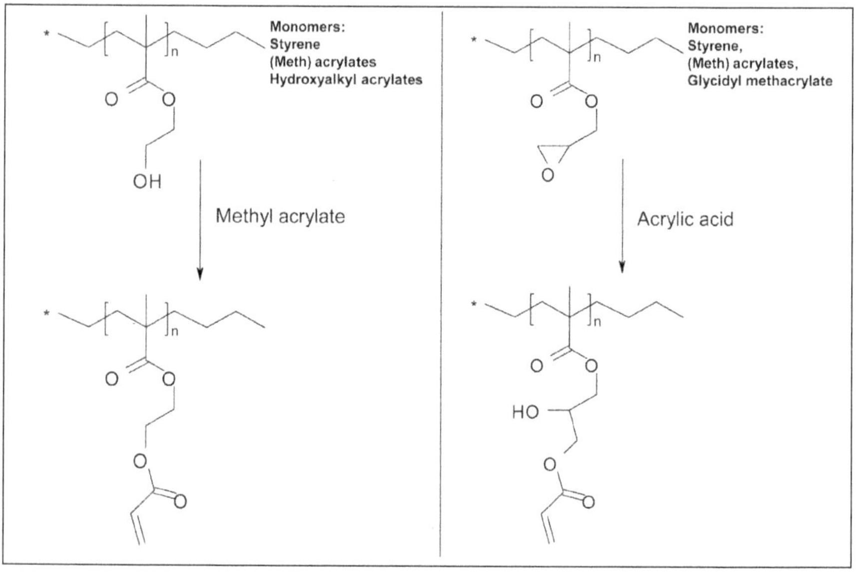

Figure 5.22: Syntheses of acrylated polyacrylates

Dendrimers and hyperbranched acrylates

Dendrimers or hyperbranched acrylates as resins for radiation curable coatings are associated with a number of unusual combinations of properties:

- High functionality
- Fast curing speed
- Low viscosity
- Low polymerization shrinkage
- Excellent hardness-elasticity ratio

The dendritic molecules are characterized by their perfect structure, starting from a core molecule and adding perfectly symmetric highly branched structures by exactly controlling the stochiometry, bearing functional groups only on the surface (from the Greek "dendron", the tree). The hyperbranched structures are built up less perfect, but are significantly easier and cheaper to synthesize, but still offering the key good properties of dendrimers (Figure 5.23).

Synthesis example: synthesis of a dendritic acrylate [33]:

A general synthesis of an acrylated dendritic oligomers based on dimethylol propionic acid (DMPA) relies on trimethylolpropane (10mmol) as the starting

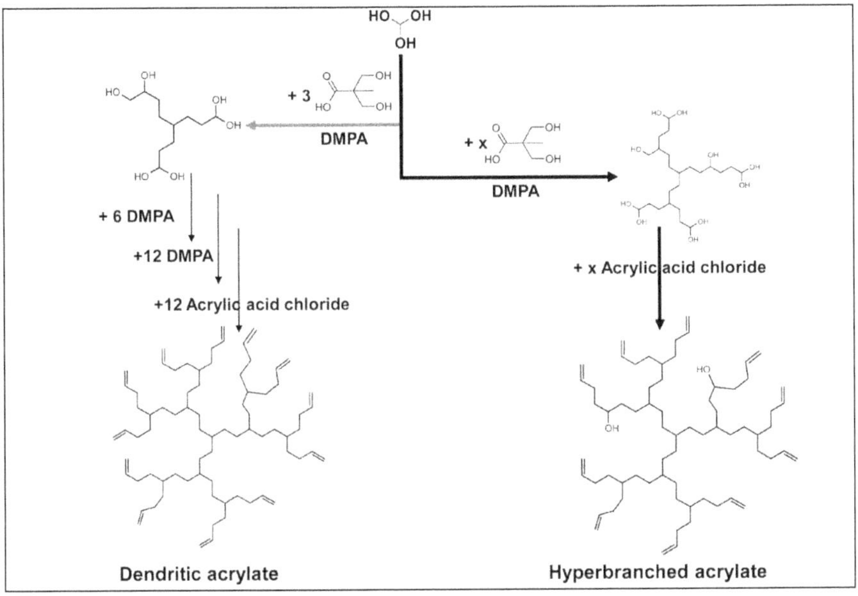

Figure 5.23: Schematic design of dendritic and hyperbranched acrylates

core alcohol, which is esterified with 30mmol dimethylol propionic acid in the presence of 0.1wt% sulfuric acid at 140°C in an argon stream. After about 30 minutes of reaction time a vacuum is applied to remove the formed water and the reaction continues for about 1 hour. This procedure gives a polyol of the first generation containing six OH-groups. To produce the second generation again one mol of DMPA per OH group of the first generation is added together with the corresponding amount of sulfuric acid and esterified in the same manner. Following the same procedure further higher generations can be synthesized. The functionalization with acrylate end groups was done in a classical reaction with acrylic acid chloride in the presence of triethylamine as the base to capture the hydrogen chloride.

Properties of hyperbranched acrylated

The general performance properties of radiation cured coatings containing as the main resins acrylate end functionalized hyperbranched or dendritic structures have been intensively investigated by several groups in recent years [34]. Almost perfect dendritic polyester acrylates based on dimethylol propionic acid have first been introduced by Perstorp under the trade name "Boltorn" [35]. Because of their ease of synthesis, hyperbranched structures can be produced much more cost effective than the perfect dendrimers. Acrylate functionalized hyperbranched structures are nowadays also offered by the industry [36].

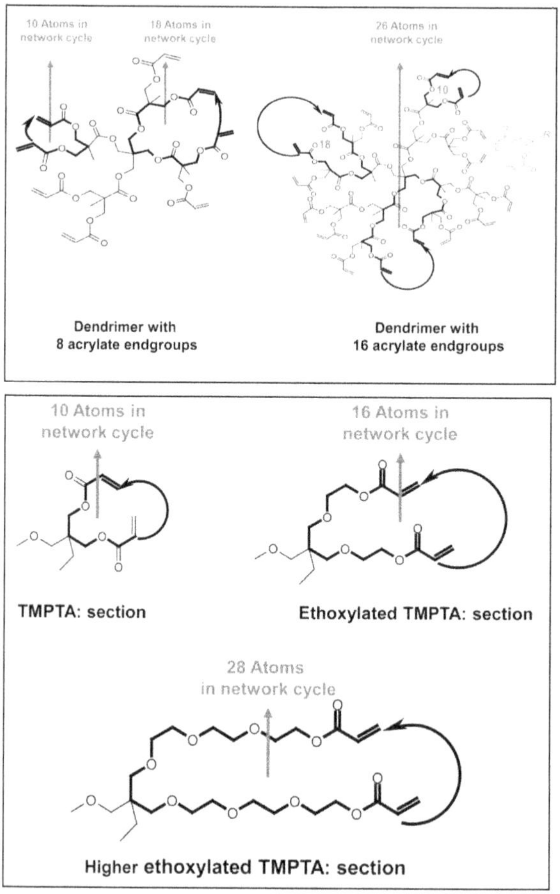

Figure 5.24: Molecular weight between crosslinks in dendritic acrylates (above) and in formulations containing classical acrylate resins and reactive diluents (bottom)

The synthesis of the hyperbranched oligomers also usually relies on a multi-functional core of, for example, ethoxylated pentaerythritol, as a starting material. However, in contrast, the condensation with the branching AB2 monomer (e.g. dimethylol propionic acid) does not rely on the exact stochiometry and is usually done in a one step reaction. From this reaction branched molecules with terminal OH groups result, but the number of OH groups is not perfectly controlled. The OH groups can subsequently be converted partially or fully to acrylate or another functional end group. Both, the core molecule and the AB_2 or A_2B_3 (e.g., reaction product of one mole of IPDI with one mol diethanolamine) branching monomers can be chosen from a variety of available raw material. Thus, it is easily possible to adjust the design of the dendritic or hyperbranched molecules to the property requirements.

The hyperbranched molecules often offer, compared to other high-functionality acrylates, advantages in terms of lower viscosity than their linear analogues. Furthermore, because of their special structure, they may provide often diametrically opposed properties, like high hardness and high flexibility in a molecule. Considering structure-property relationships, however, reveals that the dendritic or hyperbranched molecules intrinsically contain a distribution of molecular weights between crosslinks, which can be also realized by the proper choice of acrylate components in a formulation. By choosing the appropriate acrylate resins and diluents, similar network structures and a similar range of properties can be derived with formulation chemistry compared to the intrinsic molecular structure of hyperbranched acrylates.

Figure 5.24 illustrates schematically the number of atoms between crosslinks in a network structure. The smallest possible network loop is considered when the molecules crosslink intramolecular. If reactions take place with other formulation acrylates (as indicated in Figure 5.24 above) the network chain length will increase accordingly. As a result of these considerations, one can conclude, that the dendritic or hyperbranched structures always exhibit a distribution of molecular weights between crosslinks. As a consequence, a distribution of hard and soft phases is intrinsically present in such a molecule. However, due to the high functionality of these molecules, they are often compared with other highly functional acrylates (e.g., dipentaerythritol hexaacrylate) which provide a more uniform network structure and do not contain hard and soft phases per se.

In any case, hyperbranched acrylate structures in general exhibit lower viscosities compared to linear structures, and another big advantage is the possibility to introduce different functionalities in the same molecule in order to adjust further properties such as adhesion, compatibility, crystallinity and others. Compared to the mentioned classical formulations, often lower volume shrinkage may be achieved. Since, at the same crosslinking density, due to the branching and the higher molecular weight of the molecules, the volume fraction of the polymerizable acrylate groups has to be significantly lower.

Melamine acrylates

In the distant past, acrylated melamine resins have been introduced, which contains both, radical polymerizable acrylate groups as well as N-methylmethylol groups, capable of thermal condensation reactions [37]. They are synthesized by the transetherification reaction of hexa-N-methyl-methylol melamine (HMMM) with hydroxyalkyl acrylates, see Figure 5.25 [38].

Such products are still commercially available, for instance the mono- to pentafunctional melamine acrylates from Bomar Specialties (XMA-220, monofunctional, BMA-222, trifunctional and BMA-224, pentafunctional). The products are relatively low in viscosity (2000 to 5000mPas), have good weathering resistance, high scratch resistance and glass transition temperatures in the range of -20°C to 130°C.

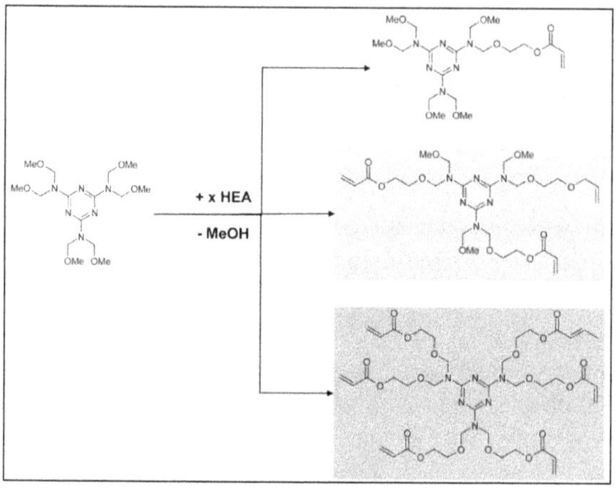

Figure 5.25: Melamine acrylates

Silicone acrylates

The term silicone acrylates, as used here, covers both the mono-functional tris (alkoxy) silyl acrylates, which are employed especially as an adhesion promoter, for example, in UV curable optical fibre coatings, as well as the classical organopolysiloxanes that can be functionalized with acrylate groups, either terminally or in the silicone chains (Figure 5.26). Main applications of the acrylated polysiloxanes are release coatings for pressure sensitive adhesives, as the polysilicones have a low surface tension. Another preferred application of polysiloxanes is coating additives to improve the surface characteristics (wetting, flow, slip) and to prevent foaming. Although silicone additives are usually not functionalized, the acrylate modified polysiloxanes may improve the compatibility with the UV curable formulation and contribute to reduced bleeding.

Acrylated silicones can also be used in Dual Cure applications, by a combined curing via photochemically induced acrylate curing and thermally induced reactions of the silicone units. Such silicone reactions may either be at a moisture

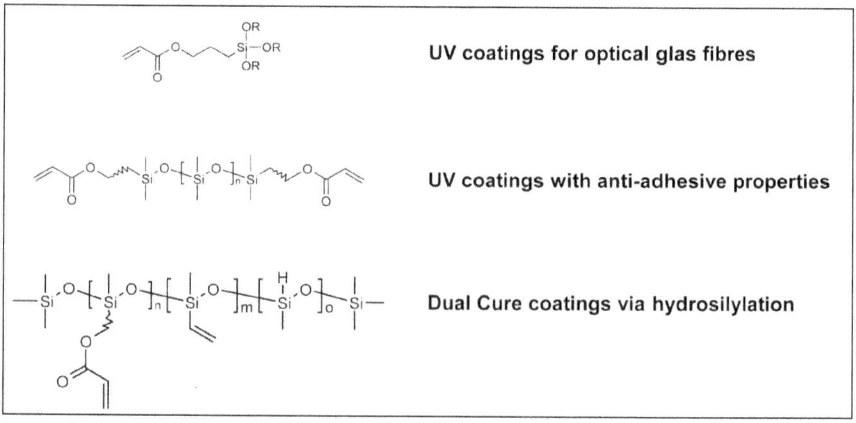

Figure 5.26: Exemplified structures of silicon acrylates

induced condensation of trialkoxy silanes or a platinum catalyzed hydrosilylation of Si-H units with vinyl or acrylate groups [39].

Acrylates for UV powder coatings

Powder coatings are one of the most environmentally friendly coating technologies because they are used as 100% systems without any solvent and the "overspray" can be relatively easily recycled. However, thermal energy is needed to melt the resins in order to form a film and cure it. Thermal curable powder coatings sometimes suffer from the fact, that the melting process and film formation is not completed before the crosslinking reaction has already begun. The results are often a poor quality coating surface, e.g., orange peel appearance.). UV curable powder coatings typically allow a decoupling of flow and crosslinking reaction, since the melting is not affected by the crosslinking. Favourably, the crosslinking can be triggered by exposure to UV light when a good coating flow is achieved. Compared to traditional thermally curable powder coatings, UV powder coatings require significantly reduced drying/curing temperatures (90 to 130°C compared to 130 to 200°C), shorter drying/curing times (1 to 2min vs. > 5min) and thus, significantly lower energy demand [40].

In the literature, a vast number of resin systems for UV curable powder coatings are described, from which the acrylate-based products are described here:

- Unsaturated polyesters in combination with Urethane acrylates [41]
- Acrylated oligomers in combination with crystalline urethane acrylates [42, 43]
- Methacrylate functionalized polyesters [44]

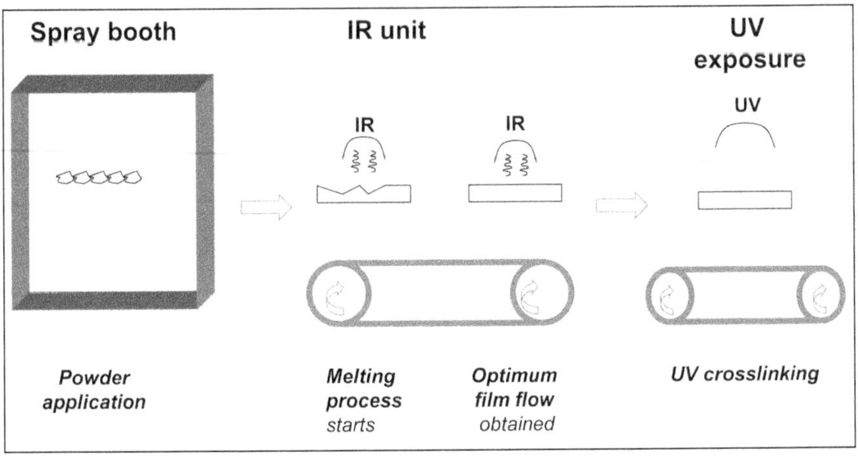

Figure 5.27: Scheme of the UV powder curing process

Unsaturated polyesters based on maleic acid copolymers have been extensively studied in combination with vinylether functionalized urethane acrylates (DSM, NL). Acrylated oligomers were based on dendritic or hyperbranched structures in order to introduce different functionalities in a molecule. For example, hydroxyl functionalized dendritic structures were modified with caprolactone units to introduce crystallization, fully or partially functionalized with acrylate groups. By careful selection of the components, the melting points and glass transition temperatures of the resins can be varied within wide ranges to the desired target values. For example, resins containing short caprolactone chains exhibit glass transition temperatures of minus 56°C, whereas, longer caprolactone units resulted in resins with melting points of about 50°C. These latter, low melting semi-crystalline resins may be used for UV curing powder coating applications at low temperatures [45].

Isocyanato acrylates

Isocyanato acrylates are molecules which contain in the same molecule both a radiation curable acrylate and an isocyanate modality, which can react with various functional groups via thermal induced addition reactions. A well available molecule of this chemical structure is isocyanatoethyl methacrylate. Synthetically easy accessible products are the under-stochiometric adducts of hydroxyalkyl acrylates to polyisocyanates. Whereas, such products are relatively high in viscosity, a recently developed polymeric isocyanato acrylate based on allophanate structures is characterized in particular by low viscosity [46].

These products may be used in different applications and different ways, for example, as a chemical reactive adhesive to functional surfaces capable of reacting with isocyanates. Furthermore, in Dual Cure coating applications where the thermally induced isocyanate reaction precedes independently from the light induced acrylate polymerization, for example, by moisture curing or reacting with other isocyanate-reactive groups, such as

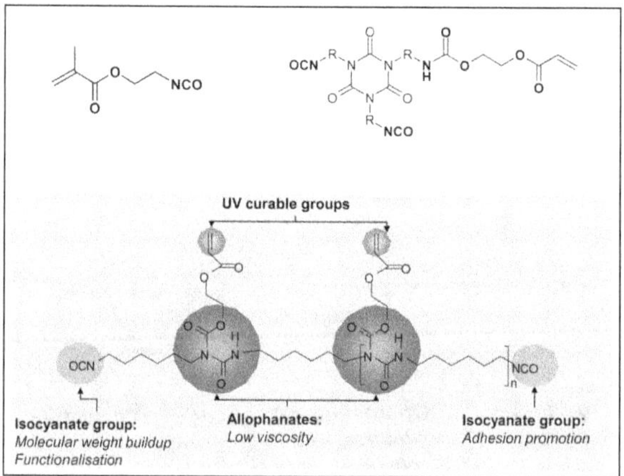

Figure 5.28 Chemical structures of isocyanato acrylates

polyols or polyamines. The isocyanato acrylates may also be used as building blocks for the synthesis of polyurethanes containing unsaturated acrylate groups in the side chain or as terminal groups.

Other specialty acrylates

Fluorinated building blocks are employed in urethane acrylate molecules in order to introduce low surface tension and low refractive indices into the resin (perfluoro-polyether (meth) acrylates [47]). The content of the fluorine components is adjusted in order to meet the desired properties. In high-quality applications (optics, photonics), in which, for instance, the lowest possible refractive index and a high weathering resistance are required, high fluoride levels are incorporated into the resin, while, where modified acrylates are used as additives to reduce the surface tension, only relatively small amounts of fluoride blocks are applied. Commercially available fluorinated methacrylates may exhibit refractive indices as low as 1.345 (MD700 "Fluorolink", Solvay).

Another class of specialty resins is the recently introduced self initiating acrylate oligomers, which are obtained by a Michael analogous reaction of beta-keto esters with polyfunctional acrylates [48]. These special acrylic resins do not need an added photoinitiator for the UV initiated polymerization, since upon irradiation the keto-esters undergo a bond cleavage with elimination of an acetyl radicals and the formation of a polymer radical, which can initiate a subsequent radical polymerization and crosslinking of the acrylates.

$CH_2{=}CH\text{-}CO\text{-}\quad \text{-}O\text{-}CF_2\,(O\text{-}CF_2)_m\text{-}(O\text{-}CF_2\text{-}CF_2)_n\text{-}OCF_2\text{-}O\text{-}\quad \text{-}CO\text{-}CH{=}CH_2$

Perfluor-polyether (PFPE) acrylate

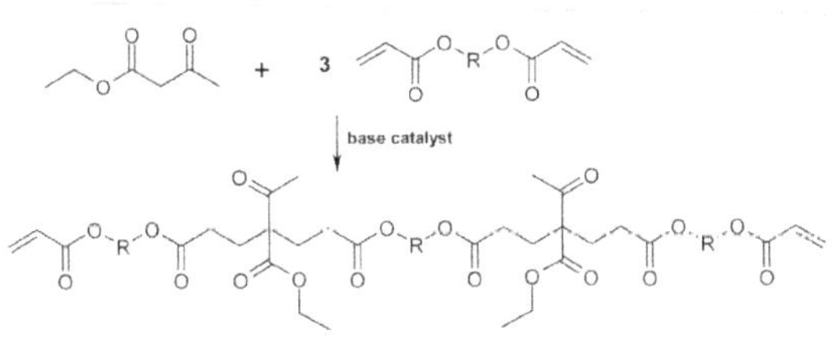

Self initiating acrylate

Figure 5.29: Scheme of the chemical structures of specialty acrylates

5.4.4 UV curable acrylate functionalized dispersions

Although the 100% solids radiation-curable acrylate coatings offer a lot of benefits, in certain applications, they may hit their limits. Thus, for example, in spray application, very low viscosities, in the range of 200mPas, are required. With 100% solid acrylate formulations, such low viscosities can only be obtained with a very high content of very low molecular weight reactive diluents. Such high levels of stenomeric acrylates are often undesirable because these diluents are still very volatile and will evaporate or when applied on porous substrates like wood, may diffuse into the pores and remain uncured and exude during storage or use. A different reason for using waterbased systems is the application of open porous coatings (translucent paints), which are permeable for water vapours. In order to realize an open porous structure, a relatively large volume shrinkage is required, which, can preferably be achieved by evaporation of solvents or water. Because solvents are generally undesirable because of VOC issues, UV curable emulsions and dispersions have been developed in order to expand the application areas of the eco-efficient UV curable acrylate coatings [49].

The general benefits of waterborne UV curable systems over 100% solids UV coatings for these mentioned applications, mostly spray applications on wood substrates, are lower extractables and migratables, better matting behaviour, better adhesion and good hardness-flexibility ratios. Commercially available are both emulsions produced by emulsification of the classical acrylate resins, for example polyester acrylates, polyether acrylates or epoxy acrylates and dispersions. These emulsions are manufactured by means of using polymeric protective colloids.

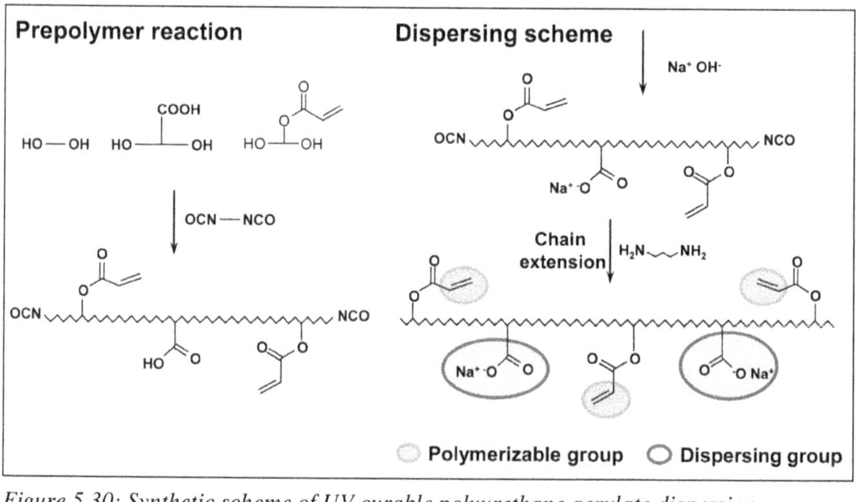

Figure 5.30: Synthetic scheme of UV curable polyurethane acrylate dispersion

Nowadays, the self emulsifiable polyurethane acrylate dispersions dominate the market. The self emulsification is realized by incorporation of ionic groups, like sulphonate or carboxylate groups, into the polymer backbone of polyurethane acrylates [50].

The higher molecular weight urethane acrylates are manufactured by a multistep process. In the first reaction step, an isocyanate terminated prepolymer is prepared in solution, subsequently dispersed in water and neutralized with base. Thereafter, the so-called "chain extension" with diamines will increase the molecular weight of the molecules in the particle and subsequently monomers and solvent are removed completely by distillation (Figure 5.30). Since high molecular weights may be realized in these dispersions, they can be designed in order to obtain physically dry and non-sticky coatings after application and drying without UV exposure. These UV curable dispersions can be designed to be very well suited for food packaging applications, since low molecular weight migratables can be greatly reduced, so that very good results are obtained out of emission and extraction studies. The physically dry films exhibit, however, often reduced conversion of the acrylate double bonds, due to the reduced mobility. Measures for increasing the conversion are discussed in Chapter 5.5.2.1.

For ease of matting, significant film shrinkage is required. Despite the fact that the polymerization shrinkage of 100% UV curable systems might be quite high, it is generally not sufficient to obtain the necessary matting. Aqueous UV curable systems, where the solids content is about 40%, however, provides during drying of the film enough shrinkage in order to obtain a sufficient matting effect.

On the other hand, the polymerization shrinkage, which often leads to adhesion problems, is especially in the higher molecular weight aqueous polyurethane acrylate dispersions relatively low. The polymerization shrinkage is a function of the polymerizable acrylate groups per volume of coating. A higher molecular weight of the resins often accounts for a reduced double bond density. The use of high molecular weight 100% acrylate resins is often restricted by the resulting high viscosity. In contrast, the viscosity of the aqueous dispersions is almost independent of the molecular weight of the resin and is determined solely by the particle size and the solid content of the dispersion.

The mechanical properties such as hardness and flexibility are determined by the chemical structure of the resins and the crosslinking density. Since it is easily possible to design the high molecular weight resins of the dispersions in such a way that they contain both, hard phases by the degree of crosslinking density and soft phases by the selection of segments with low glass transition temperature. Such aqueous polyurethane dispersions offer a well balanced hardness-flexibility ratio, exhibiting at sufficiently high hardness pretty good flexibility [51].

Such UV curable polyurethane dispersions are therefore preferably applied by spray applications in the wood and furniture industries. In addition to the already established applications of 100% UV curable formulations on plastic substrates, like printing and roller coating applications on PVC flooring and in UV inks, recently, more and more new applications of water-based spray formulations on plastics substrates are emerging [52]. These new applications on plastics concern both, very hard coatings as an alternative to solvent-based 2K systems for the coating of mobile phones, for instance, as well as so-called "soft touch" coating materials, namely coatings having a pleasant feel upon touching of, for example, automotive interior parts.

Furthermore, UV curable polyurethane acrylate dispersions have been evaluated intensively even for their suitability to be used for outdoor applications by accelerated weathering studies [53]. It was found that, in particular, the UV curable polyurethane dispersions based on cycloaliphatic structures, are already relatively stable without the use of light-stabilizers. In the presence of UV absorbers and HALS radical scavengers they offer an equally good level of weatherability as the well known 2K polyurethane automotive clear coat, which were used as a benchmark.

5.4.5 Influence of chemical structure on formulation properties

The properties of the coating films do not only depend on the chemical composition of the formulation, but also on the application method and flow properties, and the rheology of the formulation. Rheology is the science which deals with the flow and deformation properties of materials. The first part plays an important role in the application and film formation and is largely determined by the viscosity (measure of the flowability) of the formulation. The deformation properties of the films deal with the mechanical properties of the cured films and evaluate the main factors that may lead to a mechanical failure of the films. The evaluations are performed in order to determine the suitability of the examined materials for the corresponding application and window of application. The structural influence on the mechanical properties of radiation cured acrylate films is discussed further in Chapter 5.5.2.

5.4.5.1 Viscosity

Viscosity is a measure of the resistance of a material to flow when being deformed and predominantly associated with a shear stress. The shear viscosity is measured in Pa · s (Pascal · seconds). When the applied force is the gravitational force only, it is called kinematic viscosity (shear viscosity divided by the density of the liquid). The viscosity of the formulation plays a major role in processing the formulations. It has to be adjusted to the application method, for instance, for spray applications, it has to be relatively low (0.1 to 0.5Pas), whereas for roller coating applications it may be up to about 5Pas. The processing viscosity in solvent-based paints can be relatively easily adjusted via the solvent content. In radiation-curable, solvent-free (100%) coatings, the reactive diluent is responsible for the diluting properties. The dilution behaviour of acrylate based reactive diluents has already been discussed in Chapter 5.4.2 and demonstrated exemplarily for a polyester acrylate resin (see also Figure 5.11).

The viscosity of liquids is closely related to the availability of free volume and therefore all measures that reduce the free volume increase the viscosity. A few general factors influencing the viscosity are discussed briefly. In general, the viscosity increases with the strength of the interaction of the chain segments (polarity, hydrogen bonding) and with the increase of molecular weight. In resins with broad molecular weight distribution, the higher molecular weight fractions contribute disproportionately high to the increase in viscosity. Furthermore, a high glass transition temperature of the formulation, which is determined mainly by the resin and reactive diluent chemical structure, molecular weight, the resin concentration, and the interaction between resin and thinner, increases the viscosity.

Besides the impact on the flow behaviour of the formulation, the viscosity also influences the behaviour during film formation (history, film thickness, wetting of the substrate, flow behaviour on vertical surfaces, settling of fillers) and the final properties (adhesion).

Recommendations to related literature

- Wicks, Z.W., Jones, F.N., Pappas, S.P., „**Organic Coatings** – Science and Technology", Vol. 2, chapter XIX, "**Flow**", S. 1-28, John Wiley&Sons, Inc., (1994)
- Glöckner, P., Jung, T., Struck, S., Studer, K., **Radiation Curing**, Vincentz,, S. 48-52 (2009)

The general trends of resin viscosities can therefore be deduced from the chemical structures and the interactions of the chain segments. Due to hydrogen bridging in urethane and epoxy acrylates, the viscosities are rather high, whereas polyester acrylates and polyether acrylates exhibit medium to low viscosities. Since the molecular weights are generally in the range between 500g/mol and 5000g/mol, entanglements of the chain segments do not matter at all. The absolute viscosity of neat urethane acrylate resins with molecular weights around 5000g/mol, may, depending on the structure, exceed 500Pas. Urethane and epoxy acrylates are therefore offered commercially almost entirely diluted with reactive thinners.

In contrast, the viscosity of aqueous UV curable dispersions is not determined by the molecular weight and the interactions of the chain segments of the resins, but almost exclusively by the solids content, the particle size and the particle interactions. Therefore, UV curable aqueous dispersions are often used for spray applications, with the additional advantage that the film properties of the resin matrix can be adjusted according to the requirements without considering the influence of the chosen components on the formulation viscosity.

5.4.5.2 Reactivity

Besides viscosity, reactivity of the formulations of radiation curable coatings plays an important role for the selection of components. Especially in the printing ink processes the selection of the raw materials is often determined by reactivity rather than mechanical properties in order to allow for high production rates to gain economical competitiveness. However, the reactivity in this meaning is often associated with the rate at which the liquid coating is transferred into a solid paint film rather than with the pure kinetic rate of polymerization. The reactivity is therefore often determined by wipe tests with solvents or by determination of hardness (often fingernail hardness). The development of hardness is closely related to the increase in glass transition temperature during the radical polymerization and thus depends on the chemical structure of the resins and reactive diluents, the double bond conversion and crosslinking density (see Chapter 5.5.1.4). Generally, the higher the functionality of the components, the higher the reactivity of the final formulation. However, as discussed already in the section about monoacrylates (see also Figure 5.13), monofunctional acrylates may increase the conversion of the resins and thus result in an earlier solidification, resulting in an observed higher "reactivity".

5.4.5.3 Surface tension – interface tension

Due to the cohesive forces liquids tend to reduce their surface tension. The more polar the liquid, the more pronounced is the effort to form a spherical structure. In order to spread a liquid onto a surface a specific force has to be applied. The surface tension of a coating formulation thus has an influence on the flow properties of the coating on the surface, as well as on the wetting behaviour, which in turn has a strong influence on adhesion properties. The wetting depends on the viscosity and surface tension (measured in mN/m = dyn/cm) of the formulation.

The size of the surface tension depends on the strength of the interaction of the molecules, thus strongly on the polarity of the molecules (n-pentane has a surface tension of 16mN/m; ethylene glycol of 48mN/m and water of 73mN/m). The surface tension of

Table 5.5: Surface tension of selected acrylates and polymer substrates (Source: Sartomer[54])

Surface tension [dyn/cm = mN/m]							
of acrylate						of substrate	
low		medium		high			
Isooctyl-	28	DPGDA	32.5	PEG600 DA	43.7	PTFE	22.5
Lauryl-	30.3	TPGDA	33.3	E20 TMPTA	41.8	PVC	38.4
Isobornyl-	31.7	HDDA	35.7	E15 TMPTA	41.5	PE	36.1
		THFA	36.1	E9 TMPTA	40.2	PS	43.5
		TMPTA	36.1	E3 TMPTA	39.6	PC	46.7
				POEA	39.2	PET	47
						PMMA	49
						PVDC	49

radiation-curable coating materials are sometimes given in the brochures and have a range of about 25dyn/cm to 45dyn/cm for classical acrylates[54] (Table 5.5). Generally, the non-polar hydrocarbon-rich acrylics are on the low end, the more polar acrylates exhibit higher surface tensions. With the degree of ethoxylation, the surface tension increases, as shown in Table 5.5 for the series TMPTA with zero, three, nine, 15 or 20 ethylene oxide units, respectively. A good wetting behaviour is obtained when the surface tension of the formulation is lower than that of the substrate.

Regarding substrates, polar surfaces (glass, cleaned steel), but also polar plastics, such as polyamides and PET, are relatively easy to wet with conventional coating formulations, while non-polar plastics such as PE, PP and PC are often difficult to wet. Here, the lowest surface tension compounds have to be used in the coating formulation. Fluorine and silicon groups offer the lowest surface tension; however, they are often too expensive when incorporated into resins. Thus, preferably, fluorinated or silicone-containing low molecular weight surfactants are commonly used as levelling and wetting additives.

5.5 Structure and properties of coating films

5.5.1 Network characteristics

5.5.1.1 Network formation

Radiation curable coatings rely on the photo-induced excitation of photoinitiators and the subsequent formation of radicals which finally initiate a radical polymerization. The initiated polymerization follows the same laws of free-radical

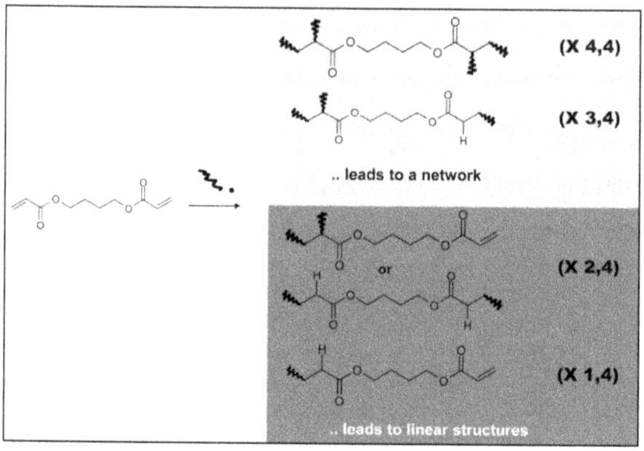

Figure 5.31: Linkage possibilities of a diacrylate in the radical polymerization on the example of hexanediol diacrylate

polymerization as already been outlined in Figure 5.5, no matter whether the polymerization proceeds in a vessel or in a film. After the first addition of an acrylate group to the photochemically produced radical, the polymerization proceeds via the classical addition and termination reactions. In contrast to the syntheses of polyacrylates in a vessel (dispersions, solution polymers), where almost exclusively mono-functional acrylates are employed, radiation-curable coatings consist predominantly of polyfunctional acrylates. With the use of di-or polyfunctional acrylates rapid crosslinks are formed, if the addition reactions lead to at least three connection points. Hexanediol diacrylate is a difunctional monomer; however, in terms of network formation the two acrylate functionalities can form

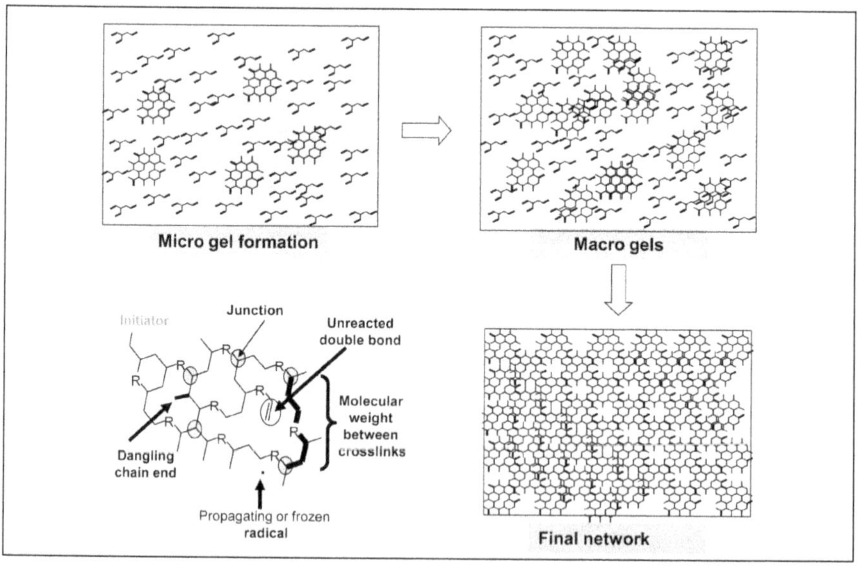

Figure 5.32: Section of a microgel/network structure (bottom, left) and growing together of microgels during polymerization

four connection centres. If the addition reaction leads only up to two other connections, a linear structures results. However, all addition reactions resulting in three or four connection points create crosslinks and therefore a network structure (Figure 5.31).

Kloosterboer [55] studied the formation of network structures of UV cured acrylates extensively and has shown that already after a very short reaction time high molecular weight chains are forming microgels. Due to an often high concentration of photoinitiators and thus initiating radicals, the microgels are more or less homogeneously distributed within the film. During the further polymerization, these microgels grow together in order to form macrogels by further addition reactions to the unreacted double bonds or by grafting to a network chain, until the final network structure results (Figure 5.32). The final network still may contain several irregularities, like dangling ends, unreacted double bonds or still radicals.

5.5.1.2 Functionality

A property spectrum of typical standard acrylic resins of the different resin classes has already been described in Chapter 5.4.3.1 (Figure 5.14). In the following, the structure-property relationships of the resins are discussed in greater detail and in particular in regard to the influence of the network structure. Since the radiation curable coating formulations are based mainly on polyfunctional acrylates, the photo-induced polymerization leads to, even at low overall conversions, gel structures. Depending on the composition and functionality of the formulation, more or less crosslinked coating structures result. The general effect of the variation of the functionality of the acrylate raw materials on the chemical and mechanical properties is illustrated in Figure 5.33.

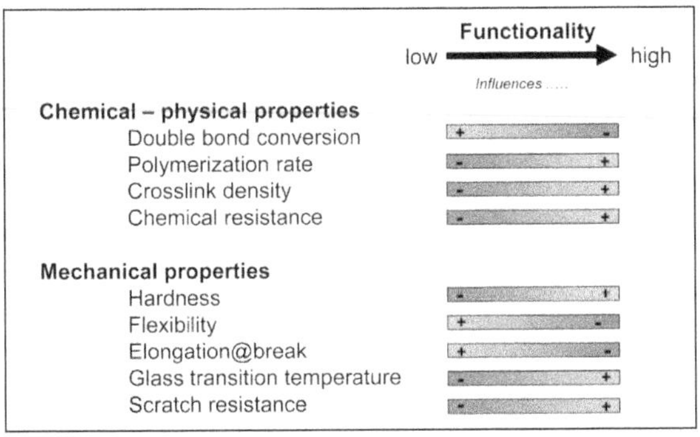

Figure 5.33: Impact of the functionality of the acrylate components on the chemical and mechanical properties

By increasing the functionality usually several properties increase, like the polymerization rate (due to a higher monomer concentration), the crosslinking density (more conjunction points) and the glass transition temperature (mobility is reduced), whereas, for instance, the polymerization conversion (due to glazing) decreases. The mechanical properties are also influenced by increasing the functionality, for example, hardness, scratch and chemical resistant are improved, while as flexibility and elongation are reduced. The functionality of the components alone, however, is not the determining factor, but rather the obtained glass transition temperature governed by the chemical structure of the components and the obtained crosslinking density. Thus, the chemical structure and the functionality of the components, their concentration and volume share, curing conversion and kinetic constants (growth, transfer, termination) determine the glass transition temperature of the formed coating network. As we will see later, most of the chemical resistance and mechanical properties are dependent on the absolute value and width of the glass transition region.

5.5.1.3 Crosslink density and molecular weight between crosslinks

The main network parameters for the characterization of networks of UV cured films were recently reported by *Meichsner* [56]. A network can be defined by the conjunction points (crosslinks) and the molecular weight between crosslinks. A **low crosslink density network** can be characterized very well by the determination of the average molecular weights between crosslinks and the concentration of the conjunction points in the coating film, and expressed in terms of crosslink density. Both parameters can be calculated by using a theory developed by *Miller* and *Macosco* [57]. Factors to be determined are the functionality, the concentration and the mole fraction of the functional groups of the crosslinker molecules, the final conversion of the functional groups and the probability of the propagation reaction, which is defined as the ratio of the growth rate constant by the sum of the rate constants of growth, transfer and termination. The theory of Macosco and Miller assumes that all functional groups are equally reactive and react independently of one another, and that no cyclization reactions occurred. Therefore, the calculations result in a "relative crosslinking density".

The parameters can also be determined experimentally by using the dynamical mechanical analysis (DMA) and by the determination of the dynamic storage modulus (E`) in the elastic region, thus above the glass transition temperature of the cured coating:

Formula 5.1: $$E' = 3\,\nu_e \cdot R{*}T = 3\,X_c \cdot \rho \cdot R \cdot T = 3 \cdot (1/M_c) \cdot \rho \cdot R \cdot T$$

in which ν_e is the number of elastically effective chains per cubic centimeter, X_c is the crosslink density expressed as molality, M_c is the molecular weight between crosslinks (cycles and dangling ends are not considered) ρ is the density of the coating film, R is the gas constant and T is the absolute Temperature.

From this formula, the relevant network characterizing parameters can be derived:

Crosslink density:

Formula 5.2: $X_c = E' / 3 \cdot \rho \cdot R \cdot T$ in [mol/g]

Molecular weight between crosslinks:

Formula 5.3: $M_c = 1 / X_c$ in [g/mol]

However, it has been shown by Meichsner et al. [56] that Formula 5.2 holds not true for **highly crosslinked networks**. They found considerable deviations from the linear behaviour and derived an **exponential correlation** between E-modulus and crosslink density (X_c). This deviating behaviour may be attributed to the short molecular weights between crosslinks, where the theory of rubber elasticity, which reflects the entropic elastic behaviour of the chain segments between the X-links, may not be valid anymore. Instead, energy elastic deformations may become dominant, which can be described by deformations of valence bonds or bond angles. They found the following correlation for modulus and X-link density:

Formula 5.4: $X_c = (\ln E' - b) / m,$

where m = slope and b = axis intercept

in the graphic correlation of ln E′ as a function of X_c.

5.5.1.4 Glass transition temperatures in highly crosslinked coatings

The mechanical properties of coating films depend strongly on the glass transition temperature of the film. Since most of the mechanical tests are done at room temperature, it is important, whether the T_g of the coating film is above or below room temperature. Thus, if T_g is below room temperature, the film is in a rubber state; whereas, it is in a glassy state if T_g is above room temperature.

The state of the glass temperature of a cured film depends essentially on the chemical structure of the components and the structure of the network.

Similar to the dependence of the storage module on crosslink density, also the dependence of the glass transition temperature on the crosslinking density is a linear behaviour only at low crosslinked coating films (Fox [58])), while highly crosslinked films exhibit an exponential relationship as shown by Stutz [59] according to Formula 5.5:

Formula 5.5: $T_g = [T_g^{\infty} - K_1(1\text{-}p)] \cdot [1 + K_2\,(X_c/(1\text{-}X_c))]$,

where T_g^{∞} is the true backbone T_g at infinite molecular weight, without end groups or crosslinks. K_1 characterizes the influence of end groups, thus reflecting the degree of cure (p), and K_2 is another constant accounting for the influence of crosslinks.

Thus, the T_g of the linear polymer is lowered by end groups if the degree of cure is not 100%, and increased due to the crosslinking reaction. This formula also explains how T_g changes during cure. As the cure conversion increases, the term lowering T_g, gets smaller and with increasing crosslink density, the term $X_c/(1\text{-}X_c)$ increases T_g exponentially.

5.5.1.5 Brittle-ductile transitions in networks

Depending on the chemical structure, in linear polymers a specific mobility of chain segments is often already observed well below the glass transition temperature. The temperature at which this mobility starts was defined as the brittle-ductile transition temperature (T_b) [60]. Below T_b the polymer is brittle, between T_b and T_g hard and ductile or tough, and above T_g soft. A polymer considered as a hard material with high impact-resistance is polycarbonate, which is due to the particularly large range of toughness of the bisphenol-A based polycarbonate with a T_b of -200°C and a T_g of + 160°C. In respect to the brittle ductile transition,

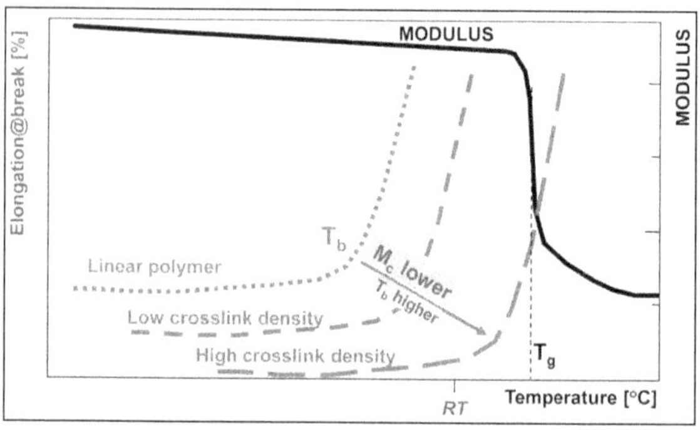

Figure 5.34: Influence of crosslinking on the brittle-ductile transition

only a few crosslinked coatings were investigated. From the published data, however, one may conclude that with increasing crosslink density the mobility of chain segments is reduced and thus the brittle-ductile transition is shifted to ever higher transition temperatures, until ultimately, in highly crosslinked coatings only one transition exists, the glass transition, shown schematically in Figure 5.34 [61].

The effects of the crosslink density (or molecular weight between crosslinks) and of the glass transition temperature, resulting from the chemical structure and the crosslinking reaction, on the mechanical properties of coating films will be discussed in subsequent chapters in more detail.

5.5.1.6 Oxygen inhibition

In the radical polymerization the growing chain radicals react with oxygen by several orders of magnitude faster than they add a new acrylate monomer. Since, during photo-polymerization, the oxygen concentration is highest at the surface of a coating, the oxygen inhibition reaction therefore retards the network formation significantly, especially at the surface. Oxygen terminates the polymerization reaction, leading to a lower conversion and thus a lower network density at the surface. This will affect many mechanical properties (scratch resistance, hardness, etc.) and the chemical resistance behaviour negatively.

To overcome the oxygen inhibition, a number of physical and chemical methods have been developed. The physical methods are very effective.

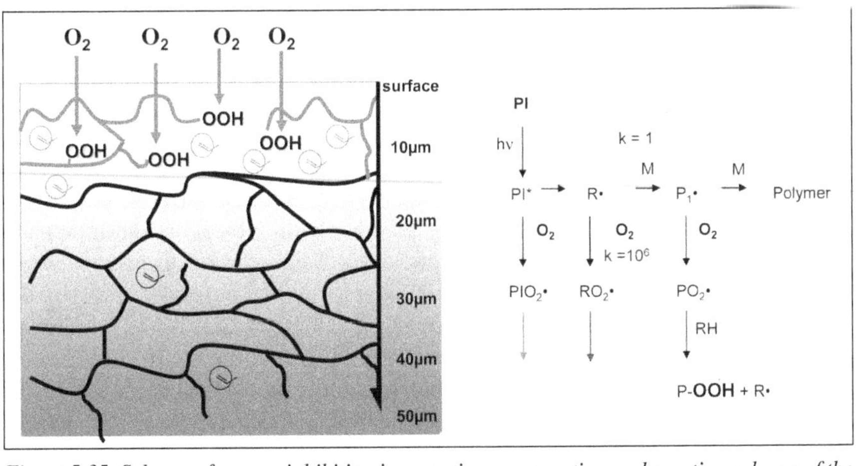

Figure 5.35: Scheme of oxygen inhibition in a coating cross section and reaction scheme of the reaction of oxygen with excited photoinitiators (quenching), primary and polymer radicals

The most widely used method is the "over-exposure" method (high energy density or high radiation intensity) [62]. A more complex and costly method is inerting the atmosphere during exposure by using an inert gas (nitrogen, carbon dioxide) [63, 64] or a cover layer, like a foil or a wax (as an oxygen barrier). Inerting is obviously the most effective way to eliminate the oxygen inhibition. In applications that do not run continuously, such as repair coatings in the automotive sector, also the recently developed "pulse inertization" could be used beneficially [65].

For three-dimensional UV exposure, inerting with carbon dioxide, which is heavier than air, is ideally suited, since it is easily possible to fill a basin with carbon dioxide and immerse the to be cured parts into it and expose it subsequently in this inert atmosphere. This process has become known as the "Larolux" process [66].

The chemical methods to overcome oxygen inhibition range from the use of high photoinitiator concentrations (disadvantage: price, migration), amines as synergists for scavenging oxygen (drawback: yellowing, odour), the use of acrylates with ethoxylated structures, to high formulation viscosities and to specific dyes as oxygen scavengers [67]. Another interesting method for radiation curing under inert conditions was recently presented as the UV "Plasma Cure" process. This method can be described in such a way that the parts to be exposed are not placed under an UV lamp, but "in a UV lamp" where they are cured. The pilot plasma chamber can accommodate an entire car body and UV cures three-dimensional parts practically without shadows under inert conditions in the vacuum (< 0.1 mbar) [68].

However, the effects of oxygen inhibition, namely the inadequate cure and remaining stickiness of the surface, can on the other hand be fully utilized by improving the interlayer adhesion in multilayer systems or to use remaining unsaturated acrylate groups for a further surface functionalization, for example, to bind crosslinked PMMA-polymer particles to a UV curable matrix [69].

5.5.2 Coating films: structure – property relationships

Coatings generally have two functions, mainly they should fashion an attractive appearance of the painted objects and on the other hand protect the substrate from any damaging impacts. The impacts on the coating film can be diverse and cover different times and rates of impact stress. Examples are the impact of gravel on the hood of a car, the penetration of red wine into the kitchen table or the weathering effect on the painted wood window. The study of structure-property relations, in particular, the viscoelastic behaviour of polymer films should lead to the identification of molecular characteristics that can be correlated to performance properties. While one often understands how structural parameters, such as chemical structure of the components, crosslink density and molecular weight between crosslinks affect

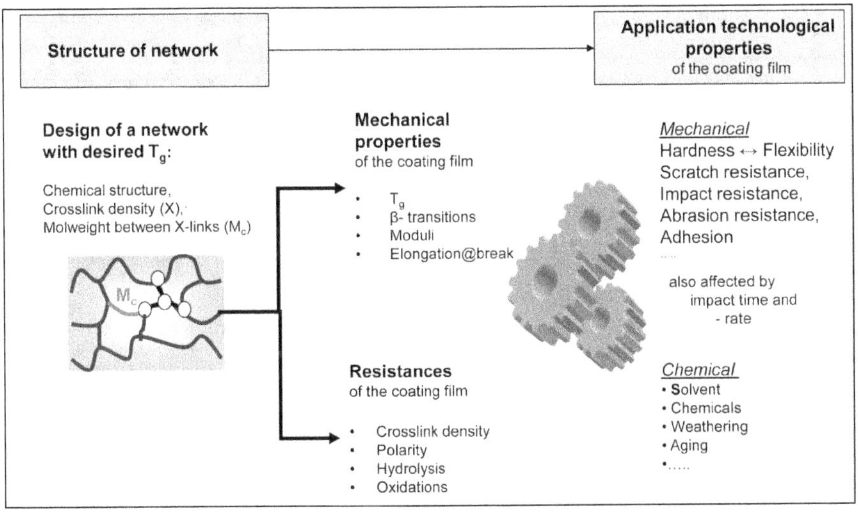

Figure 5.36: From the design of network structures to application technical properties

the mechanical properties, like glass transition temperature, elastic modulus and elongation, one is often unable to draw simple correlations with application technical service properties, such as scratch resistance, impact resistance, or abrasion resistance. Often the demands on the application technical properties are complex and diametrically opposed (high hardness and high flexibility), that the optimization of one property is inevitably realizable only at the expense of another property. The gain of knowledge how structural parameters affect the mechanical properties and chemical resistance characteristics are crucial. Since a correlation of the molecular parameters with the final major application performance characteristics can significantly speed up product development, it is still a topic of intensive research.

In order to set and adjust the mechanical application properties by the design of the network, it obviously plays a major role, whether the designed components react completely that the desired structure is achieved. Therefore, there are a number of studies on the influence of the functionality of acrylates on the curing conversion of radiation cured coatings and thus on the nature of the network structure.

5.5.2.1 Influence of chemical structure on curing conversion

By comparing the polymerization behaviour of acrylates with different functionalities it is almost generally observed that in both the homopolymerization and by using the acrylates as reactive diluents, an increase in the functionality results in an increase in polymerization rate, a decrease of the double bond conversions, and an increase of hardness and in contrast, a decrease in flexibility [70].

Table 5.6: Influence of monomer functionality on polymerization rate, double bond conversion, hardness and flexibility in three different formulations

Monomer (M)	Homopolymerization				Epoxy acrylate + M				Polyurethane acrylate + M			
	Rp	U	Hard-ness	Flex	Rp	U	Hard-ness	Flex	Rp	U	Hard-ness	Flex
EDGA (1)	1	98	n.a.	n.a.	8	95	60	0	8	98	40	0
TPGDA (2)	4	90	n.a.	n.a.	23	80	280	7	27	90	120	1
HDDA (2)	3	72	n.a.	n.a.	20	77	300	10	20	84	130	2
TMPTA (3)	10	45	n.a.	n.a.					100	64	270	5

Rp = polymerization rate; U = conversion (%); hardness = Persoz hardness (s), Flex = Mandrel flexibility (= lowest bending radius without crack; the bigger the bending radius the lower the flexibility)

The curing conversion can be increased by various measures, most notably by raising the curing temperature [71] (Figure 5.37). During exposure, the glass transition temperature of the coating film increases with increasing cure conversion. Depending on the chemical structure and the crosslink density, often a vitrification of the coating film occurs, which reduces the mobility of the unreacted double bonds significantly and prevents further conversion. Exposure at higher temperatures thus allows an enhancement of mobility, enabling further increase in conversion

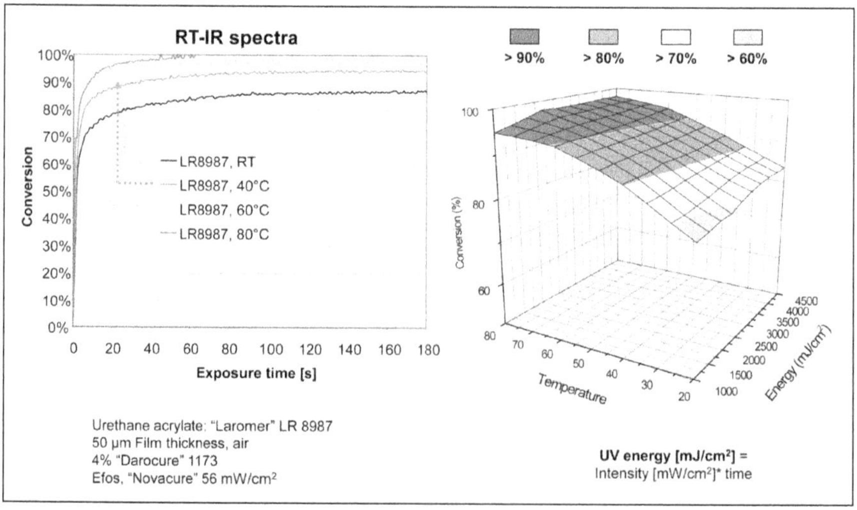

Figure 5.37: Increase of conversion of acrylates in the UV curing process by increase of exposure temperature and/or energy density

and further increase of the glass transition temperature, which is also reflected in improved mechanical properties, e.g., higher hardness and scratch resistance.

Waterborne UV curable dispersions generally exhibit significantly lower polymerization rates and thus also a lower conversion compared to 100% UV coatings under identical curing conditions. Since the aqueous polyurethane dispersions are usually physically drying, they are after drying and before UV curing already in a solid state, in which the mobility of polymer radicals is clearly limited. The mobility and therefore cure conversion can be increased by different measures, for example by adding a reactive plasticizer (reactive diluent), or by increasing the humidity or temperature [72].

5.5.2.2 Glass transition temperature: influence on hardness and flexibility

The preferred combination of properties in many coating films, such as floor coatings on wood, would be a high hardness, for example, in order to avoid marks left by high heels and increased flexibility in order to resist crack formation by the natural expansion of the wood tiles. The practice however shows rather a different picture, most coatings are either very hard or very flexible. In the property spectrum of the different UV curable acrylate resins one has already seen, that the acrylate resins are either more or less brittle or flexible. This parameter value is relatively strong for UV curable coatings based on 100% formulations, since the requirement of a low viscosity of the formulation almost excludes using high molecular weight (flexibility) resins. Exception are waterborne UV curable formulations, which contain high molecular weight resins, since the viscosity of the dispersion is not a function of the resin molecular weight, but of the particle size and concentration of the dispersions [73]. It has been shown that the range of the glass transition temperature of the cured coating determines the mechanical behaviour of the coating in terms of hardness and flexibility.

Plotting the pendulum hardness and flexibility (Erichsen) of the different "Laromer" UV curable resins against the glass temperature of the coatings obtained under identical curing conditions, reveals a nonlinear behaviour with a sudden change at the test temperature (room temperature). All coatings with glass temperatures below RT are soft and there is a significant increase in hardness for coatings with glass transition temperatures above RT. The opposite behaviour is observed in terms of flexibility, which dropped, when T_g rises above room temperature. It showed also that the mechanical properties do not depend on the product class, but on the final glass transition temperature of the cured film. Thus, while the most widely used epoxy acrylates (based on bisphenol-A diglycidyl ether) are hard and relatively brittle, there are also soft epoxy acrylates available which are, for

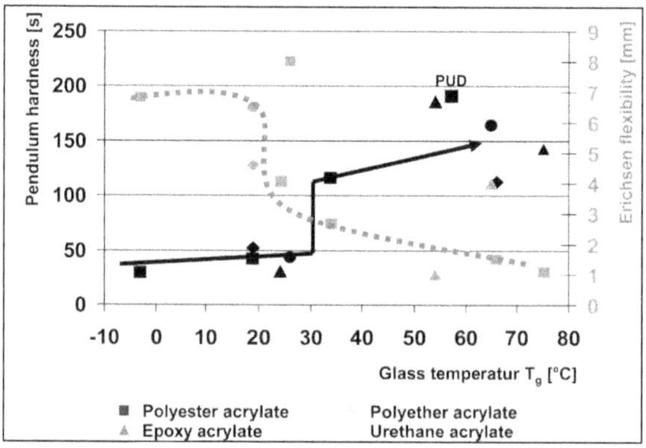

Figure 5.38: Hardness and flexibility as a function of the film glass transition temperature

example, based on aliphatic glycidyl ether. Similarly, resins of the various classes, like polyester acrylates, polyether acrylates or urethane acrylates are represented either in the hard or soft coating regions, depending on the chemical structure, crosslink density and final T_g (Figure 5.38).

In Figure 5.38 it is attracting attention that the polyurethane dispersion (PUD) has a high hardness corresponding to the high glass transition temperature of the film, however, at such a high hardness a much lower flexibility had been expected. It had been reported however, that these polyurethane dispersions have a high molecular weight between crosslinks (Mc) which provides them with a high flexibility. Because the viscosity is not the limiting element, such polyurethane dispersions can form hard phases by high spatial crosslink density of double bonds and/or hydrogen bonding of urethanes and contain as well, high molecular weight flexible soft phases and therefore exhibit ultimately a balanced hardness-flexibility ratio.

5.5.2.3 Scratch resistance: influence of the crosslink density

High scratch resistance coatings are required in many applications, for instance in floor coatings and automotive topcoats. The studies on the scratch resistance of UV curable coatings for automotive applications show the dominating influence of crosslinking density, but also that **one** mechanical parameter (e.g., modulus, elastic or plastic deformation, elongation, slip) is not sufficient to describe the scratch resistance. In model systems, a variation of the crosslink density was based either on a soft urethane acrylate or on a hard urethane, the crosslink density adjusted by the variation of the reactive diluent functionality. It appeared that high gloss retention after the scratching could be obtained with both the hard and the soft resins, solely dependent on the crosslinking density, see Figure 5.39 [74].

In a comprehensive study *Kutschera* et al. [75] compared the scratch resistance of several different automotive clear coats including classic heat-curable (1K, 2K) and UV curable coatings. They found, that considering the mechanical parameters, there is no good correlation of the scratch resistance with a single mechanical parameter, but a plot of the plastic deformation against the force leading to the first crack (mN) gave a good correlation with field tests in car wash units (Figure 5.40). These results led to the conclusion that such clear coats which exhibit high elastic restoring forces, i.e., show little plastic flow, and simultaneously withstand high impact during scratching before cracking occurs, yield the most scratch-resistant coatings. Highly crosslinked UV cured urethane acrylates have shown the highest scratch resistances and offer tremendous potential for future eco-efficient coatings in automotive applications.

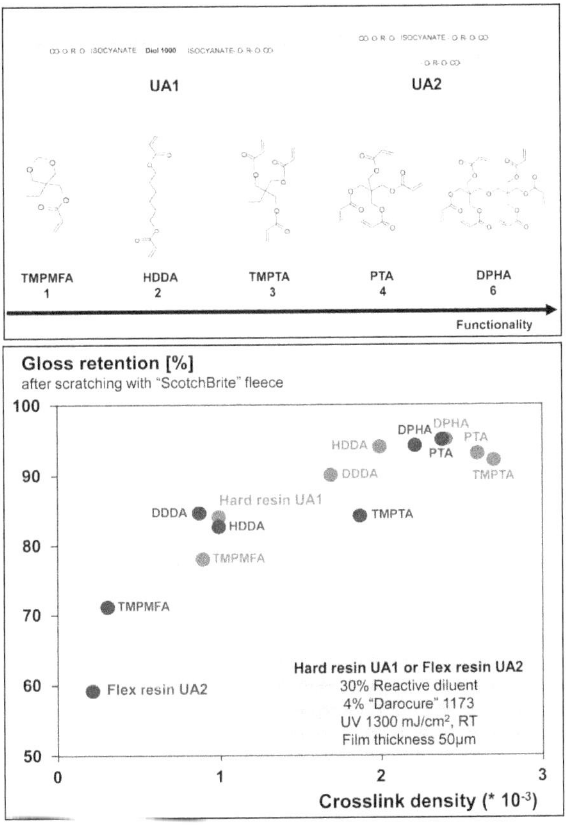

Figure 5.39: Variation of the crosslink density with hard or soft Urethane acrylate resins and reactive diluents of different functionality (above) and correlation of scratch resistance (gloss retention after scratching) with crosslink density (bottom)

Since both scratch resistance and chemical stability feature a surface property, a high crosslinking density at the surface of UV cured coatings is an essential prerequisite. Here, in particular, the strong oxygen inhibition observed in UV induced free radical polymerizations counteracts noticeably. The severe impact of the exposure atmosphere on the gloss retention after scratching with a "ScotchBrite" fleece is demonstrated in Figure 5.41.

In air atmosphere, a high scratch resistance (high gloss retention) is obtained only for the formulations with the high (hexa)-functional diluent when

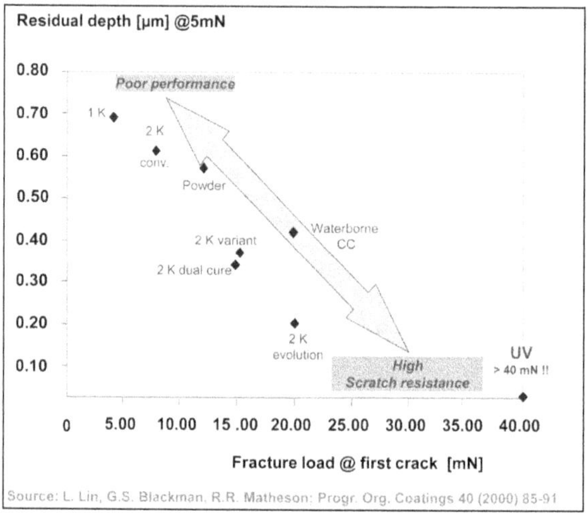

Figure 5.40: Correlation of plastic deformation after impact with a diamond tip (residual depth) with the fracture load at first crack in coating films of different formulation chemistries (according to Lin et al. [76])

exposed to high energy and high temperature. If, however, exposure is done under an inert atmosphere, already the medium functional "Laromer" 8987 (containing a difunctional reactive diluent (HDDA) exhibits high scratch resistance. As automotive coatings also have to offer good flexibility to ensure a high impact resistance, the exposure under an inert atmosphere is therefore almost inevitable since the formulations with the highly functional diluents turned out to be too brittle.

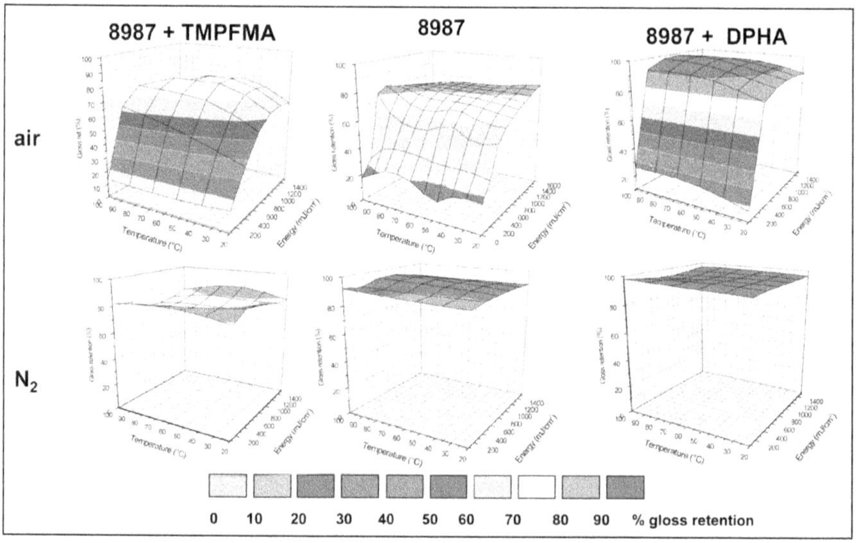

Figure 5.41: Scratch resistance (gloss retention after scratching) of coating formulations of different crosslink densities (trifunctional resin (8987) diluted with monofunctional (TMPFMA) and hexafunctional (DPHA) reactive diluent, respectively, after exposure under air and inert atmosphere (nitrogen)

5.5.2.4 Photochemical yellowing

The yellowing of coating films in the course of product life is a known phenomenon and is dependent on the chemical structure of the components used, such as resins or additives. In radiation cured coatings another effect comes along that is often referred to as the initial yellowing. This yellowing is created directly after exposure and is largely reversible, thus, it will disappear within several hours after exposure [77]. The phenomenon of initial yellowing is associated with the generation of radicals in the coating and therefore occurs both in UV curable coatings containing photoinitiators, as well as in systems which are cured with electron beams. A strong influence on the extent of the initial yellowing has to do with the structure of the acrylic resins. Although no clear correlation can be drawn with the glass transition temperature of the cured coating film, it is evident, that in particular, the coating with a high glass transition temperature exhibits the strongest initial yellowing. In addition, other factors such as energy density, exposure atmosphere, coating thickness, and photoinitiator concentration, affect the magnitude and the decay of the initial yellowing.

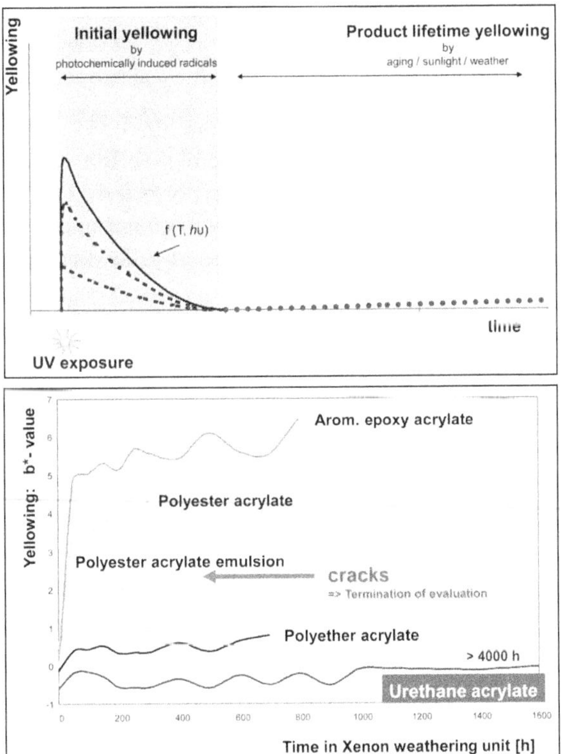

Figure 5.42: Initial yellowing (above) and yellowing during product lifetime (bottom)

The theory that frozen radicals are the cause of the yellowing is supported by the subsequent annealing of the coating film which accelerates the decay of the initial yellowing; since they may combine faster at higher temperatures where they are more mobile.

In contrast, the yellowing during product lifetime is more a function of exposure to light and oxygen and thus induced rearrangements and oxidation reactions. This yellowing is usually not reversible. In particular, aromatic compounds, such as aromatic epoxy acrylates based on bisphenol-A,

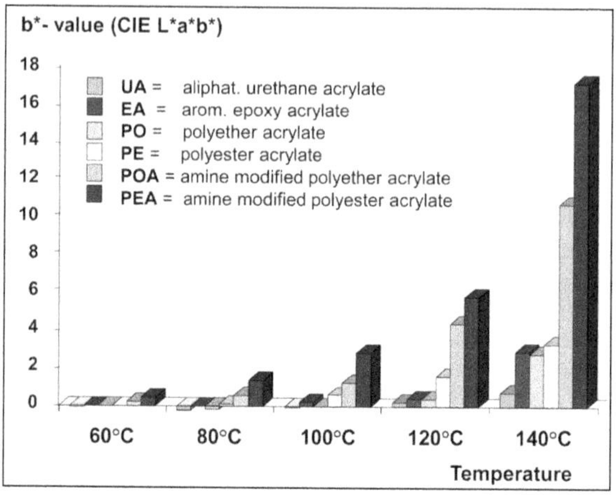

Figure 5.43: Thermal yellowing of acrylate resin classes under air and nitrogen

tend to yellow by the Photo Fries rearrangement of phenolic groups. In a screening to simulate natural UV exposure conditions with a "Suntest" CPS+ (Atlas) the main standard acrylate classes used in radiation curable coatings were evaluated upon their yellowing behaviour. Within all the tested resins, only the aromatic epoxy acrylates exhibit strong yellowing (Figure 5.42) upon pure UV exposure.

5.5.2.5 Thermal yellowing

Evaluating the purely thermal storage of UV cured coatings reveals that the thermal yellowing reactions occur mainly in amine-containing coatings, such as amine-modified polyethers or polyesters. This can be explained clearly by a thermal oxidation reaction, since if stored under an inert atmosphere; the same coatings exhibit much lower thermal yellowing (Figure 5.43).

5.5.2.6 Weathering stability

Selection criteria for resins for exterior applications have to consider the exposure of the coatings to sunlight, moisture, oxygen and temperature fluctuations. The following acrylate resin classes are out of question (Figure 5.44):

- polyether acrylates because of the ease of oxidation and concomitant embrittlement,
- some polyester acrylates, due to hydrolysis,
- aromatic epoxy and amine-modified acrylates because of the tendency to yellow.

So far, the most promising class of acrylate resins with the highest potential for use in exterior applications is the polyurethane acrylates (UA) [78].

These selection criteria apply similarly to the reactive diluents. Here especially, the acrylates with a hydrocarbon backbone, such as HDDA or decandiol diacrylate are suitable.

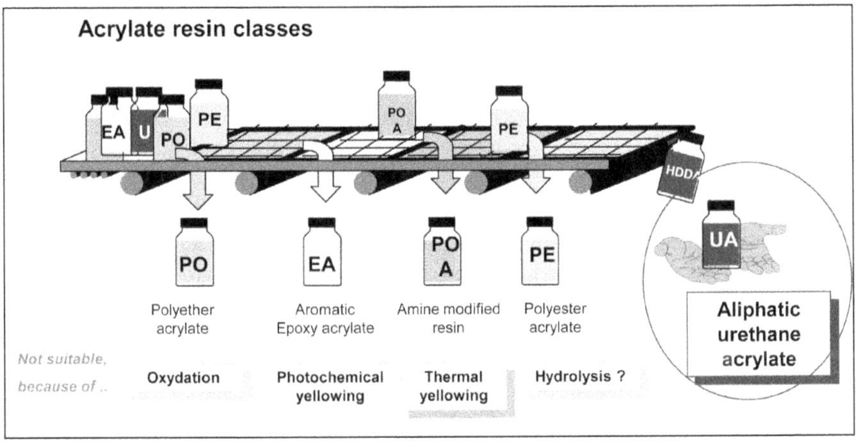

Figure 5.44: Selection criteria for resin classes to be used in radiation curable formulations for exterior applications

5.5.2.7 Performance-temperature-energy (PTE) diagrams

Radiation curing is carried out usually at room temperature, but especially in coatings where a high glass transition temperature is required (e.g., automotive clear coats), the increase of T_g during polymerization leads to vitrification and, in consequence, to a limitation of cure conversion. Higher cure conversions and therefore better properties are obtained by increasing the exposure energy and temperature. The properties of radiation cured coatings can be determined as a function of exposure energy density and exposure temperature. As an example, the scratch resistance of urethane acrylates is already shown in Figure 5.41. The method of recording the so-called PTE (performance, temperature, energy) diagrams can be used both for the optimization of process conditions and for determining the process window [79].

5.6 Applications and formulations

The main applications of radiation-curable acrylates are as high-gloss clear coatings on wood, paper, glass, ceramic, metal and plastic. Their application purpose is of decorative and/or protective nature (high gloss appearance, scratch resistance, sealing of chips and electronic components, burst protection, corrosion protection). In wood applications, like in prefinished floor tiles, often the whole multilayer assembly is based on UV curable coatings, while in other furniture coatings often mixed layers with other coating systems are used.

In the meantime, UV curing technology is recognized almost everywhere as an environmentally friendly alternative to conventional thermal curing technology; while it was previously considered only as a niche application for temperature-sensitive substrates. Therefore, more and more new applications developed in areas which are dominated almost exclusively by thermally curing systems (outdoor use, metal, glass, plastic, see Figures 5.8 and 5.9). However, these new applications require additional development of appropriate coating raw materials for suitable coating application and exposure systems, as well as the optimization of the entire coating and curing process. If everything is perfectly optimized, radiation curable systems prove to be almost always the most eco-efficient coating technology, not only, but including solvent free, rapid curing process at room temperature with the easy possibility of material recycling [80].

The global market for radiation curable resin applications is about 296,000 t/a (without unsaturated polyesters, UV curable silicones, UV adhesive resins and photoinitiators). The largest share accounts with 134,000 tons to Asia (62,000 to China, 57.500 to Japan), followed by Europe with approximately 94,000 tons and the Americas with about 68,500 tons (55,000 to North America). The global market segmentation is shown in Figure 5.45.

A rough estimate from various publications of the share of the different resin classes of radiation curable raw materials yields a proportion of acrylate based resins of about 85%, of about 10% of unsaturated polyester and about 5% of cationic curing epoxy resins and others. Of the acrylate based raw materials, approximately 45% are reactive diluents (stenomeric acrylates, of which about 50% is TPGDA), about 20% epoxy acrylate resins, and 10% each are resins with polyester acrylate,

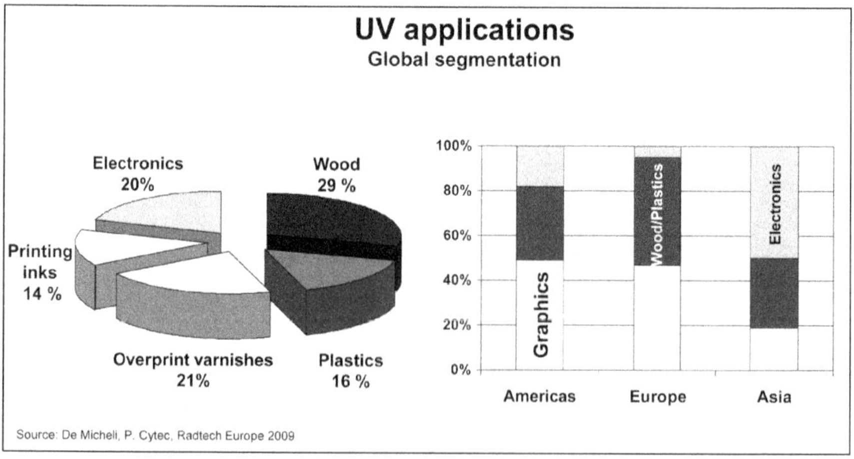

Figure 5.45: Segmentation of the resins market for UV coating applications and the regional allocation

Table 5.7: Composition and function of the raw materials of a UV curable coating formulation

Component	[%]	Function
Acrylate prepolymer	25 to 95	film formation; basic performance
Reactive diluent	0 to 60	film formation; viscosity adjustment
Photoinitiator	1 to 5	initiation of polymerization with UV exposure (not necessary at e-beam exposure)
Filler/pigments	0 to 50	grinding behavior; reduction of costs; coloration
Additives	0 to 3	defoaming; flow; slip, ...

polyether acrylate, or urethane acrylate backbone (including water-based products) and 5% of specialty resins.

The general composition of radiation-curable coatings is shown in Table 5.7. Main components are end-group functionalized acrylate prepolymers, which are responsible for film formation and the basic properties regarding mechanical and chemical resistance. Reactive diluents reduce the viscosity to the level of the required application viscosity and contribute additionally to the film forming properties. In the case of intended electron beam exposure, photoinitiators can be omitted, in the case of intended UV curing under inert gas atmosphere, only small amounts of photoinitiators are required (<1%) and for the classical exposure under air up to 5% of photoinitiators are used. Fillers are used for a variety of reasons, to reduce the price of the formulation (e.g., talc), to increase the abrasion resistance (e.g. corundum/aluminium oxide) or to apply colour (colour pigments). In addition, there are a number of additives available in order to be used for defoaming, to improve levelling, reducing friction, improving scratch resistance, etc.

However, according to the respective requirements of the intended application, the specific optimized formulation may deviate significantly from the general composition given in Table 5.7.

5.6.1 Graphic applications

The market segment of the graphics industry is the world's largest application for radiation-curable coatings. The applications in the graphics industry may be subdivided into

- UV overprint varnishes
- UV printing inks

The market for formulations for graphical applications is about 60,000 tons per year, of which about 31,000 tons are overprint varnishes and 29,000 tons printing inks.

The users of radiation curable printing inks and varnishes additionally expect a number of significant advantages over the known properties (rapid curing, no or low VOC, low energy demand), for instance:

- Reduced cleaning times, since the inks do not dry during downtime
- Powder application can be omitted, since the cured films are immediately non-sticky
- Improved properties in respect to chemical resistance and abrasion resistance

On the other hand, there are only a few disadvantages (cost of printing inks are higher, paper recycling is difficult because of the higher crosslink density) and some technical challenges to be solved (pigmentation limits the thickness and slows down the cure, low viscosity gravure inks are difficult to formulate without use of solvent).

5.6.1.1 UV overprint varnishes

Overprint varnishes are applied as a clear coat layer, for example on packaging, magazines, postcards or book covers, in order to produce high gloss surfaces and to provide surface protection. High quality packaging, for instance, in the cosmetics sector accounts for about half of the market volume in the graphics market segments, followed by labels (adhesive coating), glossy magazine covers and book or catalogue covers. Overprint varnishes are usually printed right after the ink in the same printing machine. UV curable overprint varnishes are predominantly used in offset printing, followed by flexographic printing. The printing process will be described briefly later. Benefiting from the trend to reduce solvent-based coatings, the employed amount of UV curable overprint varnishes has risen steadily in recent years and accounts for a market share of over 25%. Dominating in the market are polyacrylate dispersions (65%), the remainder being solvent-based coatings based on nitrocellulose and 2K systems [81].

UV coatings are at a much higher property level compared to dispersion varnishes. The retarded growth of UV curable varnishes however is due to a controversial discussion, which has taken place for a long time within the printing industry, in particular, the low molecular weight reactive diluent in 100% UV curable formulations may be susceptible to penetration into the porous substrates and thus elude UV curing. By slow diffusion of the acrylates into the coating layer during the lifetime of the varnish it may lead to odour problems of the coated printed products. By a suitable selection of raw materials (in particular resins and photoinitiators, avoidance of low molecular weight monomers) and adjusting the corresponding process parameters, these disadvantages can be eliminated nowadays. Thus, in the meantime, several UV curable systems exhibiting low migration and low odour have been approved for food packaging.

In the formulations for UV overprint varnishes, almost exclusively, acrylates are used as a resin class, whereof epoxy acrylates (EA) are dominating and thereof pure EA resins contribute to over 50%, blends of EA and polyester acrylates with about 20% and blends of EA and urethane acrylates with 20%, respectively.

UV overprint varnish	UV overprint varnish **Suggested starting *formulation***
Requirement profile liquid coating • low price • high reactivity, appropriate viscosity • good wetting of the substrate or printing ink • no odour, monomer-free	**52% Epoxy acrylate** *(resistance)* e.g.: "Laromer" LR 8986
Requirement profile coating film • high gloss • good abrasion resistance • no yellowing	**35% Reactive diluent** *(viscosity)* e.g.: TPGDA **5% Amine synergist** *(reactivity)* e.g.: Ebecryl® P 115
•Application • in coating unit of a printing machine, • thickness about 5 to10µm	**7% Photoinitiator** *(reactivity)* e.g.: benzophenone
..results in requirements to the formulation • Resins (epoxy acrylates, polyester acrylates, amine modified polyether acrylates) • high photoinitiator content • high content of flow and wetting additives	**1% Additives** *(wetting, flow)* e.g.: "EFKA" 3033

Figure 5.46: Requirements and a suggested formulation for an UV curable overprint varnish

A typical example of the requirements on overprint varnishes and a suggestion of a starting formulation are given in Figure 5.46. A classic example of an application of such a varnish is the coating of cosmetic packaging. The overprint varnish provides a high gloss and the necessary resistance against scratches, water or soap to the coated substrate.

The basic resins used in the formulation are epoxy acrylates, for example, aromatic epoxy acrylates like "Laromer" 8986, which may be diluted with reactive diluents like di- or tripropylene glycol diacrylate. In order to increase the reactivity amine synergists are often employed, examples are amine-modified diacrylates or low volatile amines such as triethanolamine or diethanolamine. As photoinitiator, often the relatively cheap benzophenone is used. Further ingredients are various additives, used in order to improve the flow, improve the lubricity, decrease the foaming or improve bonding or embossing behaviour.

In order to meet the demands for increasingly high reactivity in combination with high gloss and low odour, more and more UV curing units which operate under inert atmosphere have recently been installed [82]. These systems affect the UV exposure of the liquid coating in a nitrogen atmosphere. In this process the oxygen inhibition is completely eliminated and also requires much lower quantities of photoinitiator. The low amount of photoinitiator needed affects both positively, the cost effectiveness as well as the odour of the formulation and the resulting film.

5.6.1.2 UV printing inks

The printing processes in which UV inks and varnishes are mainly applied are offset printing (sheet- and roll) and flexographic printing process (Figure 5.47). Furthermore, they are used in traditional screen printing and recently also in ink-jet printing [83].

The offset printing process dominates the newsprint market, the printing of books and packaging. Offset printing is a lithographic printing process where the differentiation of printing and non-printing elements is achieved by a differentiation in polarity. The non-printing areas of the lithographic printing plate are hydrophilic, while the image on the plate is hydrophobic (oleophilic). The image to be printed thus attracts the ink from ink rollers, while the non-printing area is covered by a water-based film which keeps the non-printing areas ink-free. Printed and non-print pattern are on one level. The process involves the printing generally not directly from the printing plate to paper, but transfers the pattern first onto a blanket cylinder and subsequently to the paper. It is therefore an indirect printing process reflected in the name "offset" printing. Depending on the job size, single sheets or web rolls are printed. The printing process is divided into different types of commercial offset processes, the main being sheet fed and web offset. The printing speed is generally very high.

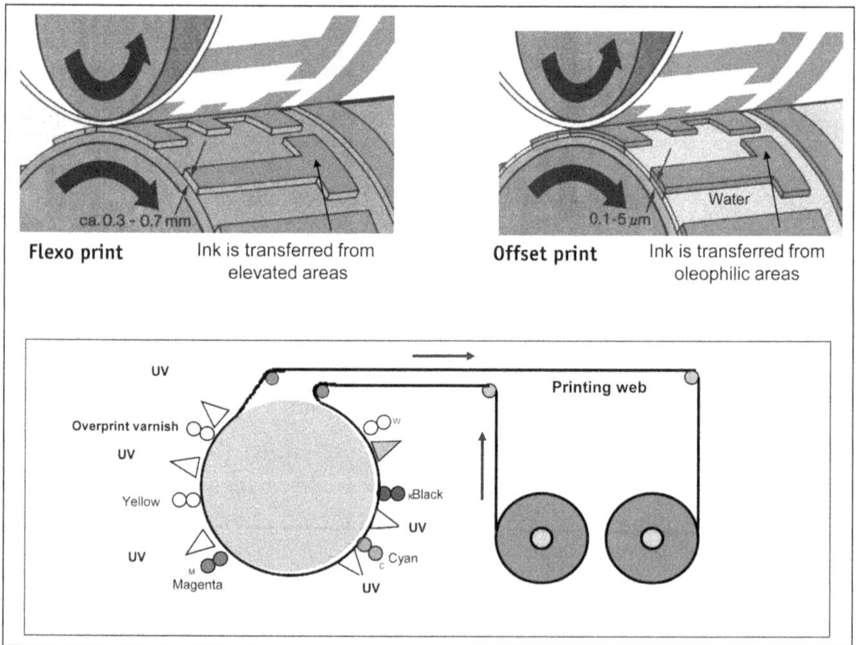

Figure 5.47: Scheme of the ink transfer in flexographic printing (top left), in offset printing (top, right) and complete four colour flexo printing with UV curable inks

The main substrates are paper, cardboard, films (PP, PE, PVC, PET) and composite materials. Packaging **applications** are for example, high-quality brochures, stickers and labels, listed securities, bank notes, credit cards, phone cards, and ID cards. Special UV equipment of the press is needed for printing with UV inks, where in addition to the UV unit with the UV lamps, the ink rollers and blankets have to have a special rubber mixture which offers enough chemical resistance against monomeric acrylates, which can cause swelling. Offset printing inks require relatively high viscosities in the range of 10 Pa·s to 1000 Pa·s

Flexographic printing utilizes a flexible printing plate in order to print on flexible substrates, such as corrugated boxes, milk and beverage cartons, and flexible plastics like shopping bags, hygiene bags, self adhesive labels, envelopes or wallpaper. It is a variant of the letterpress method, which had been introduced by Gutenberg in the form of lead type pattern. Today the printing patterns are produced by a photochemical method using a negative mask which is imaged into a flexible photopolymer printing plate. The imaged areas are polymerized and crosslinked by exposure, whereas the non-exposed parts can be washed away with a solvent or base. The resulting elevated patterns are used for the ink transfer to the substrate. The flexographic process has an advantage over offset printing in that it may use a greater variety of inks, waterbased inks can also be employed.

The formulations for radical curing UV flexo printing inks also contain oligomeric acrylate resins, reactive diluents (e.g. monomers), photoinitiators, synergists, pigments and additives. The viscosity of the ink, in contrast to offset printing inks is much lower in viscosity, in the range of 200mPas (@20°C). Since classical UV inks represent 100% liquid systems which do not dry upon application, they have to be UV "dried" (cured) This means that in a four colour printing process a UV curing step has to be executed after each colour is applied (black, cyan, magenta, yellow, see Figure 5.47).

A very innovative new development has recently been introduced by Sun Chemical (DIC group) with the "WetFlex" process which allows a wet-on-wet flexographic printing. In this process, the four colours are based on radiation curable special inks, while printed wet on wet printing is without intermediate curing. Subsequently, all the layers are cured and crosslinked simultaneously by means of electron beam curing. For this application, special water-based, radiation curable acrylate printing inks have been developed which are physically dry already after printing.

The **inkjet** method is very well suited for small lot printing jobs. Inkjet printers are spread throughout the office and home use market, the print quality however, suffers often from slow drying and insufficient durability. The UV curing process for inkjet printing and UV curable inkjet inks are therefore developed intensively in order to achieve faster drying and better stability [84].

In parallel, however, an intensive engineering of inkjet printers with attached UV unit has to occur, which will first be introduced for large-format printers, used for the printing of banners, signs, posters or advertisement boards. The inkjet inks must exhibit a very low viscosity (about 10 mPas/20 ° C), which accounts for the composition of the inkjet inks contains up to 95% of solvent, monomer or water and just the remaining 5% are the effective ink consisting of pigment, resins, additives and humectants. For UV inkjet applications the preferred formulations are based on aqueous UV curable dispersions, since 100% UV curable systems would consist almost exclusively of low viscosity acrylate monomers.

The biggest challenges for radiation curable printing inks and overprint varnishes are still associated with the adhesion capability to smooth substrates, particularly on plastics (polyethylene and polypropylene), and metals (cans, cups, caps).

5.6.2 Wood coatings

The predominant applications of UV coating systems in Europe are in the fabrication of industrially coated wood substrates, such as prefinished hardwood floor, furniture coating (e.g., kitchen cabinets, bedroom and living room furniture, table tops), doors, chairs and panels [85]. The prefinished hardwood floor, for example, has to fulfil a variety of requirements that can only be achieved by a combination of different coating layers. For an assembly of a hardwood floor based on different UV curable layers, several solutions are possible, which are dependent on the existent production line and the substrate quality. However, some general requirements apply for all options to be considered, which are the demands for good adhesion to the substrate, good interlayer adhesion, high abrasion resistance in everyday use and good resistance to "household" chemicals (such as detergents, red wine, coffee, mustard). To protect, for example, the parquet tiles against such chemicals, the topcoat especially has to exhibit good chemical resistance. Good chemical and scratch resistance is usually offered by resins based on aromatic epoxy acrylates.

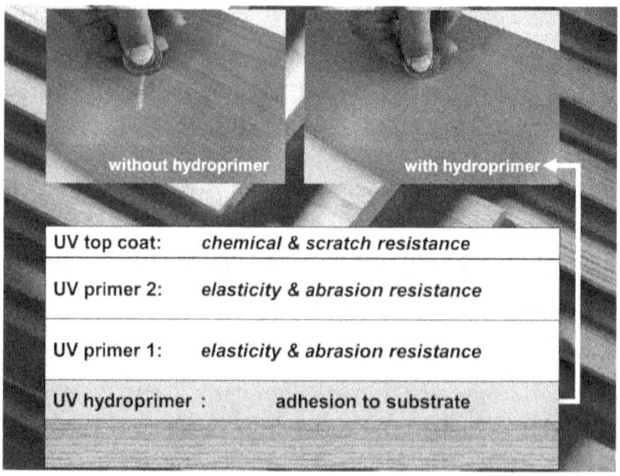

Figure 5.48: Design of the UV curable coating layers in a prefinished hardwood flooring

The high quality prefinished hardwood floorings

often consist of usually two, or sometimes three layers of abrasion resistant primer applied on top of the aqueous adhesive primer (hydroprimer). The lowermost primer coats on top of the hydroprimer are formulated with abrasive fillers and just gelled, in order to guarantee a high abrasion resistance during the lifetime of the floor. The last primer layer is free of such fillers and thus allows to create a smooth finish after UV curing by a final sanding process (the sanding of layers containing aluminium oxide or aluminium silicate would cause problems by deterioration of the sanding belts). Finally, the topcoat is applied in a thin layer to provide high gloss and good resistance properties. The process has been shown schematically in Figure 5.3 (page 283).

Table 5.8: Suggested formulation of an aqueous adhesive primer (hydroprimer)

Guide formulation: hydroprimer	
50 parts	100% UV resin, water-borne: e.g.: aliph. epoxy acrylate or polar polyether acrylate
47 parts	polyester acrylate emulsion, solid content 50%
2 parts	photoinitiator
1 part	additives (wetting, defoaming, ...)

Good **adhesion to the wooden substrate** is almost exclusively provided by using an aqueous adhesive primer (hydroprimer, see Table 5.8). The adhesion promoting effect, which is proven by a good coin-test, is achieved by a swelling and/or erection (grain raising) of the wooden fibres. This process increases the surface area and therefore offers more opportunities for penetration of the coating resins into the wood surface and providing better mechanical contact. The hydroprimer works even when applied in a thickness of about 10µm (10g/m²) at a concentration of about 20%. At that thickness and concentration the water penetrates instantaneous into the wood and therefore a specific thermal drying device for the hydroprimer can be omitted in the UV curing line.

Good interlayer adhesion is achieved either by incomplete curing of the underlying layer, or by complete curing and subsequent sanding. The possibility of

Table 5.9: Suggested formulation of a UV curable topcoat for prefinished hardwood flooring [86]

Product	Product example	Recipe [wt.-%]
Polyether acrylate	„Laromer" PO 43F (BASF SE)	74.7
Polyester acrylate	„Laromer" PE 44F (BASF SE)	8.3
Silica gel	„Syloid" 621 (Grace)	8.3
Matting agent	„Acematt" OK 607 (Degussa)	4.1
Talcum	Micro-Talc AT1 (Norwegian Talc)	2.5
Photoinitiator	„Darocure" 1173 (BASF SE)	2.1

incomplete curing (gelling) is very effective and makes sense from an economic perspective, as UV lamps of lower power or less photoinitiator may be used. Gelling will also always be used when a sanding of the coating layer is not possible or difficult (e.g. in the case of using abrasive pigments). Often an intermediate sanding of a cured coating layer is preferred, since it does not only provide better adhesion to the underlying layer, but also eliminates roughness and improves the appearance. An intermediate sanding step is therefore usually used at least before the application of the topcoat (formulation example in Table 5.9).

As mentioned already above, one of the main functions of a coating is to protect the substrate against impact or damage during use. This requires that the coating stays intact during a long lifetime and does not deteriorate easily. The lifetime durability is often tested by accelerated tests. To simulate the abrasion resistance, the abrasion behaviour is tested with a "Taber" rotary platform abrader. For parquet tile applications it has been found that fillers must be added in order to meet the required abrasion resistance. The best results are achieved with highly abrasive filler materials such as corundum (aluminium oxide). In some cases, however, even a higher layer thickness might be sufficient to yield the desired abrasion resistance result.

In furniture coatings often mixed structures consisting of layers of non-UV curable and UV curable coatings are employed. Thus, on highly porous surfaces, classical dispersion coatings are often applied as a hydroprimer, because they "seal" the surface and thus "prevent" the penetration of the low molecular components contained in UV coatings into the substrate. However, the reverse design, first applying a 100% UV coating and then a water-based coating is used, if the much better grain enhancement (German = "Anfeuerung") of the 100% UV coatings compared to dispersions has to be fully utilized. Such a process requires only a thin layer of UV coating, a short UV curing line and the topcoat dispersion will dry physically. Furthermore, combinations of UV coatings with nitrocellulose lacquers and polyurethane coatings are also employed.

5.6.3 Electronics

In the electronics and microelectronics industry radiation curable systems are used in order to pattern materials (photoresists), as protective coatings for the encapsulation of IC chips, for fixing of modules (transistors, diodes, etc.), or for labelling (printing) purposes. The applications of UV curable materials in Asia are dominated by the electronic applications. Such applications are for instance the production of printed circuit boards (PCB), where they are employed as photoresists, solder masks or labelling inks. On the other hand, such radiation curable coatings are used as hard protective top layers on displays (liquid crystal displays, plasma TV and PC's) as well as on electronic devices (optical discs, mobile phones and PC housings). The former applications require in most cases a patterning process, for example, to generate circuits, transistors, or connections. The formation of such pattern can be done either by photolithographic techniques, where a liquid or solid photoresist layer is patterned with the help of a mask, or by a printing process. The most widely used printing process for electronic applications is screen printing, a printing method where the ink is pressed through a patterned fine-mesh sieve onto a substrate.

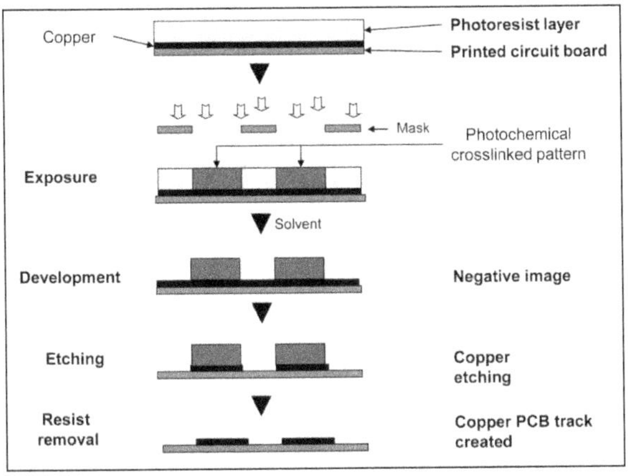

Figure 5.49: Negative working photoresists and scheme of the formation of printed circuit board tracks

Almost all acrylate coatings for electronic applications are based on epoxy acrylate resins. The photoresists based on acrylate technology are negative-working (Figure 5.49), which means that the transparent areas of the mask expose the photoresist and crosslink these patterns, which become insoluble, while the unexposed area can be removed with a solvent or a basic solution. The obtained pattern functions in the next step to protect the underlying copper in the etching process in order to produce copper tracks used on circuit boards. After etching, the patterned protective photoresist layer is removed, preferably with alkaline solutions. Therefore, in these formulations, the employed epoxy acrylates contain carboxylic acid groups or carboxylic acid group containing monomers, such as carboxyethyl acrylate, are used. Similarly, photoresists are used as solder resists, where the patterns, which do not have to be soldered, are protected by the photoresist pattern. In the meantime, due to the low thermal stress evolved during room temperature UV curing, only a few heat-curable coating systems are used in these applications, while UV curable coatings are almost exclusively applied. Furthermore, UV curable coatings are used as adhesives for the assembly of components such as transistors, diodes, etc. Due to the required high resolution, photopolymerizable negative photoresists are no longer used for the production of integrated circuits (IC). However, for casting and sealing of such integrated circuits UV curable urethane acrylates are employed [87].

UV curable coatings are also used to apply hard topcoats onto electronic devices. In China, mainly laptops, cell phones, digital cameras, PDAs and DVD players are coated with hard surface protection layers, while in Taiwan, Korea and Japan mainly mp3 players, PCs and video cameras are coated [88]. For such hard coats, mostly highly functional monomers (dipentaerythritol hexaacrylate) and high-functional urethane acrylates are employed. For plastic applications, more and more water-based UV systems come into use.

5.6.4 Other industrial applications

5.6.4.1 UV adhesives

The classification of adhesives is usually done either by the chemical basis (acrylates, polyurethanes, silicones), the **curing mechanism** (physical drying, for example of solvent-based adhesives or dispersions, or reactive adhesives, such as super glue (1K), epoxy (2K) or radiation-curing adhesives), or the function, like laminating adhesives, pressure sensitive adhesives, etc.

As with coatings, also within the application of adhesives, the reduction of solvents and rapid curing are the main reasons for the growth of UV technology. The requirements for a good adhesive are on the one hand, good adhesion to the substrate and on the other hand, good cohesion within the adhesive. Usually the cohesion increases with increasing molecular weight, but an increase of molecular weight also affects an associated increase in viscosity, which accounts for an increased required amount of solvents. In order to lower the viscosity for the application, one solution is the application of low molecular weight reactive systems (for example, unsaturated acrylates), and increase the molecular weight after the application in order to increase the cohesive forces by exposure to radiation. UV curable acrylic resins may only be used where the adhesive can be cured by at least one transparent layer (glass, PC, PMMA) or when applied directly on the surface. In contrast, photochemically initiated cationic curable systems cure much slower than free radical curable acrylate systems, so that the exposure can be performed on the surface and just generate the initiating species. With this method, for example, laminating adhesives which do not contain a transparent layer can thus be joined and stuck together, since the cationic polymerization reaction proceeds well in the dark.

The UV technology is mainly promoted in the application of sensitive adhesives (tapes), laminating adhesives (labels), CD-or DVD bonding and to a minor extent of adhesive structural bonding [89]. Radiation curable oligomers for use as laminating adhesives [90] are offered by few manufacturers for both, curing with UV or with electron beams [91]. In addition to the low viscosity unsaturated acrylates,

UV curable hot melt pressure sensitive adhesives are also applied as 100% systems at elevated temperature (about 130°C) and exposed at higher temperatures by UV exposure in order to further increases the molecular weight and improve the cohesion properties. These UV curable formulations consist of acrylate copolymers based on n-butyl acrylate and 2-ethylhexyl acrylate and an acrylated benzophenone as copolymerized photoinitiator [92, 93]. During UV exposure the benzophenone units are excited and abstract hydrogen atoms preferably from ethylhexyl acrylate and crosslink by recombination of radicals. By this reaction the molecular weight increases as a function of exposure energy density and allows to fine tune the level of adhesive and cohesive forces. These systems are commercially available under the trade name "acResin" by BASF SE.

UV curable acrylate based adhesives are employed in electronics applications for mounting devices onto printed circuit boards. The so-called chip bonder acrylics are based on a Dual Cure mechanism with first UV and then temperature curing. With a short flash of UV light, a quick fixing of the components on the circuit board is achieved, so that a dispread of the adhesive in the pure thermal curing can be avoided. Therefore the bulk of the adhesive is located under the surface mounted device (SMD) and then cured by heat. Acrylate adhesives are stable and allow for fast curing. As an alternative to the acrylates also epoxides are used in this application [94].

5.6.4.2 Optical glass fibres

Optical glass fibres for telecommunications applications consist of several layers on the light guiding fibre core (Figure 5.50). The fibre core is typically made of doped silica with a high refractive index, coated with an undoped silica cladding of lower refractive index. The fibres are then coated first with a soft UV curable coating during the fibre spinning process.

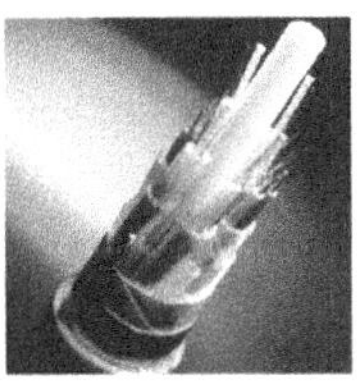

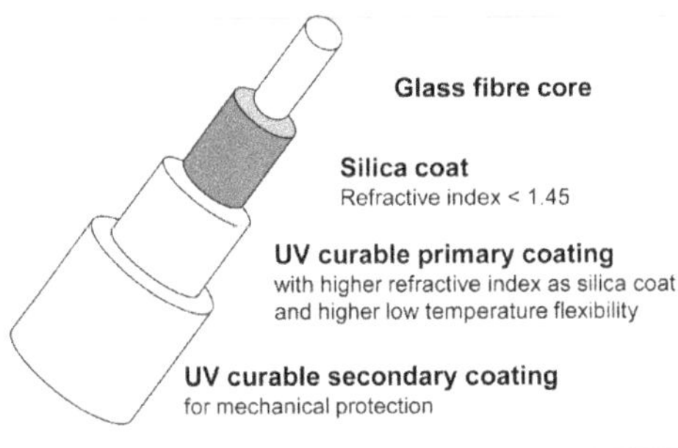

Figure 5.50: Layout of an optical glass fibre

The coating has to exhibit a higher refractive index (> 1.45) than the cladding in order to improve the signal to noise ratio of the optical fibre. Finally, the fibre is coated with a hard UV coating. The UV coatings are exposed in-situ right after application with UV light. The thickness of the fibre core is about 50 microns, the cladding about 75 microns, the first UV coat of about 60 microns and the protective final layer of about 65 microns.

Although other coating systems, like thermal curing silicone elastomers or polyamides are in use, UV curable coatings are dominating since they allow high-speed fibre spinning. The UV curable coatings consist mainly of urethane acrylates, epoxy acrylates or silicone acrylates. In order to minimize micro-cracks, especially at low temperatures, the soft primary coating has to be flexible at temperatures down to minus 60°C. Urethane acrylate resins, often blended with mono-functional acrylates such as phenoxyethyl acrylate or isobornyl acrylate, are the dominant resins in these applications because they easily provide the required high flexibility. The primary coating is applied as a flexible soft UV material in a thickness of about 50 microns in order to avoid micro-cracks during fibre bending. The cured coatings have to provide elongation at break values exceeding 100%. Besides the urethane acrylates, resins offering such high elongations are found in UV curable silicones, polybutadienes or polyether urethane acrylates. In contrast, the secondary protective layer is a hard coat which has to ensure mechanical protection and resistance to solvents, moisture, acids and bases. The formulations are often blends of urethane acrylates and epoxy acrylates, applied also in a thickness of about 50 microns. The design of the protective coating is less demanding, thus the components can be selected from the classical range of available resins in combination with reactive diluents. Elongation at break of up to 10 to 20% is sufficient. Besides coating of the glass fibres with the described two layers of UV curable materials, sometimes UV curable coatings are furthermore used for embedding the fibres in the cable harness and as UV curable inks for coding the fibres.

5.6.4.3 Stereolithography

For rapid prototyping the process of stereolithography has been developed by Chuck Hull (3D Systems). In this process, a laser is focused onto the surface of a photopolymer. The laser beam traces a pattern on the surface of the liquid resin. During the concomitant exposure the UV laser cures and solidifies the pattern traced on the resin, while the so formed shape adheres to the layer below. The so-generated form is dropped down at the laser beam thickness into the liquid photopolymer and the pattern wrote again and again according to the CAD design of the to-be build device, until the complete three-dimensional prototype of a

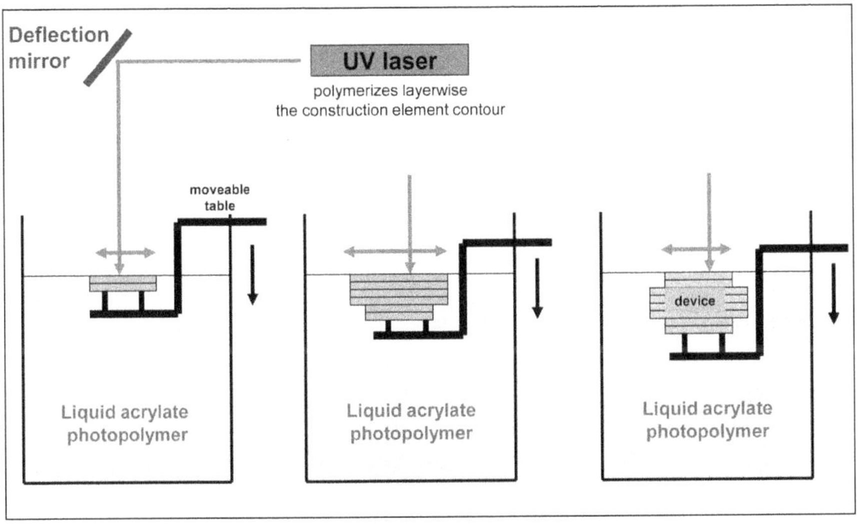

Figure 5.51: Layout of the stereolithographic process

device is generated. In order to avoid swelling of the just polymerized layer during dipping into the liquid photopolymer, alternative methods were also developed. One method makes use of re-applying always a new liquid layer with a wiper system ("Zephyr" Coating System).

The photopolymerizable materials used are specially designed acrylic type resins based on the chemistry used in classical radiation curing coatings [95]. Since acrylates often show relatively high polymerization shrinkage, latterly cationically curable epoxy resins have been enforced, which are cured by using an acid-generating photoinitiator.

5.6.4.4 Dental materials

Photocurable methacrylate based resins are used in dental applications mainly as an alternative to amalgam filling [96]. The resins used are almost all based on methacrylate chemistry, mainly due to higher T_g and lower toxicity (skin irritation). The early formulations were based on methacrylate resins (bisphenol-A reacted with glycidyl methacrylate) and diluents (triethylenglycol-dimethacrylate) in combination with modified silica glass and photoinitiators (camphorquinone). Bis-GMA exhibits much lower volume shrinkage over the previously used methyl methacrylate. The inorganic glass fillers are often functionalized with methacryloylpropyl trimethoxysilane in order to ensure the chemical linking of the filler to the polymer matrix. The main disadvantage of such composites is still their polymerization shrinkage; a considerable advantage against amalgam and other alternatives is the possibility of composites to match the tooth colour.

As an alternative to the glass-filled resins, the Ormocers (organically modified ceramics) have been established, which are based on sol-gel chemistry, where an inorganic network is formed via siloxane condensation reactions and a photochemically induced organic network based on methacrylate crosslinkers.

Recommendations of related literature

- Albert, P., Dermann, K., Rentsch, H., „Amalgam und die Alternativen", **Chemie in unserer Zeit**, Vol. 34 (5), S. 300-305 (2000)
- Fraunhofer, J.A., "Dental Materials at a Glance", Wiley&Sons, (2009)
- Fouassier, J.P, Rabek, J.F., Radiation Curing in Polymer Science and Technolgy, Elsevier 1993, Vol. IV, Chapter 13: Photocuring of Polymeric Dental Materials, S. 387-466

5.6.5 UV coatings for exterior applications

UV curable coatings are by now also used in exterior applications. Since the dogma has been overcome, that UV curing will not work if UV stabilizers, which are necessary for exterior applications, are present in UV curable coatings [97], a lot of outdoors applications have evolved. For a weather-resistant, low-yellowing formulation, the proper selection of resins, photoinitiators and monomers is crucial.

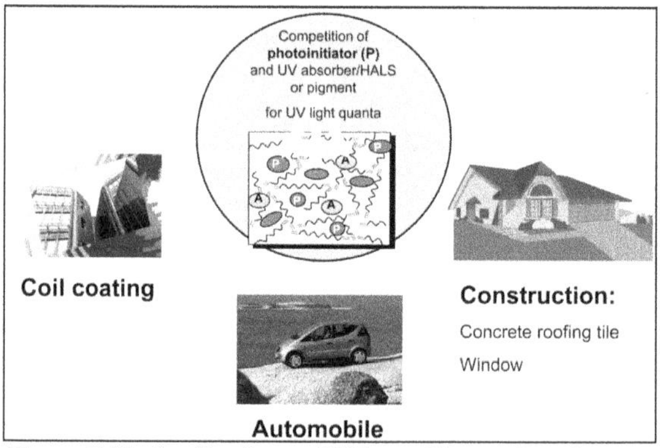

Figure 5.52: Applications of UV curable coatings in outdoor areas: UV absorber and HALS radical scavengers are necessary for lifetime stability, but may interfere with the UV curing process

Almost all exterior coatings contain light stabilizers, namely UV absorbers or absorbing pigments and HALS radical scavengers in order to protect both, the clear coat and the underlying layers from degradation reactions caused by sunlight exposure. The influence of UV absorbers and HALS radical scavengers on the UV curing reaction was investigated by

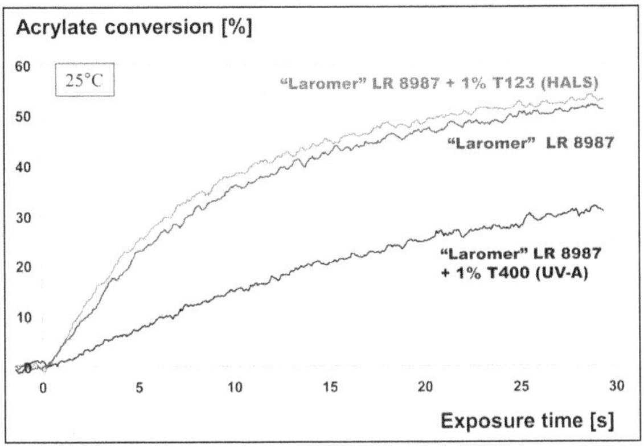

Figure 5.53: Conversion of an UV cured urethane acrylate resin ("Laromer" LR 8987) in the absence and presence of an UV absorber (UV-A) and HALS radical scavenger, respectively

Decker [98]. It had been found that the addition of UV absorbers such as "Tinuvin" T400 reduces the conversion of the acrylate double bonds significantly, while the addition of the HALS amine radical scavengers had no influence at all on the conversion (Figure 5.53). This, at first glance, is a surprising result. However, it can be explained by the fact that the used HALS amine is present not as a radical in the initial coating, but the efficient free radical scavengers are formed during exterior use upon oxidization by the combined action of light and oxygen.

With the Fourier transform infra red (FT-IR) method developed by *Decker* [99] the tracking of the decrease (urethane C-NH vibration or CH bonds) or increase (CO and OH) of chemical bonds during weathering simulation test (QUV test) can be done easily. The effect of the stabilizers is shown in Figure 5.54 by plotting the remaining urethane groups of the coating as a function of aging time. Appropriate stabilized UV cured coatings exhibited excellent weathering stabilities, which were at least comparable or exceeded the already known excellent exterior stabilities of the 2K polyurethane automotive coatings (Figure 5.55). Superior results of the outdoor stability of UV cured clear coats were also confirmed in field tests (Jacksonville, FL) [100].

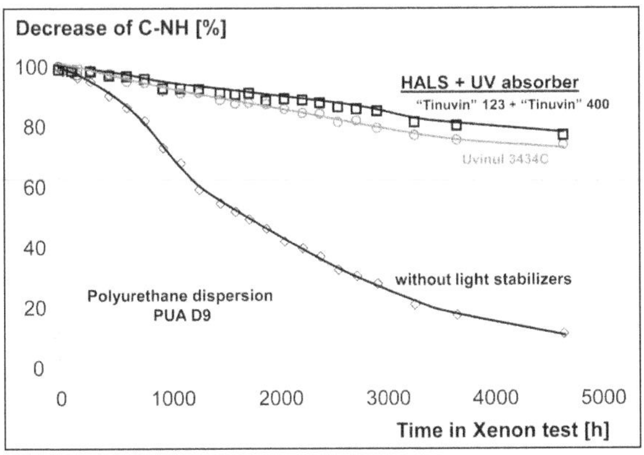

Figure 5.54: Weathering stability in Xenon QUV-A test of unstabilized urethane acrylate dispersion compared to the dispersion containing light stabilizers

In particular, urethane acrylates have demonstrated their enormous potential for the use as UV curable coatings in outdoor applications. In model formulations it has been demonstrated that their weathering resistance is at least comparable to the currently dominating 2K PU coatings in the automotive sector [101]. Since automotive coatings have to provide the most challenging requirements, such UV curable urethane acrylates will be appropriate also for other applications, such as window coatings, the coating of tiles, coating of concrete roof tiles or floor tiles, etc.

5.6.5.1 UV curable coatings for automotive applications

Since now for almost 15 years, the coating of automobile headlights was the pioneering application of UV curable coatings in automotive parts. UV curable coatings are used for two different purposes, on the one hand, the polycarbonate lenses are coated with a hard, scratch-resistant UV coating, and on the other hand, the reflector housing was coated with an UV primer that provides an ultra smooth surface for subsequent metallization [102, 103]. The UV curable coatings are used in the application for the headlights because of the favourable property spectrum of UV coatings, particularly the high scratch resistance. In other automotive applications which are under development, for instance, the use of UV curable coatings in the refinish industry, another aspect comes to the forefront. The fast curing of the coated parts is an enormous advantage as the period in which the coating is still tacky and where it is sensitive to pick up dirt particles, can be reduced significantly. Therefore considerable efforts are ongoing in the development of UV curable coatings for repair applications [104]. It is not only the chemical formulation which has to be optimized in respect to the property spectrum and especially the

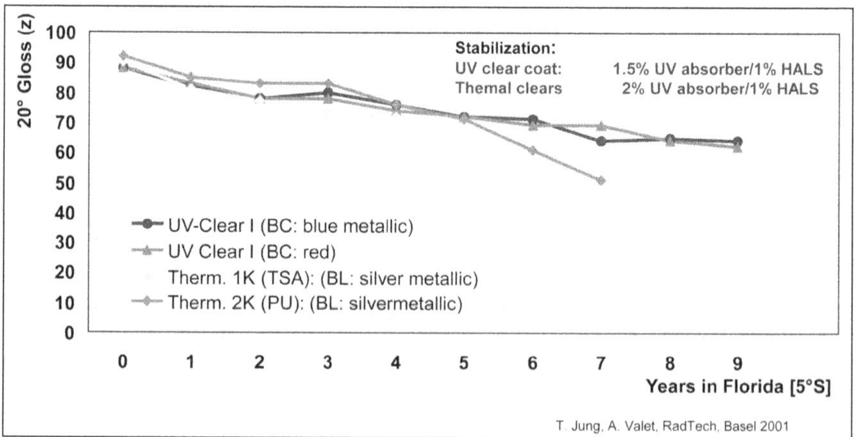

Figure 5.55: Weathering stability of model clear coats containing light stabilizers tested in Florida climate

tolerance to the oxygen inhibition effect, but also the exposure equipment has to be designed in terms of safety aspects (UV-A lamps), ease of use and efficiency (flashlights) as well as the flexibility to adapt to the component contours.

The appropriateness of UV clear coatings based on urethane acrylates for weather-resistant applications has already been demonstrated [105] (Figure 5.55) in automotive specific tests.

Furthermore, monomer free aliphatic urethane acrylate blends of a hard and a soft urethane acrylate resin have been used in a formulation designed for exposure with a UV-A lamp and could demonstrate their suitability in practical repair applications. After the usual processing, like sanding and polishing, the level of properties, such as resistance and optical impression (gloss, haze) were absolutely comparable to the benchmark repair based on 2K polyurethane repair systems [106].

Further uses of UV curable coatings already realized in automotive applications are UV curable sealers for SMC (sheet molding compounds) parts, applied in order to prevent the outgassing of air from the composite materials during the classical thermal cured painting.

UV clear coatings for plastic mounting parts for both interior and exterior components, as well as clear coatings for the coating of wheel rims and wheel covers and ultimately for OEM coating of the whole automotive body are currently intensively under development [107, 108].

5.6.5.2 UV curable coatings for construction applications

The industrial coating of window frames is done either with solvent based or aqueous coatings. For repair coating of window frames by the do-it yourself application water-borne paints have been more and more accepted. For the industrial coating of window frames also UV curable coatings are in development, since they dry (harden) faster and via an additional crosslinking they exhibit better (longer) weathering stability.

For the coating of concrete floors Quaker Chemical developed an UV curable coating system called “Rapid Shield”, in which the UV curable coating is applied by a roller or brush on the floor and is then cured with special mobile lamps [109] (see Figure 5.59). Furthermore, for surface protection of concrete stone products, such as paving stones, acrylic emulsions are usually applied, however, also in this application UV curable coatings are advancing [110]. For the coating and UV curing of these applications, special equipment is needed (e.g. BM Anlagenbau, Hameln). Last, but not least, another example of the use of UV curable systems in construction applications, in the repair of concrete construction elements, UV curable aqueous dispersions are employed [111].

Another application of acrylate based radiation curable coatings is the top coating of prefinished facade panels. The company Trespa, for example, offers under the brand name "Trespa Meteon" facade panels, where the pigmented surface coating was cured with electron beam radiation.

5.6.6 UV curing within alternative coating technologies

5.6.6.1 UV powder coatings

Powder coatings are one of the most environmentally friendly coatings because they are applied without any solvent and excess powder can be easily recycled. The main disadvantage of thermally cured powder coatings, however is, that the melting process starts usually at high temperatures (> 100°C), and at these conditions the subsequent film formation interferes with the simultaneous thermal curing process. This often leads to the surface defects recognized in powder coatings, the so-called orange skin texture. In the case of UV curable powder coatings, it was expected to separate the film forming process from the curing reaction, since at the point where the film forming process is completed and an optimum surface is obtained, the curing reaction can be triggered by switching on the UV lamps.

The first applications of UV curable powder coatings started with the coating of electric motors (Baldor Electric, 1997), engine coolers (Valeo Engine, 1999) and fibreboards (MDF, Decorative Veneer, 2001), however, until now, the share of UV curable powder coatings of the total powder market is still negligible low.

UV curable powder coating formulations consist of unsaturated resins (unsaturated or acrylated polyesters), if necessary crystalline crosslinkers (vinyl ethers or acrylated urethanes), and additives (photoinitiators, flow additives, pigments). The formulation of the powder coating is done by mixing of the components in an extruder above T_g and below the melting temperature, cooling down and grinding the mixture to a powder. The particle size of the powder is generally in the range of 40 to 60 microns. Tailor-made properties for example for metal applications could be achieved with a combination of amorphous and crystalline urethane acrylate resins [112]. In order to develop powder coatings that melt at lower temperatures (<100°C) and thus can be used also on temperature sensitive substrates such as wood, the quadrant between good process ability (powder has to be storage stable at 40°C), and good film forming properties (melting and levelling <100°C) had to be solved. In this respect crystalline UV curable reactive diluents have been proven as the components with high potential [113].

Although UV curable powder coatings enable the decoupling of the melting process (flow) from concurrent crosslinking (as in the case of thermal curing), they always

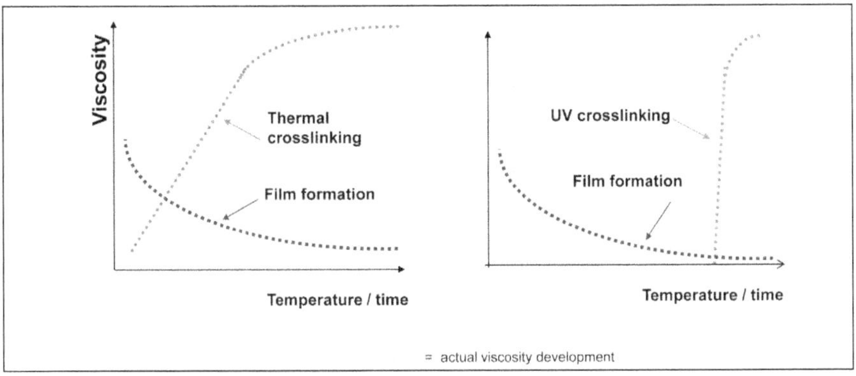

Figure 5.56: Melt diagram showing the viscosity dependence of thermally curable and UV curable powder coatings on the baking temperature and time

compete with 100% UV coating (better flow) and classical coatings, like high-solids or low temperature curing powder coatings. Up to now, they were applied mainly if the user had already installed a powder paint application line and wants to speed up the curing reaction or for a specific technical reason, such as when different temperature-sensitive materials are present in a construction element.

The RadTech Europe organization [114] has installed a "UV Powder Working Group", which is dealing with all relevant issues related to UV powder coatings, for example:

- Powder applications
- Comparison of conventional powder with 100% UV and solvent based coatings
- General advantages of the UV powder technology
- Limitations
- Economy of UV powders

Application and curing lines for the processing of UV curable powder coatings for the coating of MDF (medium density fibreboard) substrates have already been realized based on different concepts [115].

5.6.6.2 Dual Cure systems (UV and non-UV curing)

Major challenges for the further spread of UV curing systems represent three-dimensional substrates, especially when they contain undercuts into which the UV light, which is necessary for the curing, hardly can penetrate. Furthermore, in coatings which need heavy pigmentation in order to achieve opaque layers, the UV curing often is problematic because the pigments absorb the light and

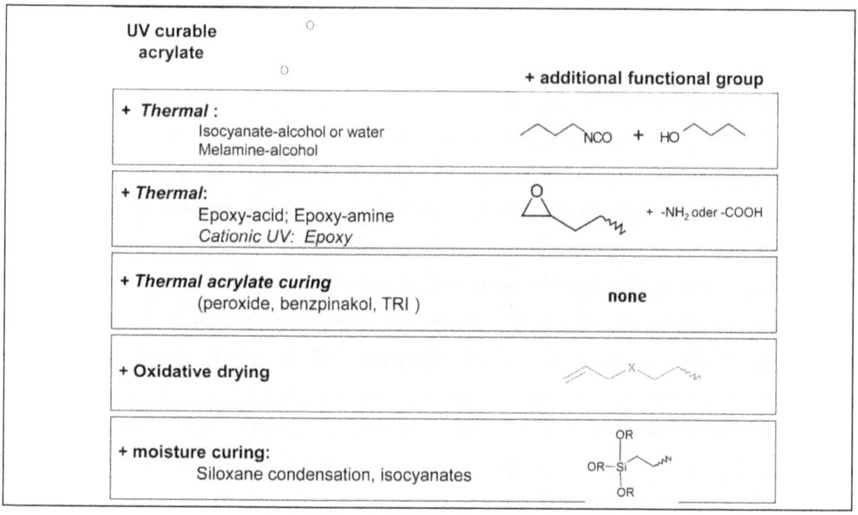

Figure 5.57: Dual Cure coating systems reflecting the combination of radiation curing functionalities with non-radiation curing mechanism

therefore limit curing in thicker layers. This also holds true if thicker layers than the few microns which are applied in printing inks are necessary. The problem of curing in the so-called shadow areas can be solved sustainably for many applications by the suitable combination of two curing mechanisms, where one mechanism is independent from curing by radiation. The many possible combinations of UV curing with different other non-UV curing mechanisms are exemplified in Figure 5.57.

The most common method is the combination of radiation curing with thermal crosslinking. Thermal crosslinking methods are available in many variations. There are a number of possible combinations of radiation curing with thermal hardening, such as the common reaction of hydroxyl group containing unsaturated acrylic esters with isocyanate resins or melamine crosslinkers. These OH group containing acrylates can be applied in two-component systems in combination with polyisocyanates or as one-component coatings in combination with melamine resins, so that upon radiation and thermal crosslinking a homogeneous network of radically polymerized acrylates and thermally crosslinked groups results. Therefore curing even of the coating in shadow areas or in highly pigmented coating layers is possible.

Other applicable systems can be based on epoxy groups introduced into acrylic esters, so that a cationic epoxy curing takes place induced via thermal latent acid donors as the UV light-independent mechanism, or the epoxy acrylates can be crosslinked with carboxylic acid or amino-group containing resins.

Other possibilities are the combination of acrylates with oxidative drying groups, which cure slowly with oxygen analogue to alkyd resins.

In some cases it may already be sufficient, that the shadow areas or pigmented coating systems only dry physically. Thus, for example, radiation-curable, aqueous dispersions often exhibit physical drying without radiation curing. The resins of polyurethane dispersions, for example, exhibit a much higher molecular weight than traditional 100% UV resins, and may be designed to offer glass transition temperatures of the uncured coatings above room temperature. They can be crosslinked in the use areas by radiation curing in order to offer the designated performance spectrum and solidify in shadow areas, in which often the demands are only related to low emission, low migration and low odour.

A further favourable curing mechanism is moisture cure. Capable of curing by moisture, are for instance, siloxane or isocyanate groups. Compounds containing both, unsaturated acrylates and siloxanes or free isocyanate groups are commercially available. A particularly useful polymer having free isocyanate and unsaturated acrylic ester functions has been introduced recently (Figure 5.28, "Laromer" LR 9000). This molecule can be used either as the polyisocyanate component in the formulation of UV curable 2K systems in combination with hydroxyl-containing resins, or as a sole binder resin utilizing the moisture cure of the isocyanate groups and as a primer resin, where the isocyanate groups react with functional groups of the substrate. For example, the reaction between the NCO groups of the resin with the respective reaction partners in the substrate (OH-, HOOC-, etc.) results in excellent adhesion to the substrate.

However, all these systems have an inherent disadvantage, unless the thermal curing temperature is so high that the acrylate groups are polymerized thermally, the shadow areas rely only on the curing of the thermally reactive groups.

The most elegant way of Dual Cure systems is therefore the additional thermal curing of the unsaturated acrylate groups using peroxides, benzpinacol (tetraphenyl ethanediol) or other thermally latent radical formers (TRI) [116]. In these formulations both, the radiation-induced and the thermal induced polymerization utilize the same functional groups and thus, at least theoretically, should give the same network. Thus, in the radiation cured area as well as in the pure thermally cured shadow areas a complete conversion of all available reactive groups takes place.

5.6.6.3 UV film coating technology

The substitution of traditional spray coating by preformed coating films has been gained a lot of attractiveness [117]. Especially in the automotive industry, more and more different substrates (steel, aluminium, plastic) are coated by more and more different suppliers. During the assembly the colours of the different parts have to match by 100% to the desired colour. Even when using the same colour batch, by the spraying process, significantly deviating colours may result, a problem which appeared under the keyword "colour match". The hope is that a coated film which is decorated to the relevant parts can significantly mitigate the problem and ultimately substitute the whole spray painting process at the car manufacturers' sites. Along the way, there are first automotive parts produced by this technology [118]. Key to implementing this technology is a thermoplastic deformable and in the final state UV curable clear coat, which offers to the automotive part at least the same properties as the conventional 2K clear coats. To implement this approach, plastic mounted parts were the first to be selected, because the film technology could more easily be implemented by this special method (Figure 5.58). As shown in Figure 5.58 a thermoplastic carrier foil (e.g. ABS) is coated first with the colour-producing basecoat and then coated with a UV curable clear coat and a cover sheet. The film can be cut according to the component dimensions and is then thermally deep drawn into the final shape. Thereafter, the clear coat is crosslinked through the protective foil (inert conditions) with UV light and finally the thus prepared film is injection moulded with the same thermoplastic (ABS) in order to construct the final mounted part, for example, a roof module.

The final clear coat on automotive parts must meet the high requirements for scratch, chemical and weathering resistance, which are mainly accessible via a high crosslinking density. A pre-applied clear coat with such a high crosslinking density however, would only tolerate deformations well below 50%. However, in the manufacture of complex components, such as mirror housing, elongations of more than 100% are required. This can be achieved with UV curable urethane acrylate coating films, which behave like thermoplastic deformable polyurethanes in the unexposed film. After the part is moulded into its final shape the UV exposure converts the thermoplastic film into a highly crosslinked coating. To ensure the required weathering resistance urethane acrylates have proven appropriate as resins. The formulations for the clear coat must be adjusted in respect to the glass transition temperature in order to exhibit a tack-free film of the unexposed film. In

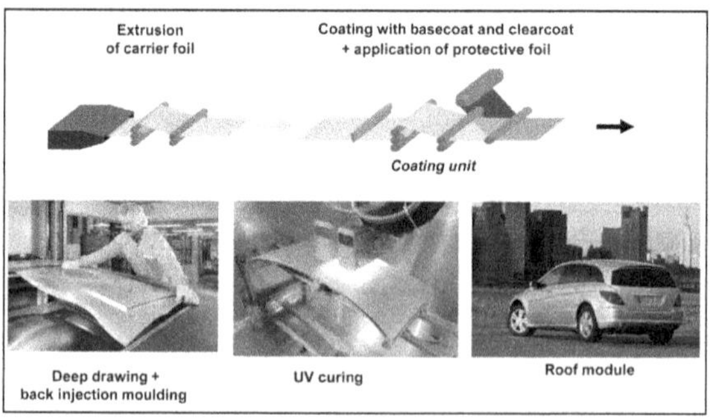

Figure 5.58: Scheme of the process to manufacture automotive plastic mounting parts by UV curable foil technology

order to provide still enough mobility of the crosslinkable groups during UV exposure the irradiation conditions (temperature, energy) have to be optimized precisely. This film technology opens up a number of new applications where the surface properties (scratch resistance, chemical resistance) of the currently applied thermoplastic films are not satisfactory, like in automotive surfaces, computer cases, mobile phone housings, industrial or furniture applications.

5.6.7 Opportunities for UV coatings in the "new" applications

Based on the advantages of radiation curable coatings with regard to eco-efficiency, fast curing at room temperature, and the ability to pattern surfaces by selective exposure, there are a number of new or well-known applications which are being explored intensively. Applications in which UV coating technologies can be used preferably include [119]:

- Coil coating
- Primer and topcoats for metalized plastic parts
- Floor coatings
- Printed electronic circuits

With the coating of metal sheets (coil coating), an application came into focus again, which had been already under investigation in the early days of radiation curing technology. The radiation curing method is generally ideally suited because the coating is cured on two-dimensional metal bands at high cure speed. In order to limit the baking oven length at a curing speed of about 180m/min, the conventional coating is now cured at temperatures up to 240°C. Thus, the substitution of thermal curing by a radiation curing process at considerably lower temperatures, ideally at room temperature, would result in enormous energy savings. A breakthrough in the use of radiation curable systems so far has failed mainly because of technical problems, like lack of adhesion to the metal due to high volume shrinkage or incomplete curing due to high pigmentation of the topcoat layer. Pioneers in the development of radiation cured coatings for coil coating processes using electron beam exposure can be found in Japan, where in particular household appliances and kitchen furniture is manufactured with electron beam cured topcoats and adhesives [120]. The traditional coil coating manufacturing lines are almost all designed for a two-coat application. In the first step the primer is coated and cured at temperatures of up to 240° C, cooled down and then, the metal coil is recoated with the topcoat and heated up again to 240° C to cure the top layer. In applications where a third functional layer would be desirable, for instance, to provide higher scratch resistance, the integration of an UV curing unit is very beneficial. It can be easily implemented because it needs only very limited space and does not need the heating cooling cycle. The easiest way to incorporate an UV curable layer would be the primer. Newly developed UV curable primers based on flexible urethane acrylates exhibiting good adhesion to the metallic substrates have already been reported to meet the basic requirements [121]. Furthermore it has been demonstrated with model formulations that UV curable acrylate based formulations can meet the high performance standards of metal coatings with respect to mechanical properties, adhesion and corrosion protection [122].

Chrome-plated plastic parts are represented in many daily use applications (household fittings, automobile interior parts, electronic components). Their conventional production process however, is expensive. Therefore, the conventional process is more and more replaced by the physical vapour deposition of very thin metal layers (0.1 micron; PVD). UV curable coatings offer ideal opportunities in order to run the process fast and efficient. Furthermore, the curing of the planarization layer on the plastic part under the thin metal layer and the clear coat to protect the metal layer can be done at such low temperatures that the plastic part is not warped. Acrylate based formulations similar to those already used in the coating of automobile headlamps, will also have great potential in these applications.

In a special issue of the Journal of Surface Technology (JOT), a series of "new" applications for UV systems are described [123]. In addition to detailed cost comparisons of UV curable coating processes with conventional coating technologies, new concepts for the installation of three-dimensional curing lines including material recycling of overspray are discussed. The examples given include a number of applications on wood, plastic and metal, such as scratch-resistant coatings on guitar bodies, plastic packaging for cosmetics, steel pipes for oil and gas transport, pipes, diesel fuel injection pumps and vacuum cleaner housings. With the installation of a first automated UV coating line for truck axles in the Mercedes factory in Gaggenau, the suitability of UV curing technology for the curing of large complex parts has been shown.

Conventional coatings used for the coating of concrete floors are nowadays based on epoxy resins, urethane systems or redox initiated formulations based on methylmethacrylates. These systems however, are very slow in curing speed and often require hours or days until they are fully cured. Fast curing UV coatings would be beneficial in these applications, since the floors can be entered immediately after curing. In addition to the high performance requirements of the considered coatings, also appropriate coating and radiation curing devices have to be developed. The first prototypes of such an UV curing equipment has been presented recently by the Quaker company with the "RapidShield" system [124].

The propagation of electronic devices such as flexible displays, electronic paper (e-paper), flexible photovoltaic cells, transistors, flexible batteries, RFID labels

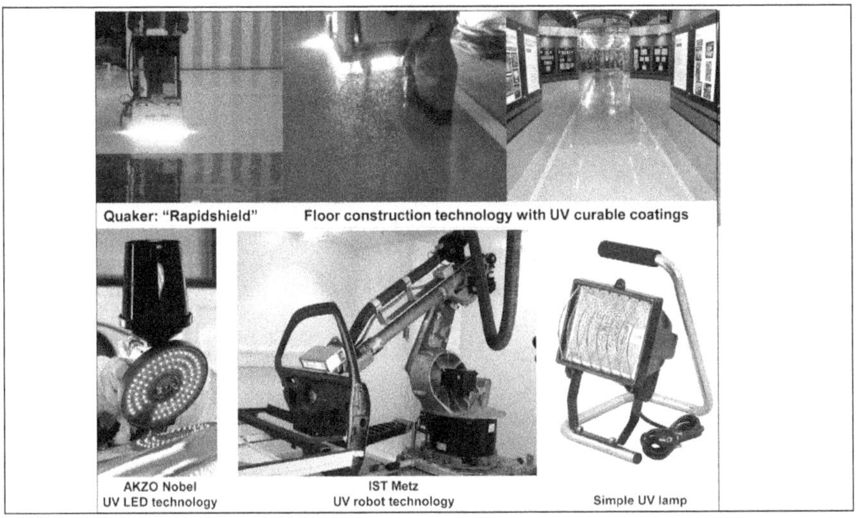

Figure 5.59: Innovative curing equipment developments for new applications, like floor coatings (Quaker), spray applications with concomitant LED exposure (AKZO), contour exposure with robots (IST) and simple modified lamps as cost efficient alternative (e.g.: for exposure under inert atmosphere)

will grow significantly with the implementation of printed electronic circuits on inexpensive plastic substrates, associated with low-cost processing and rapid manufacturing. Here too, UV curing systems will prevail because of the advantages of curing at low temperatures (on heat-sensitive films/substrates), the possibility of easy application in the printing processes and the rapid curing process as key technology. In these applications, also unsaturated acrylate resins will be used, however, only a few uses will be opened up for the standard acrylates, since these systems require specifically tailored resins in order to comply with the specific property requirements.

5.7 Literature

Recommendations of related literature

General Coatings Technology

- Wicks, Z.W., Jones, F.N., Pappas, S.P., **Organic Coatings,** 3rd Edition, John Wiley & Sons (2007)
- Goldschmidt, A., Streitberger, H.J., **BASF Handbuch Lackiertechnik**, Vincentz Verlag, Hannover (2002)
- Müller, B., Poth, U., **Coatings Formulation**: *An international textbook,* 2nd Edition, Vincentz Network (2011)
- Kittel, H., **Lehrbuch der Lacke und Beschichtungen**, Bd. 1-10, S. Hirzel Verlag, Stuttgart (2008)

Radiation Curing

- Fouassier, J.P., Allonas, X., "Basics and applications of photopolymerization reactions", Vol. 1-3, Research Signpost, Kerala, India (2010)
- Glöckner, P., Jung, T., Struck, S., Studer, K., **Radiation Curing**, Vincentz Network (2009)
- Schwalm, R., **UV Coatings**, Basics, Recent Developments and New Applications, Elsevier (2007)
- Müller, B., Poth, U., **Coatings Formulation**: *An international textbook,* 2nd Edition, p. 237-249 (Radiation Curing), Vincentz Network, Hannover (2011)
- Garratt, P., **Strahlenhärtung**, Vincentz, Hannover, (1996)
- Fouassier, J.P, Rabek, J.F., Radiation Curing in Polymer Science and Technology, Elsevier 1993; Vol. I: Fundamentals and Methods; Vol. II: Photoinitiating Systems; Vol. III: Polymerisation Mechanism; Vol. IV: Practical Aspects and Applications
- Holman, R, UV&EB Curing, Formulation for Printing Inks, Coatings and Paints, SITA, London, (1984)

References

[1] a) Glöckner, P., Jung, T., Struck, S., Studer, K., Radiation Curing, Vincentz, (2009)
b) Schwalm, R., UV Coatings, Basics, Recent Developments and New Applications, Elsevier, 2007
c) Garratt, P., Strahlenhärtung, Vincentz, Hannover, 1996
d) Fouassier, J.P., Rabek, J.F., Radiation Curing in Polymer Science and Technology, Vol I-IV, Elsevier 1993

e) Holman, R, UV&EB Curing Formulation for Printing Inks, Coatings and Paints, SITA, London, 1984

[2] Beck, E., in Kittel, H., Lehrbuch der Lacke und Beschichtungen, Chapter 2.3.2.11.1.2

[3] http://www.radtech-europe.com/files_content/rtevdmaprotokolmay05.pdf

[4] a) Biehler, M., Menzel, K., Saling, P., UV Spektrum, Fusion UV, Winter 2003
b) Schwalm, R., UV Coatings, Elsevier, 2007, p. 13–14

[5] Harbournc, D. , „Overview and Update of the radiation curing market and technology in North America", RadTech Asia, **(2005)**, Proceedings, p. 3–22

[6] Luo, V., „Perspectives on radiation curing industry in Asia", RadTech Asia, (**2005**), Proceedings, p. 24–25

[7] Burger, P., besser lackieren, 15/**2000**, p.1.
Schneider, M., Klein, W., Schröder, C., RadTech Europe, (**1999**), p. 711–716.

[8] Lapin, S., ACS Symposium Series, Vol. 417, Kap. 26, p. 363–381 (1990)

[9] Schwalm, R., Binder, H., Funhoff, D., Lokai, M., Schrof, W., and Weiguny, S., RadTech Europe 1999, Conference Proceedings, p. 103–109

[10] Hoyle, C., Clark, T., Lee, T.Y., Roper, T., Pan, P., Wei, H., Zhou, H., Lichtenhan, J., RadTech Europe 2005, Conference Proceedings, Vol. I, p. 289–293

[11] Fouassier, J.P., Rabek, J.F. (Ed.), Radiation Curing in Polymer Science and Technology, Vol. III, Elsevier, 1993, Peeters, S., p. 181–183

[12] Albert, P., Dermann, K., Rentsch, H., Amalgam und die Alternativen, Chemie in unserer Zeit, 34 (5), 300–305 (2000)

[13] Huemmer, T.F., Industrial Finishing, 46 (5), 34 (1970)

[14] http://www.cefic.be/files/Publications/Labelling%20Guide%20German.pdf

[15] Jansen, J.F.G.A., Dias, A.A., Dorschu, M., Coussens, B., Macromolecules 2003, 36, 3861–3873

[16] Fouassier, J.P. and Rabek, J.F. (Ed.), Radiation Curing in Polymer Science and Technology, Vol. III, Elsevier Applied Science, London, 1993, Chapter 2, Decker, C., p. 33–64

[17] Van Holen, J., Stone, V., van den Bergen, H., RadTech Europe, 2005, Conference Proceedings, Vol. I, p. 237–242

[18] Kilambi, H., Reddy, S.K., Schneidewind, L., Lee, T.Y., Stansbury, J.W., and Bowman, C.N., Macromolecules, 2007, 40(17), 6112–6118

[19] Osaka Organic Chemicals, DOL series

[20] Rounds, N., Plamondon, J., Kohr, A., Radiation Curing Conference Proceedings (1986)

[21] DE 3843930; DE3843938 (Henkel KG)

[22] DE 12634 (1983)

[23] WO 907483 (Henkel KG)

[24] DE 2838691 (UCB, BE)

[25] GB 1006587 (1965)

[26] Harka, U.D., Pfundhelle, D., Muehlber, A.J., Webster, D.C., RadTech US, Baltimore, (2010), online

[27] a) Meier-Westhues, U., Polyurethane, Vincenz Network, Hannover, 2007, p. 58
b) PKT Oldring (Ed.), Chemistry & Technology of UV-, EB Formulations for Coatings, Inks & Paints, Vol. 2, SITA Technology, London, (1991), p. 97–105

[28] Dvorchak, M., Henderson, K., Gambino, C., RadTech US 2010, Proceedings, online; *and* Ludewig, M., Kuczewski, H., Fischer, W., RadTech Europe 2009, Proceedings, on CD

[29] US 4,587,313 (Hitachi Chemical, 1986)

[30] Anon., Jap. Patent 58,113,280 (1983)

[31] Chisholm, M.S., Harte, H.C., Rumoroso, I., Slark, A.T., Williams, M.C., Surface Coatings Int., Part A, 06/2003, p. 221–228

[32] Fouassier, J.P. and Rabek, J.F. (Ed.), Radiation Curing in Polymer Science and Technology, Vol. IV, Elsevier Applied Science, London, 1993, Chapter 3, Huber, H.F., p. 51–85

[33] Hult, A., Malström, E., Johansson, M., Polym. Mater. Sci. Eng., 72 (1995), p. 528–529

[34] a) Malmstroem E, Hult A, Gedde UW. Relaxation processes in hyperbranched polyesters: influence of terminal groups. Polymer 1997; 38:4873–9.
b) Wei H, Lu Y, Shi W, Yuan H, Chen Y., UV curing behaviour of methacrylated hyperbranched poly(amine-ester)s. J Appl Polym Sci 2001;80:51–7.
c) Kou H, Asif A, Shi W., Photopolymerization kinetics of hyperbranched acrylated aromatic polyester. J. Appl. Polym. Sci. 2003; 89:1500–4.
d) D.K. Chattopadhyay1, K.V.S.N. Raju, Prog. Polym. Sci. 32 (2007) 376–377

[35] Shi, W.F., Ranby, B.J., J. Appl. Polym. Sci., 1996, 59, 1937 (1996)

[36] Klang, J.A., "Radiation curable hyperbranched Polyester Acrylates (Sartomer), RadTech Europe 2007, Conference Proceedings on CD

[37] Gummeson, J.J., Radiation curing of Polymeric Materials, American Chemical Siciety, Washington, DC, 1990, Chapter 14, p. 176–193

[38] Wendrinsky, J., Melamine acrylates: An interesting class of oligomers, RadTech Europe 1989, Conference papers (1989), p. 453–463

[39] Fouassier, J.P., Rabek, J.F., Radiation Curing in Polymer Science and Technology, Vol. III, Peeters, S., Chapter 6, p. 201–203

[40] Wimscheider, H., Pulverbeschichten, mo, Carl Hanser Verlag, München, 55 (2001), 4, p. 44–45

[41] EP 0702067, Hoechst (1995)

[42] Paulus, W., Binder, H., Henning, I., Rieger, J., Schrof, W., Schwalm, R., UV härtende Pulverlacke, Blocken und fließen, Farbe&Lack, 107, 10/2001, p. 46–52

[43] US 7,001,931 (BASF, 2001)

[44] Fink, D., Brindöpke, G., Eur. Coat. J., (1995), p. 606–608

[45] Johansson, M., Glauser, T., Jansson, A., Hult, A., Malström, E., Claesson, H., Progress in Organic Coatings, 48, (2003), p. 194–200

[46] EP1144476 and US 6617413 (BASF SE)

[47] Trombetta, T., Pollone, P., Zompatori, A., Vezzulli, G., RadTech Europe 2007, Conference Proceedings (CD)

[48] Gould, M.L., Narayan-Sarathy, S., Hammond, T.E., Rechter, R.B., RadTech Europe **2005**, Conference Proceedings, Vol. I, p. 245–251

[49] Beck, E., Keil, E., Lokai, M., Farbe&Lack, 98, (3), (1992), p. 165–170

[50] Meier-Westhues, U., Polyurethane, Vincenz Network, Hannover, 2007, p. 84

[51] Schwalm, R., Häußling, L., Reich, W., Beck, E., Enenkel, P., Menzel, K., Keil, E., Tuning the mechanical properties of UV coatings towards hard and flexible systems, Prog. Org. Coat., 32 (1–4), (1997), p. 191–196

[52] Weikard, J., Lühmann, E., Sommer, S., The New Frontier: Waterborne UV Coatings for Plastics, RadTech Europe, 2007, Wien, Conference Proceedings auf CD
[53] Decker, C., Masson, F., Schwalm, R., Polymer Degradation and Stability, 83, p. 309–320 (2004)
[54] Sartomer Application Bulletin, Surface Tension of Sartomer Monomers
[55] Kloosterboer, J.G., Network formation by chain crosslinking photopolymerization and its application in electronics, Advances in Polymer Science 84, Springer Verlag, Berlin, (1988), p. 37–49
[56] Meichsner, G., Deuter, F., Groß, T., Beck, E., and Menzel, K., Netzwerkgrößen UV-gehärteter Lackfilme, Farbe&Lack, 103, 8/97, p. 45–50
[57] Miller, D.R., and Macosco, C.W., Makromolecules 9 (1976), p. 206 and p. 199
[58] Fox, T.G., and Loshaek, S., J. Polym Sci., 15, (1955), p. 371
[59] Stutz, H., Illers, K.H., Mertens, J., J. Polym. Sci., Part B: Polym. Phys. 28, (1990), p. 1483
[60] Wu, S.J., J. Appl. Polym. Sci., 20, p. 327 (1976)
[61] Schwalm, R., Crosslinking effect on mechanical properties of UV curable coatings, PPCJ, October 1999, p. 18–22
[62] Awokola, M., Lenhard, W., Löffler, H., Flosbach, C., Frese, P., UV crosslinking of acryloyl functional polymers in the presence of oxygen, Prog. Org. Coat., 44, (2002), 211–216
[63] Wight, F.R., Oxygen inhibition of acrylic photopolymerization, Journal of Polymer Science: Polymer Letters Edition, 16, (1978), 121–127
[64] a) Beck, E., Proceedings RadTech Europe, Basel, p. 643 (2001)
b) Müller, R., Proceedings RadTech, Europe, Basel, s. 149 (2001)
[65] a) Pieke, S., Heering, W., RadTech Europe, Conference Proceedings, 2007
b) EP 1967284 (Verfahren zur UV Strahlungshärtung, 2008)
[66] a) Beck, E., UV curing for everybody, RadTech Europe, Proceedings, p. 643–648 (2001)
b) Studer, K., Decker, C., Beck, E., Schwalm, R., Prog. Org. Coatings, 48, p. 92–111 (2003)
[67] a) Schwalm, R., UV Coatings, Basics, Recent Developments and New Applications, Elsevier, p. 179–185 (2007)
b) Davidson, R.S., The role of amines in UV Curing, in Fouassier, J.P., Rabek, J.F., Radiation Curing in Polymer Science and Technology, Vol. III, p. 153–176 (1993)
c) Pappas, S.P. (Ed.), UV Curing: Science and Technology, Vol. II, p. 125 (1985)
d) Decker, C., Makromol. Chem., 180, p. 2027 (1979)
[68] Jung, T., Simmendinger, P., Tobisch, W., UV Plasma Curing", RadTech Europe 2005, Conference Proceedings, Vol. I, p. 171–175
[69] Wu, Y., Nie, J, The effective utilization of oxygen in the process of photopolymerization, RadTech US, Baltimore (2010), online
[70] Fouassier, J.P. and Rabek, J.F. (Ed.), Radiation Curing in Polymer Science and Technology, Vol. III, Elsevier Applied Science, London, 1993, Chapter 2, Decker, C., p. 34–38
[71] Schwalm, R., Schrof, W., Gruber, N., European Coatings Journal, 02/08, p. 26–30 (2008)
[72] Decker, C., Masson, F., Schwalm, R., How to speed up the UV curing of waterbased acrylic coatings, Journal, Coatings Technology Research, Vol. 1, (2), p. 127–136, April 2004

[73] Schwalm, R., Häußling, L., Reich, W., Enenkel, P., Beck, E., Menzel, K., Fine tuning the mechanical properties of UV coatings, Modern Paint Coating, 87 (9), p. 42–52 (1997)
[74] Schwalm, R., Scratch resistant UV curable coatings for automotive application, in Basics and Applications of Photopolymerization Reactions, Vol. 1: 239–253 (2010), Research Signpost, Fouassier, J.P., Alones, X., (Eds)
[75] Kutschera, M., Sander, R., Hermann, P., Weckenmann, U., Poppe, A., J. Coat. Techn. Res., 3(2) (2006)
[76] Lin, L., Blackman, G.S., Matheson, R.R., Prog. Org. Coatings, 40, p. 85–91 (2000)
[77] Koeniger, R., Studer, K., European Coatings Journal, (1–2), p. 26–58 (2001)
[78] Königer, R. Strahlungshärtbare Lackrohstoffe für Außenanwendungen; Farbe&Lack 105 4/99 (233–240)
[79] Schwalm, R., Schrof, W., Gruber, N., Graphic optimisation, ECJ, 02/08, p. 26–30 (2008)
[80] Schwalm, R., UV Coatings-Basics, Recent Developments, and New Applications, Elsevier, 2007, p. 13 und p. 258–264
[81] Becker, M., Weilburgeer Graphics, Die aktuelle Wochenschau, http://www.aktuelle-wochenschau.de/2007/woche45/woche45.html
[82] Menzel, K.; Bankowsky, H-H; Enenkel, P.; Lokai, M., Jüngste Untersuchungen über die UV-Inert-Technologie, Surface Coatings International, Part B: Coatings Transactions, Volume 84, Number 1, 1 February 2001 , pp. 63–66(4)
[83] Gieseke, W., Siegwerk Druckfarben AG, Die Aktuelle Wochenschau, http://www.aktuelle-wochenschau.de/2007/woche35/woche35.html
[84] Stowe, R.W., „Key factors of UV curing of inkjet printing", IS&T`s NIP19: International Conference on Digital Printing Technolgies, Final Program and Proceedings, New Orleans, LA, US, 2003, p. 268–271
[85] Bankowsky, H.H., Enenkel, P., Lokai, M., Menzel, K., „Strahlungshärtung in der Holzlackierung", Farbe&Lack, 106. Jahrgang, 10/2000, p. 50–56
[86] Müller, B., Poth, U., Lackformulierungen und Lackrezepturen, Vincentz Verlag (2003), p. 251
[87] www.dymax.com/de/pdf/literature/lit019eudt_light_curable_adhesives_coatings_encapsulants_for_electronic_assembly_sg.pdf
[88] Frost&Sullivan: http://www.allbusiness.com/asia/843153-1.html
[89] Kowalik, T., UV light – Challenges in Structural Bonding, RadTech Europe, Conference Proceedings 2009 on CD
[90] Sciangola, D., UV curable laminating adhesives, Sartomer, www.sartomer.com
[91] Gould, M., Electron beam laminating adhesives study, Rahn, RadTech Europe Conference Proceedings 2009, on CD
[92] Meyer-Roscher, B., Barwich, J., Düsterwald, U., Zosel, A., Radiation curable Acrylic Systems in Handbook of Pressure Sensitive Adhesive Technology (ED. D. Satas), 3rd Edition (1999)
[93] Czech, Z., Butwin, A., Gluch, U., Madejska, E., Coating International, 3-2010, p. 26–28
[94] Loctite Produkte für die Elektronikindustrie; http://loctite.fast.de/wwdh/de/i183ch11.htm
[95] Wouters, M., tenCate, T., Maalderink, H., Boot, R., RadTech Europe 2007, Conference Proceedings, on CD

[96] Albert, P., Dermann, K., Rentsch, H., „Amalgam und die Alternativen", Chemie in unserer Zeit, Vol. 34 (5), p. 300–305 (2000)
[97] Decker et al., J. of Polymer Science, Part A Polymer Chemistry Vol. 29, 739–747 (1991)
[98] Decker, C.und Biry, S., Progr. Org. Coatings, 29, p. 81–87 (1996)
[99] Decker, C., Zahouily, K., Conference Proceedings, RadTech 1998, Chicago, IL, p. 396 ff
[100] Frigge, E., Farbe&Lack, 106 (7), p. 78–80 (2000)
[101] A. Valet, Progress in Organic Coatings, 35, p. 223–233 (1999)
[102] Starzmann, O., UV Developments in 3D-UV Curing for Automotive Parts, RadTech2005, Conference Proceedings, p. 231; http://www.ist-uv.com/fileadmin/user_upload/UV-Einsatzmoeglichkeiten/pdf/02_Automotive_GB.pdf
[103] Weber, E., Abell, P., UV curable basecoats and clearcoats for automotive lighting, RadTech Report November/December 2003, p. 59–65
[104] Stropp, J.P., Wolff, U., Kernaghan, S., Löffler, H., Osterhold, M., Thomas, H., UV curing systems for automotive refinish applications, Progress in Organic Coatings, 55, p. 201–205 (2006)
[105] Jung, T., Valet, A., UV curable clearcoats, RadTech Report November/December 2001, p. 30–35
[106] Jeffries, M., Gambino, C., Pierce, J., JCT Coatings Tech., April 2008, p. 32–36
[107] Joessel, K., Automotive Applications of UV cured coatings, Fusion, www.pfonline.com/articles/article_print1.cfm
[108] Mills, P., Robotic UV curing for automotive exterior applications: a cost effective and technically viable alternative for UV curing, SURCAR 2005,
[109] www.rapidshield.com; ConcreteDecor, Vol. 9, No. 2, p. 42–44, March/April 2009
[110] http://www.ha-be-hameln.de/_mediafiles/367-oberflaechenschutz-bm.pdf
[111] Haering Elastocryl D 1180 (Haering GmbH, Untergruppenbach)
[112] Wenning, A., Tailor-made UV-curable powder clear coatings for metal applications, Macromol. Symp. 187, p. 597–603 (2002)
[113] Paulus, W., Binder, H., Henning, I., Rieger, J., Schrof, W., Schwalm, R., Blocken und fließen, Farbe&Lack, 107, 10/2002, p. 46–52
[114] http://www.radtech-europe.com
[115] Santandrea, S., RadTech Europe 2005, Conference Proceedings, Vol. I, p. 45–50
[116] Studer, K., Nesvadba, P., Jung, T., Benkhoff, J., Powell, K., Lordelot, C., Progress in Organic Coatings, 61 (2–4), February 2008, p. 119–125
[117] Gruber, N., Beck, E., Hoben, C., Lafuente, O., Menzel, K., Schwalm, R., UV Film Technology as an Alternative to Conventional Coating Applications, RadTech Europe 2007, Wien, Conference Proceedings on CD
[118] Am Anfang ist der Lack, DaimlerChrysler HighTech Report, 1.2005, p. 10–15
[119] Skinner, D., Making light work for a living, European Coatings Journal, 10/2009, p. 18–22
[120] Garrat, P.G., Strahlenhärtung, Vincentz, Hannover, p. 149 (1996))
[121] EP2160423, BASF SE (2007)
[122] Roose, P., Fallais, I., Vandermiers, C., Olivier, M.-G., Poelman, M., Progress in Organic Coatings, 64, p. 163–170 (2009)
[123] www.jot-oberflaeche.de; JOT Spezial UV Lackierung, in Kooperation mit Sturm-Gruppe und Lankwitzer Lackfabrik
[124] Quaker Chemical Corp., http://www.rapidshield.com/

Ulrich Poth, born in 1941, underwent training at Herberts (now DuPont Performance Coatings) where he worked on developing electro-insulation coatings and resins, and powder coatings. He studied at the University of Applied Sciences, Niederrhein, Krefeld/Germany, graduating as an engineer. In 1971 he joined BASF Coatings, Münster/Germany where he worked in research & development, and on the preparation of resins for industrial coatings applications, with special emphasis on resins for automotive coatings. He ultimately held a managerial position. In 1995, he assumed responsibility development and production operations of automotive topcoats and clear coats in the OEM business unit of BASF Coatings. Although he retired in 2002, he still works on projects, provides consulting services, and is a lecturer.

Dr. Reinhold Schwalm, born in 1954, studied chemistry at the Universität of Marburg/Germany and received his Ph.D in 1983. He joined BASF Polymer Research in 1985, coming from a post-doc year at the IBM Research Lab in San Jose, California/USA. His research activities can be characterized by the design and synthesis of functional organic materials with emphasis on materials for coating applications. Since 1989 he has been manager of different research groups dealing with photoresists; polymer printing plates; coating resins, crosslinkers and dispersing agents for pigments. The recent research activities focus on UV curing technology, raw materials for polyurethane coatings and surface modifications. He is (co)-author of more than 60 journal publications. He is author of a reference book about "UV coatings" and (co)-inventor on more than 110 issued patents.

Dr. Manfred Schwartz, born in 1960, began his career in 1987 with BASF, where he worked in the polymer research laboratory. From 1992 to 1994, he was responsible for paint development at a producer in South America and, from 1995 until 2001, he was a BASF marketing manager for emulsions for exterior application. As a visiting professor at the Federal University of Pernambuco in Recife/PE Brazil, he conducted research into polymer additives from 2002 to 2009. He has been a consultant to the chemical industry in Brazil since 2010.

Poth/Schwalm/Schwartz: Acrylic Resins

ISBN: 978-3-86630-857-2

Acknowledgement

The authors would like to thank their numerous colleagues and assistants in BASF SE and BASF Coatings, whose varied input and previous publications led to the writing of the present book. Overall, their success in the working field described herein was due to close cooperation in the research, development and product maintenance of resins on one hand, and paint development, application experience and customer demands on the other.

Furthermore, Dr. Reinhold Schwalm would like to thank his US colleague Kyle Sass for proofreading Chapter 5.

Poth/Schwalm/Schwartz: Acrylic Resins

ISBN: 978-3-86630-857-2

Index

Symbols

A

B

C

D

E

F

G

O

P

R

S

T

U

V

W

www.ingramcontent.com/pod-product-compliance
Lightning Source LLC
LaVergne TN
LVHW020002031026
849161LV00023B/814